Climate Change in the Arabian Peninsula

Fayma Mushtaq · Luai Muhammad Alhems ·
Majid Farooq
Editors

Climate Change in the Arabian Peninsula

Challenges and Pathways to Solutions

Springer

Editors
Fayma Mushtaq
Applied Research Centre for Environment and Marine Studies
King Fahd University of Petroleum and Minerals
Dhahran, Saudi Arabia

Luai Muhammad Alhems
Applied Research Centre for Environment and Marine Studies
King Fahd University of Petroleum and Minerals
Dhahran, Saudi Arabia

Majid Farooq
Department of Ecology, Environment and Remote Sensing
Government of Jammu and Kashmir
Srinagar, Jammu and Kashmir, India

ISBN 978-3-032-02480-0 ISBN 978-3-032-02481-7 (eBook)
https://doi.org/10.1007/978-3-032-02481-7

This Springer imprint is published by the registered company Springer Nature Switzerland AG
The registered company address is: Gewerbestrasse 11, 6330 Cham, Switzerland

Dedicated to beloved parents, whose unwavering support, guidance and encouragement have shaped every step of this journey

Preface

The Arabian Peninsula, encompassing approximately 3.2 million square kilometers, is the world's largest peninsula and occupies a uniquely strategic position at the intersection of Southwest Asia and Northeastern Africa. Bordered by the Red Sea to the west, the Arabian Sea and the Gulf of Aden to the south, Arabian Gulf to the northeast, the region holds immense geopolitical, economic and environmental significance. It is a land of remarkable contrasts, where ancient cultural heritage thrives alongside modern urban development and where expansive deserts meet dynamic coastal zones. While abundant hydrocarbon reserves underpin its role in the global energy sector, the Peninsula faces considerable environmental challenges, ranking among the most climate-vulnerable regions worldwide. The Arabian Peninsula's climate is predominantly arid, with high temperatures, scarce rainfall and limited renewable water resources. In recent decades, the impacts of climate change have become increasingly evident across the region, reflected in more frequent and intense heatwaves, shifting seasonal patterns, erratic extreme rainfall events and growing pressure on already stressed freshwater systems. These evolving climate conditions, coupled with rapid urbanization, population growth and economic transformation, present complex challenges for sustainable development. Nevertheless, within these intricacies lies significant potential to address these issues sustainably through analysis of its climate vulnerabilities and the adoption of innovative adaptive strategies.

This book, *Climate Change in the Arabian Peninsula: Challenges and Pathways to Solutions*, explores the region's complex climate dynamics and their implications for environmental sustainability, socio-economic development and policy. Combining regional case studies, detailed climate data and sectoral analyses, it examines risks and opportunities across key areas such as temperature and precipitation trends, water stress, public health, food security and ecosystem responses. The multidisciplinary insights provided aim to support researchers, students and policymakers in addressing current and future challenges. Yet, in the face of these challenges, the countries across the Arabian Peninsula have demonstrated remarkable resilience and foresight by investing in climate-smart infrastructure, advancing renewable energy initiatives and fostering regional cooperation in environmental management. These collective efforts, explored subsequent chapters, provide valuable examples for other arid and climate-vulnerable regions worldwide.

In particular, the book aligns this scientific inquiry within the broader context of the region's transformative sustainability agenda. As the largest country in the Arabian Peninsula, the Kingdom of Saudi Arabia has demonstrated a growing commitment to climate leadership through its Vision 2030 strategy and flagship initiatives such as the Saudi Green Initiative (SGI) and the Middle East Green Initiative (MGI), which emphasize renewable energy, afforestation and circular carbon economy frameworks. Alongside Saudi Arabia's efforts under Vision 2030, countries across the Arabian Peninsula are advancing their own climate agendas strengthening Nationally Determined Contributions (NDCs), promoting nature-based solutions and prioritizing resilient, low-emission development. These collective actions reflect a shared shift toward sustainability and position the region as an emerging model of climate ambition. The United Arab Emirates has launched the UAE Net Zero by 2050 Strategic Initiative and hosted the Conference of the Parties (COP28), reinforcing its leadership in clean energy deployment and international climate diplomacy. Oman's National Strategy for Adaptation and Mitigation of Climate Change, coupled with investments in green hydrogen, reflects a forward-looking approach to decarbonization. Bahrain has committed to achieving net zero emissions by 2060 and is implementing integrated coastal zone management to respond to sea-level rise. Qatar's National Environment and Climate Change Strategy, Kuwait's Vision 2035 and dedicated initiatives in Yemen, Iraq and Jordan ranging from reforestation and water security to institutional capacity-building further underscore the region's shared commitment to sustainable development and climate resilience. Across the Arabian Peninsula, national institutions have played a vital role in translating climate strategies into meaningful action. These agencies support ecosystem restoration, biodiversity conservation, climate monitoring, early warning systems and renewable energy deployment. Through their work, countries are strengthening the scientific and operational foundations needed to manage climate risks and advance sustainability goals across the region.

Recognizing the importance of region-specific examples in enhancing public understanding and informing climate action, this volume focuses on the Arabian Peninsula's evolving challenges and the diverse responses emerging across its countries. Through illustrative case studies and integrated assessments, the book seeks to engage a broad audience including students, researchers, scientists and policymakers by highlighting forward-looking solutions tailored to the region's socio-ecological realities. By focusing on these dimensions, it contributes to the development of actionable adaptation and mitigation strategies while supporting both regional and international climate policy frameworks.

The editorial team acknowledges the contributions of a diverse group of authors and experts whose rigorous work forms the backbone of this compilation. We are particularly grateful to those working on the frontlines of climate science, environmental management and policy implementation in the Arabian Peninsula. We hope that this book will serve as a meaningful resource for academic communities, governmental bodies, environmental practitioners and all stakeholders engaged in building a sustainable and climate-resilient future for the region.

Dhahran, Saudi Arabia	Fayma Mushtaq
Dhahran, Saudi Arabia	Luai Muhammad Alhems
Srinagar/Jammu and Kashmir, India	Majid Farooq

Acknowledgement

The completion of this book, *Climate Change in the Arabian Peninsula: Challenges and Pathways to Solutions*, stands as a testament to the strength of scholarly collaboration, institutional support and the unwavering encouragement of colleagues, mentors and well-wishers who believed in its purpose. While words may not fully capture our appreciation, we extend our sincere thanks to all who supported this endeavor through their goodwill, expertise and commitment. We express our profound appreciation to the panel of esteemed experts who generously dedicated their time to evaluating the submitted chapters. Their scholarly engagement and detailed assessments have significantly enhanced the academic rigor and overall quality of this publication. We are equally grateful to the anonymous reviewers who, through their impartial and constructive input, played a pivotal role in shaping the final manuscript.

Dr. Fayma Mushtaq extends her sincere gratitude to Dr. Luai Muhammad Alhems and Dr. Majid Farooq for their steadfast support, thoughtful guidance and belief in her leadership throughout this project. Their collaboration and trust were instrumental in shaping the vision and execution of this book. She also gratefully acknowledges the Applied Research Centre for Environment and Marine Studies (ARCEMS), King Fahd University of Petroleum and Minerals (KFUPM), Dhahran, Saudi Arabia, for their institutional support, infrastructure and logistical facilitation without which this project would not have been possible.

To our contributing authors, your academic integrity, responsiveness and high-quality contributions form the foundation of this work. Your dedication represents the intellectual core of this volume and we are sincerely grateful for your continued cooperation and enthusiasm throughout the editorial process. Finally, we offer our deepest gratitude for the blessings of the Almighty and the unwavering support of our families, especially our parents. Their encouragement, faith and guidance have been instrumental in sustaining our focus and perseverance. We hope their blessings remain with us as we continue to contribute meaningfully to the scientific and academic communities.

To all who accompanied us on this journey—thank you.

Disclaimer

The authors of individual chapters are solely responsible for the ideas, views, data, figures and geographical boundaries presented in their respective contributions to this book. These contents have not been reviewed or endorsed, in any form, by the publisher, the editors, or the authors of the prefaces, or other chapters. It should also be noted that the views expressed in this book do not necessarily represent those of the organizations to which the authors are affiliated.

Contents

1 **Understanding Climate Dynamics and Resilience in the Arabian Peninsula: An Introductory Perspective** 1
Fayma Mushtaq, Luai Muhammad Alhems, and Majid Farooq

2 **Geographical and Climatic Characteristics of the Arabian Peninsula: Evolution of Landforms and Climate Regimes** 21
Ruby Gorka and Rakesh Kumar

3 **Historical Climate Dynamics and Regional Variability in the Arabian Peninsula: Trends in Temperature and Precipitation Patterns** 37
Reet Kamal Tiwari, Nitin Arora, Sartajvir Singh, and Vishakha Sood

4 **Future Climate Projections Over the Arabian Peninsula: Insights from CMIP6 Models** 61
Fayma Mushtaq, Gowhar Meraj, and Mansour Almazroui

5 **Hazards and Extreme Climate Events in the Arabian Peninsula: Patterns, Drivers and Resilience Pathways** 95
Fayma Mushtaq, Akhouri Pramod Krishna, Manikandan Karuppasamy Ponnambalam, and Thadickal V. Joydas

6 **Socio-Economic Dimensions of Climate Change in the Arabian Peninsula: Challenges and Strategic Responses** 129
Zaiema Rouf and Aasimah Tanveer

7 **Biodiversity and Ecosystem Dynamics of the Arabian Peninsula: Climate-Driven Changes, Species Vulnerability and Conservation Strategies** 161
Neeta Kumari, Saahil Hembrom, and Soumya Pandey

8 **Climate Change and Public Health in the Arabian Peninsula: Epidemiological Threats, Vulnerabilities and Adaptive Pathways** 185
Mohammad Muslim and Hepa Susami

9 **Building Climate Resilience in the Hyper-Arid Arabian Peninsula: Policy Pathways, Adaptation Measures and the Water-Energy-Food Nexus** 217
Mir Khursheed Alam, Umra Rashid, and Aarifah Jan

10 **Climate Adaptation Strategies and Policy Frameworks in the Arabian Peninsula** 243
Md Arif Hasan, Fatema Hossain, and Syed Masiur Rahman

11 **Saudi Arabia's Climate Leadership: National Initiatives, Institutional Frameworks and Alignment with the SDGs** 277
Fayma Mushtaq, Luai Muhammad Alhems, and Majid Farooq

About the Editors

Fayma Mushtaq is a Postdoctoral Researcher at the Applied Research Center for Environment and Marine Studies (ARCEMS), King Fahd University of Petroleum and Minerals (KFUPM), Dhahran, Saudi Arabia. She earned her Ph.D. in Remote Sensing from the Birla Institute of Technology, Ranchi, India, in 2019 and graduated as a gold medalist with M.Sc. in Remote Sensing and GIS from Jiwaji University, Gwalior, India in 2011. Her research interests include spatial analysis and modeling, image processing, climate change vulnerability and risk assessment, water quality studies, lightning activity analysis and policy documentation. She has authored over 19 journal articles, 1 book, 5 book chapters and 10 policy documents. Dr. Mushtaq was a recipient of Department of Science and Technology (DST), Government of India, Innovation in Science Pursuit for Inspired Research (INSPIRE) and Science and Engineering Research Board National Post Doctoral fellow (SERB-NPDF) award. Dr. Mushtaq is also member of the American Geophysical Union (AGU) and the European Geosciences Union (EGU). She represented King Fahd University of Petroleum and Minerals (KFUPM) at the UNCCD COP16 in Riyadh (2024), delivering a talk on "The Role of Civil Society, Women, and Youth in Land Stewardship and Climate Action," reflecting her active role in international climate engagement.

Luai Muhammad Alhems earned his Ph.D. from Texas A&M University, College Station, TX, USA, in 2002. He currently serves as the Director of the Applied Research Center for Metrology, Standards and Testing (ARCMST) and the Applied Research Center for Environment and Marine Studies (ARCMES). Additionally, he is a Professor of Thermo-Fluid in the Department of Mechanical Engineering at King Fahd University of Petroleum and Minerals (KFUPM), Dhahran, Saudi Arabia. Widely recognized for his contributions to research, he has authored and coauthored over 180 journal articles and patents. His research focuses on gas turbines, energy systems, failure analysis, wind energy and energy conservation.

Majid Farooq is a Scientist with the Department of Ecology, Environment and Remote Sensing, Government of Jammu and Kashmir, India. He holds a Ph.D. in Geoinformatics from SGVU Jaipur India, in 2022 and has over 20 years of experience in geospatial technology and climate change research. He earned his M.Tech in Remote Sensing and GIS from SRM University, Chennai, India in 2009. With an extensive publication record, he has authored over 50 journal articles, 8 books and 10 book chapters. His research focuses on climate change modeling, vulnerability and risk assessment, spatial analysis and modeling, forest and wildlife studies, disaster management, environmental planning and policy design. Dr. Farooq's awards and honors include the National e-Governance Award (Innovative Use of Technology) by the Government of India (2012), representation of India at the UNFCCC COP-24 in Katowice, Poland (2018) and the National Geospatial Fellow Award—2024.

Chapter 1
Understanding Climate Dynamics and Resilience in the Arabian Peninsula: An Introductory Perspective

Fayma Mushtaq, Luai Muhammad Alhems, and Majid Farooq

1.1 Introduction

Climate change refers to a series of global processes that trap solar heat in the Earth's atmosphere. It is a large-scale phenomenon with far-reaching impacts, posing significant challenges to both human society and natural ecosystems (Kuniyal et al., 2021). Specifically, it involves long-term changes in the Earth's climate, leading to variability in the patterns and frequency of hydrometeorological events. While Earth's climate has naturally varied over millions of years, recent shifts are predominantly caused by human activities. These include anthropogenic activities such as fossil fuel combustion, biomass burning and deforestation (Ampofo et al., 2023; El Kasri et al., 2021) leading to increase in concentration of greenhouse gases (GHGs), such as carbon dioxide (CO_2), methane (CH_4) and nitrous oxide (N_2O) (Wei et al., 2016).

The accumulation of these gases disrupts the Earth's energy balance through a process known as radiative forcing (RF). RF quantifies the net change in the balance of incoming and outgoing energy in the Earth's atmosphere due to external disturbances. A positive RF leads to warming, while a negative RF causes cooling. This concept is crucial for understanding how GHGs drive global warming and for comparing the contributions of various agents to changes in global surface temperature (Fig. 1.1).

The intensification of GHG levels accelerates global warming and contributes to significant alterations in regional and global climate patterns over extended periods. Climate change has been directly linked to rising temperatures, one of its most visible and widely documented effects. Anthropogenic activities that release GHGs into the

F. Mushtaq (✉) · L. M. Alhems
Applied Research Centre for Environment and Marine Studies, King Fahd University of Petroleum and Minerals, Dhahran, Saudi Arabia
e-mail: fayma.mushtaq@kfupm.edu.sa

M. Farooq
Department of Ecology, Environment and Remote Sensing, Government of Jammu and Kashmir, Srinagar, Jammu and Kashmir, India

F. Mushtaq et al. (eds.), *Climate Change in the Arabian Peninsula*,
https://doi.org/10.1007/978-3-032-02481-7_1

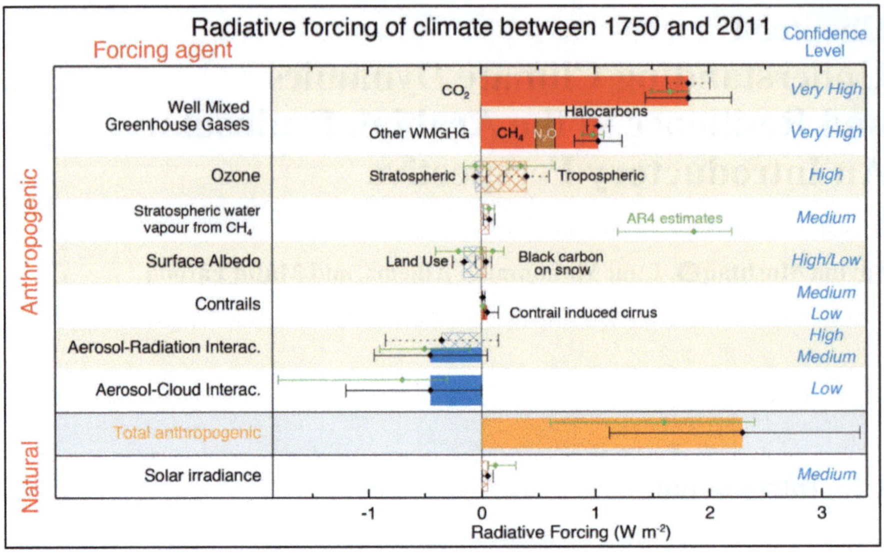

Fig. 1.1 Radiative forcing (RF) and effective radiative forcing (ERF) of climate change during the industrial era (1750–2011) are represented, where solid bars indicate ERF, hatched bars represent RF and green diamonds with associated uncertainties correspond to RF values. *Source* https://climateknowledgeportal.worldbank.org/overview

atmosphere are the primary drivers of this phenomenon, which has resulted in a 1.1 °C increase in the Earth's average surface temperature since the pre-industrial era (Kumar et al., 2023; IPCC, 2021; Singla et al., 2023) (Fig. 1.2). Significantly, 2024 recorded the highest global temperature since 1850, surpassing the 1.5 °C threshold, according to NOAA records (Fig. 1.3). This temperature rise has triggered significant consequences, including more frequent and severe weather events, melting ice caps and glaciers, rising sea levels and widespread ecological disruption (Kumar et al., 2021). If significant reductions in GHG emissions are achieved, the global temperature rise could be limited to 2 °C or less; however, without substantial efforts to curb emissions, the increase could exceed 5 °C by the end of this century (Rogelj et al., 2013).

The latest IPCC assessment report highlights that extreme hydrometeorological events are expected to become more frequent and intense in many regions, including North America, Europe and Asia, significantly increasing the risk of water-related disasters in the coming decades (IPCC, 2021). These events encompass a wide range of phenomena, such as storms, floods and droughts, each with distinct and far-reaching impacts on ecosystems and human systems (Stanković et al., 2019). Importantly, these events are often interrelated, with their relationships influenced by various factors, including geographic location and environmental conditions (Fonseca et al., 2023). For instance, extreme high temperatures can trigger meteorological droughts, which accelerate river depletion, while rapid temperature fluctuations increase the likelihood of extreme precipitation, potentially resulting in flood

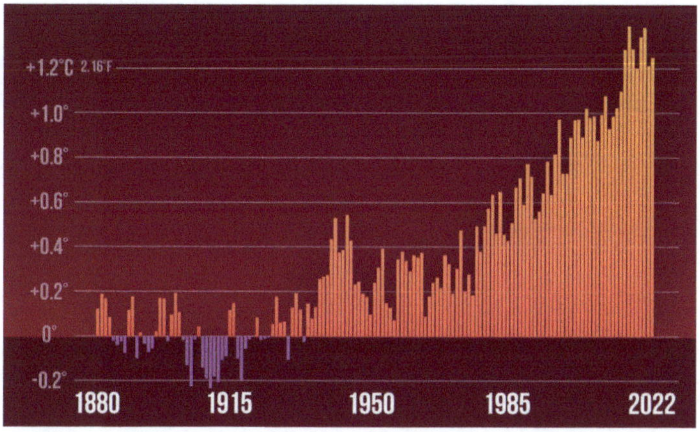

Fig. 1.2 Global average temperature anomalies; since the 1880s global average temperature has warmed approximately 1 °C. *Source* https://climateknowledgeportal.worldbank.org/overview

disasters (Feng et al., 2022). Among these hydrometeorological factors, rainfall plays a critical climatic role, as changes in its patterns can directly or indirectly affect agricultural production, water resource management, human livelihoods and ecosystem structure and function (Kumar & Gautam, 2014). The interconnected nature of these events underscores the importance of comprehensive climate risk assessment and adaptive strategies to mitigate cascading impacts.

1.2 Climatic Evolution and Current Patterns of the Arabian Peninsula

The Arabian Peninsula, the largest peninsula in the world, is located in the southwestern part of Asia and includes Saudi Arabia, Oman, the United Arab Emirates (UAE), Qatar, Bahrain, Kuwait, Yemen, Jordan and Iraq (Wikipedia, 2024). Geographically, it spans approximately 12 °N to 32 °N, from southern Yemen to northern Saudi Arabia and 34 °E to 60 °E, from northwestern Saudi Arabia to eastern Oman (Fig. 1.4). Covering an area of approximately 3 million km^2, the peninsula is characterized by diverse landscapes such as rugged mountains, wadis, alluvial plains, expansive sand dune deserts and varied coastlines (Ghazanfar, 2024). It is bordered by Jordan, Iraq and the Arabian Gulf to the north; the Arabian Sea to the south; the Arabian Gulf and the Strait of Hormuz to the east; and the Red Sea to the west. Strategically positioned at the crossroads of Africa and Asia, the Arabian Peninsula serves as a meeting point for diverse biogeographic realms, including the Palearctic, Afrotropical and Indomalayan regions. This unique geographical position is reflected in its biodiversity, which encompasses species and ecosystems influenced by both African and Asian biogeographic elements (White & Léonard, 1991).

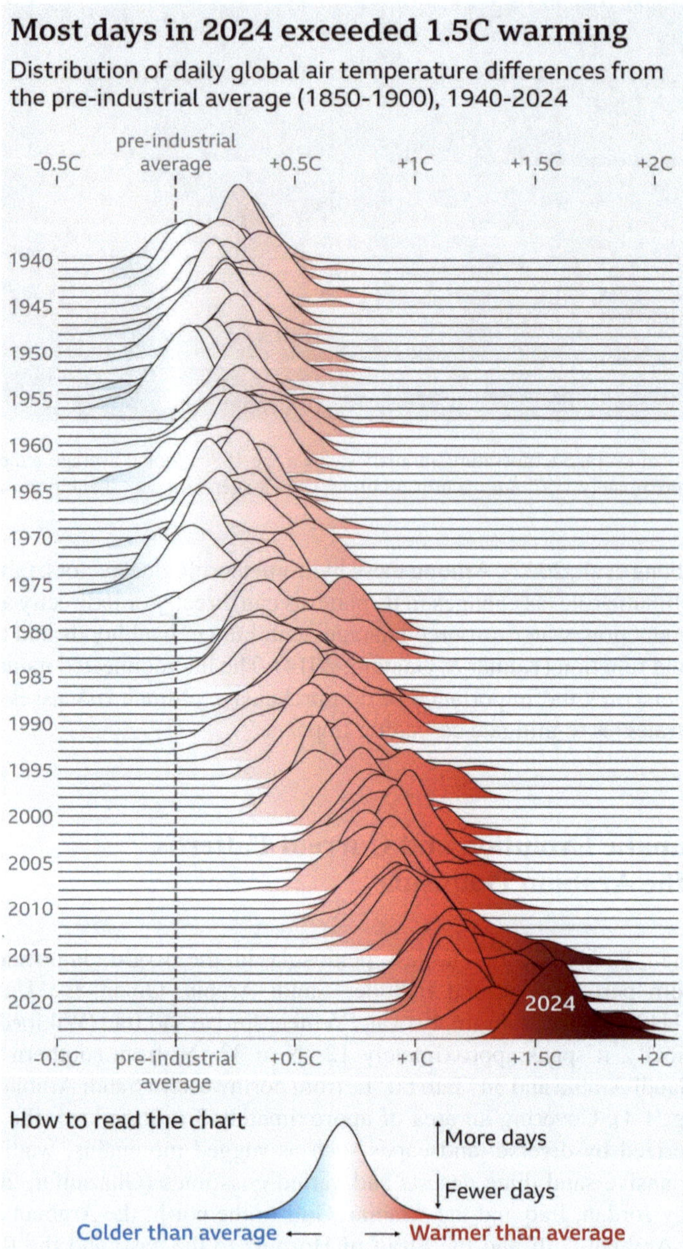

Fig. 1.3 The figure indicates that 2024 was the warmest year globally since 1850, with temperatures surpassing the 1.5 °C threshold. *Source* ERA5/C3S/ ECMWF. Accessed via BBC News

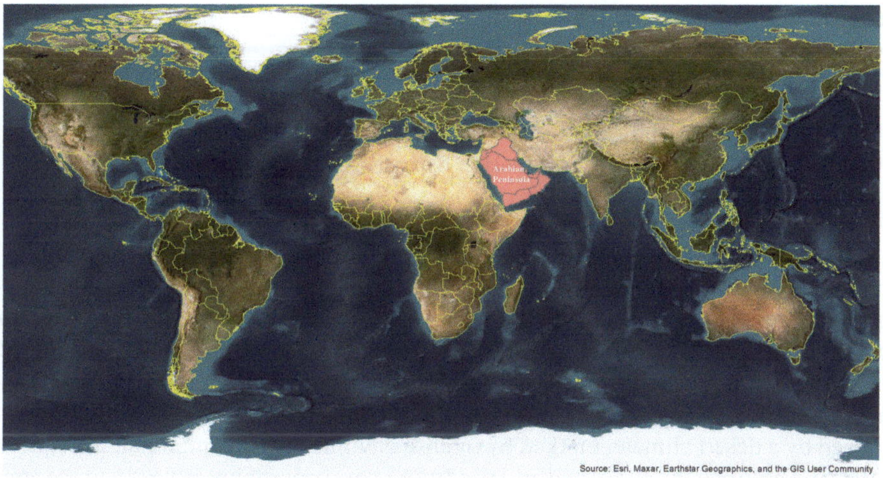

Fig. 1.4 The map illustrates the geographical position of the Arabian Peninsula on a global scale

The past climate of the Arabian Peninsula has undergone significant shifts, transitioning from a fertile and temperate landscape to one of the driest regions on Earth (Kotwicki & Al Sulaimani, 2009). During the Late Cretaceous and the Precambrian era, tectonic activities shaped the region, including the formation of the Red Sea Rift (Table 1.1). The Rubʿ al-Khali desert began forming over 2 million years ago, with alternating hyper-arid and temperate periods, evidenced by fossilized remains of animals and ancient lakes that existed between 37–17 ka and 10–5 ka. Climatic fluctuations during the Late Quaternary were driven by the Indian Ocean monsoon and westerlies, oscillating between wet and dry phases (Table 1.2). Extreme aridity during the Pleistocene resulted in significant droughts, migrations and the adaptation of irrigation systems along the Tigris and Euphrates rivers, fostering the growth of early civilizations. Historical records and paleoclimatic studies reveal a gradual desiccation of the region, driven by declining rainfall and expanding deserts, which shaped human settlement, biodiversity and resource availability. The climate gradually transitioned to arid conditions, ultimately shaping the desert landscapes observed in the present day.

Table 1.1 Key climatic events and impacts

Time period	Event	Impact
870–542 Ma	Formation of Arabian-Nubian Shield	Shaped regional topography and geological features
37–17 ka	Lake formations	Wet periods supporting biodiversity
8 ka	Cooling period	Reduced rainfall by 30%; migration to irrigated areas
4.2 ka	Aridification event	Severe cultural upheaval

Table 1.2 Drivers of past climatic variability

Driver	Timeframe	Impact
Indian Ocean monsoon	Late Quaternary	Enhanced rainfall and fluvial activity
Pleistocene Glaciation	135–75 ka	Reduced sea levels, extreme aridity
Desiccation processes	Recent millennia	Expansion of deserts, decline in cultivable land

The present climate of the Arabian Peninsula is influenced by its complex topography and dynamic atmospheric circulation, resulting in significant spatial heterogeneity (Patlakas et al., 2019a, 2019b). The region exhibits a combination of uniform climatic features and substantial variability, driven by intricate physical and chemical processes within the atmosphere, diverse physiographic characteristics and seasonally shifting atmospheric circulation patterns. The region is predominantly characterized by a desert climate, marked by intense daytime heat and minimal, infrequent rainfall. These extremely hot and arid conditions cover approximately 80% of the area (Almazroui et al., 2012). Based on the Köppen–Geiger classification, the climate of the Arabian Peninsula is categorized as a Hot Desert Climate (BWh) (Patlakas et al., 2019a, 2019b). The region experiences four distinct climatological seasons, each representing a unique climatic regime: the northeast monsoon (December-March), the spring transition (April–May), the southwest monsoon (June–September) and the fall transition (October–November) (Corps, 1990). Temperatures range from highs of up to 45 °C in desert regions to lows of around 2 °C at the highest summits in Yemen, Saudi Arabia and Oman (Patlakas et al., 2019a, 2019b). Winter temperatures in the lower hills and plains typically occur from November to January, averaging around 22 °C. Precipitation is generally highest during the spring and winter seasons. Temperature gradients across the peninsula indicate cooler conditions in the northern regions, with progressively warmer temperatures toward the south. During summer, precipitation is minimal, while autumn generally receives slightly higher rainfall compared to summer. Despite the low average annual rainfall, occasional high-intensity storm events occur intermittently. These events are most commonly observed during winter months when weather systems traveling from west to east are uplifted over the mountainous regions (De Vries et al., 2016).

Rainfall patterns across the Arabian Peninsula exhibit significant spatial variability (Mushtaq et al., 2025). Climatic data analysis from Saudi Arabia, as reported in previous studies, highlights significant trends when comparing recent data (1994–2009) with long-term records spanning over 30 years (1979–2009) (Ghazanfar, 2024). These studies indicate a substantial decline in rainfall and an increase in mean temperatures across much of the region. However, rainfall has increased in the southern parts of the Arabian Peninsula, particularly along the Red Sea coast, over the past few decades compared to levels observed in the 1980s. In contrast, inland areas show no significant changes in rainfall patterns during the same period (Almazroui et al., 2012).

1.3 Climate Change and Environmental Challenges in the Arabian Peninsula

The Arabian Peninsula, with its unique climatic and geographic characteristics, has shown increasing exposure to extreme weather events in recent years, reflecting its sensitivity to evolving climate patterns (Almazroui et al., 2012; Vinodhkumar et al., 2024). Heavy floods, prolonged droughts, have become increasingly common, underscoring the region's susceptibility to climate variability (World Bank, 2014). Frequent exposure to extreme heat further intensifies these challenges, significantly affecting the economy, society and natural environment (Hamed et al., 2024). These climatic shifts have led to extensive research into temperature dynamics and their wide-ranging consequences, highlighting their role in influencing environmental, agricultural and public health issues (Pal & Eltahir, 2016; World Bank, 2019). Adding to these pressures is the severe water stress across the Arabian Peninsula, necessitating a heavy reliance on desalinated water from the Arabian Gulf to meet urban demands. However, this growing dependence on desalination introduces additional environmental challenges, particularly for the Gulf's ecosystem, threatening marine biodiversity and water quality (Sheppard et al., 2010).

Temperature and Precipitation Dynamics

Rising temperatures in the Arabian Peninsula have had far-reaching impacts. For instance, 2010 was recorded as the hottest year in Saudi Arabia (Almazroui et al., 2012). Studies on temperature extremes reveal significant warming trends, with critical consequences for agriculture, biodiversity and energy demand (El-Nesr et al., 2010; Nasrallah & Balling, 1995; Smadi & Zghoul, 2005). Heat stress and water scarcity threaten agricultural productivity, while increased cooling demands strain energy systems. Biodiversity faces habitat loss and thermal stress and migration patterns are shifting as resources diminish, placing food security and ecosystem stability at significant risk (Alkolibi, 2002). In addition to temperature extremes, precipitation patterns exhibit significant spatial variability across the Peninsula. These patterns are influenced by diverse climatic systems such as the Mediterranean climate, the Indian monsoon and orographic effects (Bucchignani et al., 2016). While these mechanisms are of considerable importance, their complexity, combined with limitations in long-term observational data, presents challenges for accurately assessing water availability in the region (Evans, 2010). The region's increasing risk of desertification, driven by rising temperatures and declining precipitation, further threatens already scarce water resources and accelerates environmental degradation (Huang et al., 2017).

Sea Level Rise and Coastal Sensitivities in the Arabian Gulf

Sea level rise (SLR) is a global phenomenon with varying local impacts influenced by geophysical drivers and the low-lying, hyper-arid coastal areas of the Arabian Gulf are particularly vulnerable (Ba-Khamis et al., 2024). The coastline of the eastern Arabian Peninsula, encompassing populous cities and critical infrastructure in Iraq, Kuwait, Saudi Arabia, Bahrain, Qatar and the UAE, faces significant threats (Fig. 1.5). Airports, seaports, power plants, refineries and desalination plants in this region are vital to both the local economy and global trade, making these vulnerabilities especially concerning. Human-induced climate change is the primary driver of accelerated SLR, with global warming causing rising sea temperatures and the melting of the cryosphere. Thermal expansion, responsible for 41.7% of global SLR, is a major contributor to these changes (Frederikse et al., 2020). In the Arabian Gulf, regional factors exacerbate these impacts, with increasing temperatures intensifying thermal expansion and the region's overall vulnerability. Han et al. (2010) emphasize how warm climates affect areas like the Bay of Bengal and the Indian Ocean, similarly impacting the Arabian Gulf. Additionally, rising salinity levels, influenced by the growing demand for desalination to meet the needs of an increasing population, further complicate the region's challenges (Dawoud, 2012).

The countries within the Arabian Gulf region are significantly affected by SLR, leading to measurable losses in land area (Hereher, 2020) (Fig. 1.6). The Arabian Gulf has experienced a considerable increase in sea level, with an average rise of 4.3 mm per year and a margin of error of 0.4 mm, while the Indian Ocean has seen an increase of 3.3 mm per year with a similar margin of error. These rising trends are projected to intensify over time, posing significant challenges for the semi-enclosed Arabian Gulf due to its unique geophysical characteristics. The region's extremely low precipitation rates, with most rainfall concentrated along the northeastern coast, further amplify its vulnerabilities. The Arabian Gulf is also the warmest sea in the world (Hereher, 2020) and exhibits higher salinity levels compared to other regions, driven by increasing desalination activities and climatic conditions. These interconnected factors highlight the urgent need for comprehensive adaptation strategies to address the unique vulnerabilities of the Arabian Gulf to SLR (Bagherian et al., 2014).

Biodiversity and Ecosystem Vulnerability

The Arabian Peninsula's unique biogeography harbors remarkable biodiversity, with countries such as Saudi Arabia, Yemen and Oman showcasing diverse landforms, including mountains that serve as refugia for endemic species and deserts that support uniquely adapted flora and fauna (Talhouk & Abboud, 2009). However, this extraordinary biodiversity is increasingly under threat due to rising temperatures, erratic rainfall patterns and prolonged droughts, which are reshaping ecosystems, leading

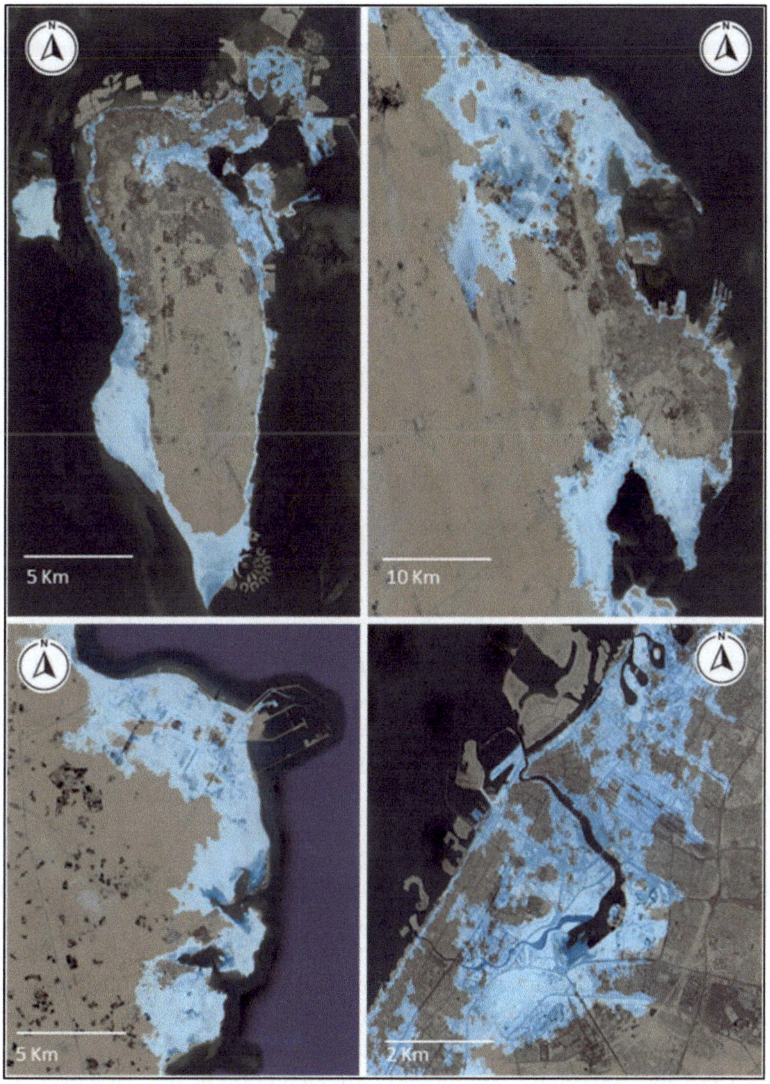

Fig. 1.5 Projected risk zones for coastal submersion in major Arabian Gulf cities including Bahrain, Ras Laffan (Qatar), Dammam and Al Khobar (Saudi Arabia) and Dubai (UAE) under a 3-m SLR scenario by the end of the twenty-first century, based on findings by Ba-Khamis et al. (2024). *Source* Ba-Khamis et al. (2024)

to habitat degradation and species decline (Ghazanfar, 2024). Human activities such as urbanization and overgrazing further disrupt this fragile ecological balance, exacerbating the vulnerability of ecosystems. Freshwater ecosystems, in particular, are at significant risk, with species inhabiting tropical, semi-arid and arid freshwater systems highly susceptible to climate change (Korkmaz et al., 2023). Fragmented

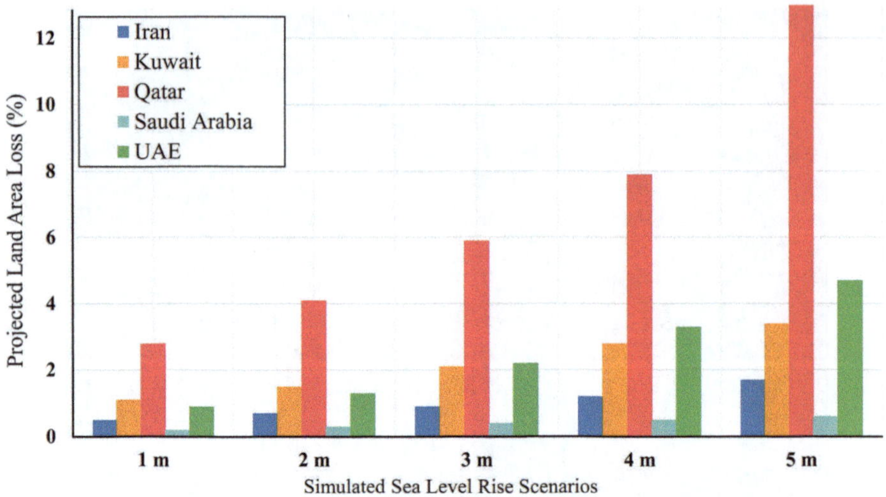

Fig. 1.6 Countries in the Arabian Peninsula potentially affected by sea level rise (SLR), highlighting projected land loss under varying scenarios based on findings by Ba-Khamis et al. (2024). *Source* Ba-Khamis et al. (2024)

habitats, increasing water scarcity due to intensive use and sensitivity to temperature fluctuations have created significant challenges for the survival of endemic freshwater fish species, highlighting the importance of targeted conservation efforts (Makki et al., 2023). Terrestrial biodiversity is similarly threatened, with approximately 80 endemic mammal species in the Afro-Arabian region projected to face severe challenges, as many are expected to lose 50–100% of their current suitable habitats under future climate scenarios (Soultan et al., 2019). Climate change is poised to drastically reshape the distribution and diversity patterns of these mammals, placing numerous species at elevated risk of extinction (Soultan et al., 2019). Marine ecosystems in the Arabian Gulf are also experiencing severe impacts, as temperature-driven climate change disrupts sustained reef-building processes through recurring temperature anomalies, changes in aragonite saturation states and other stressors (Riegl, 2003). Frequent temperature-related coral bleaching events, such as those in 1996, 1998 and 2002, have significantly affected coral health. However, signs of phenotypic adaptation have emerged, with some coral species exhibiting improved resilience following repeated stress events, offering a glimpse of hope amidst these pressing challenges.

Water Scarcity and Management Challenges

Water scarcity is a critical issue in the Arabian Peninsula, driven by climate change, population growth and unsustainable consumption patterns (Odhiambo, 2017a,

2017b). The region's arid and hyper-arid zones are characterized by minimal precipitation, high evaporation rates and limited groundwater recharge, which further exacerbate the crisis (UN Water, 2012). Climate scenarios forecast severe water shortages, increased soil salinity and reduced water runoff, placing immense strain on already limited resources (IPCC, 2007; Milly et al., 2005). In Saudi Arabia, projections suggest a decline in precipitation by up to 20–25% during the dry season by 2050, which may intensify drought conditions and contribute to rising groundwater salinity, particularly in inland regions (Ragab & Prudhomme, 2000). Similarly, in the United Arab Emirates (UAE), average annual temperatures are projected to rise by 2.3–5.9 °C by 2100 (Ajaj et al., 2019), coupled with a 45% decline in rainfall, while in Yemen, water scarcity, declining quality and advancing desertification pose severe challenges. Governments across the region have implemented strategies to address these issues, including expanding desalination facilities and constructing dams to capture seasonal floodwaters (Al-Nuaimi & Dawoud et al., 2020; Murad, 2007). For instance, the UAE has built over 113 recharge and storage dams, capturing 150 million cubic meters of water annually, while Saudi Arabia operates 232 dams designed to capture flash floodwater and enhance groundwater recharge (Saud et al., 2013). However, as climate change leads to reduced rainfall and exacerbates arid conditions, the effectiveness of these measures may diminish, highlighting the need for innovative solutions. The evolving climate and water-related challenges in the Arabian Peninsula highlight the importance of adopting sustainable management practices, strengthening policy frameworks and exploring innovative conservation strategies to support the long-term stewardship of the region's water resources and the preservation of its environment.

1.4 Significance of Climate Policies for Sustainable Development

Climate-related policies can play a pivotal role in addressing the environmental challenges of the Arabian Peninsula, which stem from its inherent aridity and increasing developmental pressures. While rapid economic growth and urban expansion have driven significant progress across the region, they have also placed pressure on groundwater reserves, ecosystems and biodiversity (Burt, 2014). Studies indicate that a substantial portion of the Peninsula is affected by overgrazing, with certain areas experiencing considerable land degradation (Gallacher & Hill, 2013). These conditions underscore the need for climate policies that are carefully tailored to the region's ecological characteristics and socio-political context. Mitigation policies should focus on reducing greenhouse gas emissions and advancing sustainable energy solutions, while adaptation strategies should aim to strengthen resilience to climate variability. Together, these approaches should form the foundation of effective climate governance, which is essential for securing natural resources and supporting long-term socio-economic sustainability across the region (Shindell et al., 2017).

Mitigation Strategies: Reducing Carbon Footprints

The economies of the Arabian Peninsula rely heavily on hydrocarbons, making the transition toward renewable energy a central pillar of effective climate change mitigation. Solar and wind energy offer tremendous potential in the region, supported by high solar irradiance and vast open landscapes. Landmark initiatives such as Saudi Arabia's NEOM and the UAE's Masdar City exemplify regional leadership in energy diversification and innovation (Madakam & Bhawsar, 2021; Sankaran & Chopra, 2020). Additionally, Saudi Arabia's Sakaka PV Solar Plant and the Dumat Al-Jandal Wind Farm have demonstrated scalable, grid-integrated renewable solutions (Ali, 2023; Almutairi & Alahmadi, 2022). Transitioning to renewable energy not only contributes to significant reductions in carbon emissions but also enhances energy security and economic diversification by reducing dependence on fossil fuels. Mitigation efforts must also extend to the transport and industrial sectors. This includes strategic investments in electric vehicle infrastructure, the expansion of efficient and affordable public transport networks and the adoption of carbon capture, utilization and storage (CCUS) technologies in energy-intensive industries (Jiang et al., 2020). Moreover, incorporating green building standards such as the Estidama Pearl Rating System in the UAE or LEED-certified developments in Saudi Arabia into urban planning and construction practices can substantially reduce energy consumption and urban emissions (Metwally, 2019; Salim & Abu Dabous, 2018). These multifaceted mitigation strategies, when aligned with national visions such as Saudi Arabia's Vision 2030 and UAE's Net Zero by 2050 initiative, play a crucial role in shaping a low-carbon and sustainable future for the region.

Adaptation Strategies: Building Resilience

Adaptation strategies hold particular importance for the Arabian Peninsula, where the impacts of climate change such as water scarcity, rising temperatures and flash flooding are becoming more pronounced. As one of the most water-stressed regions globally, the Peninsula faces mounting pressure on its already limited freshwater resources. This challenge is further intensified by unsustainable irrigation practices and the overextraction of groundwater in agricultural and urban zones. To address these vulnerabilities, adaptation policies should prioritize innovative and integrated water management solutions, including the use of advanced desalination technologies, treated wastewater reuse, rainwater harvesting and precision agriculture techniques. For example, Saudi Arabia has made significant investments in desalination infrastructure, with facilities such as the Ras Al-Khair Plant combining energy-efficient technology and water security planning (Allouhi & Almohammadi, 2024). Infrastructure planning must also be climate-informed to reduce the exposure of urban areas and essential services to hazards such as heatwaves and sudden floods. The design and construction of climate-resilient infrastructure, including heat-resistant building materials, shaded public spaces and effective drainage systems, are crucial for urban adaptation. The UAE's Masdar Institute Neighborhood, for instance, demonstrates how passive cooling designs and green architecture can reduce heat stress and enhance overall urban comfort. In addition, strengthening early warning systems and developing localized risk mapping can improve preparedness and response capabilities (Šakić Trogrlić et al., 2022). These technological interventions should be complemented by community-based adaptation initiatives that engage local populations in decision-making processes. Empowering communities through climate education, participatory planning and traditional knowledge integration helps ensure that adaptation strategies are context-specific and socially inclusive. These adaptation strategies, informed by both technological advancement and community engagement, offer a pathway to strengthening environmental resilience across the Arabian Peninsula. By integrating climate considerations into water management, infrastructure design and local planning, the region can better navigate future uncertainties while fostering inclusive, climate-aware development.

Addressing Institutional and Policy Gaps

Strong and well-coordinated institutional frameworks are widely recognized as essential for advancing effective climate governance (Farooq et al., 2021). Robust institutions can enable the successful formulation and implementation of climate strategies that align with national development priorities and global sustainability goals. To support this, enhancing institutional capacity, encouraging inter-agency coordination and promoting inclusive planning processes can contribute to more coherent and resilient policy outcomes. Regional collaboration offers an important opportunity in this context. Organizations such as the Gulf Cooperation Council (GCC) can facilitate joint climate initiatives, support harmonized regulatory standards and foster platforms for knowledge exchange and resource sharing. Some countries in the region have already taken proactive steps such as the UAE's National Climate Change Plan and Saudi Arabia's recent environmental governance reforms which provide promising models of structured institutional engagement (Alomari, 2018; Holmbukt, 2021). A cooperative regional approach is particularly valuable for addressing shared environmental challenges, including water resource management, energy transition and coastal adaptation.

Furthermore, deeper engagement with international partners can further strengthen climate efforts. Participation in global climate finance mechanisms, such as the Green Climate Fund and collaboration with multilateral institutions, academic networks and technical agencies can provide access to resources, expertise and innovative technologies (Browne, 2022). Such partnerships not only enhance national capacities but also contribute to building a resilient and sustainable future for the region through informed, forward-looking climate policies.

Socio-cultural and Economic Considerations

The socio-cultural and economic context of the Arabian Peninsula presents unique opportunities and considerations for advancing effective climate policy. The region's economic structures, shaped in part by long-standing reliance on hydrocarbon revenues, have contributed significantly to national prosperity and development. At the same time, transitioning toward more diversified and sustainable economic pathways presents both a challenge and an opportunity for long-term resilience. Societal values and cultural norms also play an important role in shaping environmental priorities and governance approaches. Enhancing public awareness, encouraging meaningful community participation and fostering shared environmental responsibility can strengthen the effectiveness of climate initiatives. Efforts to mainstream sustainability principles through environmental education, youth engagement and support for community-led initiatives can help align national climate goals with broader societal aspirations. Promoting a culture of sustainability that respects local traditions while embracing innovation and inclusive decision-making will be essential in creating a balanced and resilient development model. In this regard, climate policy can serve as a bridge integrating economic vision, environmental stewardship and community well-being across the Peninsula.

The Way Forward: An Integrated Approach

The pathway to climate resilience in the Arabian Peninsula calls for a holistic and forward-looking approach one that seamlessly integrates mitigation and adaptation strategies into cohesive and actionable policy frameworks. As climate impacts intensify, a proactive investment in renewable energy, sustainable water management and climate-resilient infrastructure will be central to safeguarding long-term socio-economic and ecological well-being. Enhancing institutional capacity and fostering regional cooperation through platforms such as the GCC can improve coordination and ensure shared environmental challenges are addressed collectively. At the same time, leveraging international partnerships, whether through global climate finance or knowledge exchange, will remain vital. For instance, Jordan has advanced solar energy development through international cooperation, while Oman's Integrated Coastal Zone Management Plan reflects an adaptive strategy that aligns ecological protection with development needs (Al-Awadhi et al., 2022; Salah et al., 2023). Community engagement remains equally essential. Building a culture of environmental responsibility through public awareness, school curricula and grassroots initiatives ensures that climate action is inclusive and grounded in local realities.

In Bahrain, public outreach on water conservation and sustainable urban planning demonstrates the role of civic engagement in advancing national goals. Similarly, Qatar's Green Schools Program provides students with hands-on sustainability experience, nurturing a future generation of environmentally conscious leaders (Hijawi et al., 2020). As the Arabian Peninsula continues to navigate the multifaceted dimensions of climate change, its commitment to integrated action rooted in science, inclusivity and innovation can set a powerful example. By aligning long-term visions with on-the-ground implementation, the region has the potential not only to enhance its own resilience but also to contribute meaningfully to global sustainability efforts.

1.5 Conclusion

The Arabian Peninsula occupies a pivotal position in the global discourse on climate change, with its distinctive geographic and climatic characteristics underscoring the importance of understanding and addressing regional vulnerabilities. Rising temperatures, shifting precipitation patterns and an increasing frequency of extreme weather events such as heatwaves, droughts and flash floods are creating complex challenges for both natural ecosystems and human systems. These multifaceted impacts intersect with socio-economic dependencies, including reliance on groundwater extraction, desalination technologies and hydrocarbon-based economies, which further influence the region's climate resilience. Responding to these evolving dynamics calls for a comprehensive and integrated approach one that embeds mitigation and adaptation measures within broader development strategies, while remaining sensitive to the region's environmental and socio-economic realities. Strengthening this approach requires continued investment in scientific research, technological innovation and human capacity building. Many countries in the region have already taken promising steps in this direction, including the expansion of climate monitoring networks, renewable energy deployment and the establishment of national visions that incorporate sustainability as a core objective. Political commitment and financial support remain key to enabling the implementation of adaptive strategies. Equally important is a participatory, bottom-up approach to planning one that encourages locally informed, context-specific solutions. Community engagement and stakeholder involvement are vital to strengthening adaptive capacity across sectors, ensuring that climate action is inclusive and attuned to on-the-ground needs. Public awareness and education also play a crucial role in building societal resilience. Enhancing communication between scientists, policymakers and the public helps bridge knowledge gaps and foster a deeper understanding of climate impacts on both environmental and human health. Tailored outreach grounded in credible information and accessible messaging can empower communities and decision-makers to respond proactively and collaboratively. Taken together, these measures underscore not only the urgency of coordinated climate action across the Arabian Peninsula but also the opportunity to shape sustainable, resilient development trajectories. The chapters that follow delve into observed climate trends, projected future scenarios, extreme heat events, ecosystem responses and actionable strategies for mitigation and adaptation. These insights aim to support informed, science-driven pathways that contribute to a more secure and sustainable future for the region and beyond.

References

Ajaj, R. M., Shahin, S. M., & Salem, M. A. (2019). The challenges of climate change and food security in the United Arab Emirates (UAE): From deep understanding to quick actions. *Current Nutrition & Food Science, 15*(5), 422–429. https://doi.org/10.2174/1573401314666180608121333

Al-Awadhi, T., Nasiri, N. A., Ahsan, R., Hereher, M., & Charabi, Y. (2022). An integrated approach to coastal zone management to control development and ensure sustainability in a rapidly increasing coastal urban environment: The Sultanate of Oman. *Environmental Justice, 15*(4), 214–227.

Ali, A. (2023). Transforming Saudi Arabia's energy landscape towards a sustainable future: Progress of solar photovoltaic energy deployment. *Sustainability, 15*(10), 8420.

Alkolibi, F. M. (2002). Possible effects of global warming on agriculture and water resources in Saudi Arabia: Impacts and responses. *Climatic Change, 54*(1), 225–245.

Allouhi, A., & Almohammadi, K. M. (2024). Towards green desalination: A multi-site analysis of hybrid renewable energy integration in Saudi Arabian RO plants. *Desalination, 592*, Article 118087.

Almazroui, M., Islam, M. N., Jones, P. D., Athar, H., & Rahman, M. A. (2012). Recent climate change in the Arabian Peninsula: Seasonal rainfall and temperature climatology of Saudi Arabia for 1979–2009. *Atmospheric Research, 111*, 29–45.

Almutairi, K., & Alahmadi, R. (2022). The feasibility of the first utility-scale wind farm in Saudi Arabia (the 400 MW Dumat Al-Jandal Project). *Journal of Clean Energy Technologies, 10*(1), 1–10.

Al-Nuaimi, H. S., & Murad, A. A. (2007). The role of dams in securing the surface water in the northern and eastern parts of the United Arab Emirates (UAE). *IAHS Publications—Series of Proceedings and Reports, 315*, 206–214.

Alomari, M. A. N. (2018). *Applying good governance practices in the Middle East in the case of environmental protection: A Saudi Arabian case study* (Doctoral dissertation, University of Leeds).

Ampofo, S., Annor, T., Aryee, J. N., Atiah, W. A., & Amekudzi, L. K. (2023). Long-term spatio-temporal variability and change in rainfall over Ghana (1960–2015). *Scientific African, 19*, Article e01588.

Bagherian, M. M., Akhoundali, A. M., Moazed, H., Jaafarzadeh, N., Ahadian, J., & Hasoonizadeh, H. (2014). Evaluation of Karun river water quality scenarios using simulation model results.

Ba-Khamis, A. N., Bilal, H., Heggy, E., Al-Kuwari, M. S., & Al-Ansari, T. (2024). On the drivers, forecasts, and uncertainties of relative sea level rise in the Eastern Arabian Peninsula: A review. *Regional Studies in Marine Science*, 103503.

Browne, K. E. (2022). Rethinking governance in international climate finance: Structural change and alternative approaches. *Wiley Interdisciplinary Reviews: Climate Change, 13*(5), Article e795.

Bucchignani, E., Mercogliano, P., Rianna, G., & Panitz, H. J. (2016). Analysis of ERA-Interim-driven COSMO-CLM simulations over Middle East-North Africa domain at different spatial resolutions. *International Journal of Climatology, 36*(9), 3346–3369.

Burt, J. A. (2014). The environmental costs of coastal urbanization in the Arabian Gulf. *City, 18*(6), 760–770. https://doi.org/10.1080/13604813.2014.962889

Corps, U. M. (1990). *A Persian Gulf Region climatology*. Department of the Navy, Headquarters United States Marine Corps, 19.

Dawoud, M. A., Alaswad, S. O., Ewea, H. A., & Dawoud, R. M. (2020). Towards sustainable desalination industry in Arab region: Challenges and opportunities. *Desalination and Water Treatment, 193*, 1–10. https://doi.org/10.5004/dwt.2020.25256

Dawoud, M. A. (2012). Environmental impacts of seawater desalination: Arabian Gulf case study. *International Journal of Environment and Sustainability, 1*(3).

De Vries, A. J., Feldstein, S. B., Riemer, M., Tyrlis, E., Sprenger, M., Baumgart, M., & Lelieveld, J. (2016). Dynamics of tropical–extratropical interactions and extreme precipitation events in Saudi Arabia in autumn, winter and spring. *Quarterly Journal of the Royal Meteorological Society, 142*(697), 1862–1880. https://doi.org/10.1002/qj.2781

El Kasri, J., Lahmili, A., Soussi, H., Jaouda, I., & Bentaher, M. (2021). Trend analysis of meteorological variables: Rainfall and temperature. *Civil Engineering Journal, 7*(11), 1868–1879.

El-Nesr, M. N., Abu-Zreig, M. M., & Alazba, A. A. (2010). Temperature trends and distribution in the Arabian Peninsula.

Evans, J. P. (2010). Global warming impact on the dominant precipitation processes in the Middle East. *Theoretical and Applied Climatology, 99*, 389–402.

Farooq, M., Singh, S. K., & Kanga, S. (2021). Mainstreaming adaptation strategies in relevant flagship schemes to overcome vulnerabilities of climate change to agriculture sector. *Research Journal of Agricultural Sciences an International Journal, 12*, 637–646.

Feng, S., Hao, Z., Zhang, X., Wu, L., Zhang, Y., & Hao, F. (2022). Climate change impacts on concurrences of hydrological droughts and high temperature extremes in a semi-arid river basin of China. *Journal of Arid Environments, 202*, Article 104768.

Fonseca, A., Fraga, H., & Santos, J. A. (2023). Exposure of Portuguese viticulture to weather extremes under climate change. *Climate Services, 30*, Article 100357.

Frederikse, T., Landerer, F., Caron, L., Adhikari, S., Parkes, D., Humphrey, V. W., & Wu, Y. H. (2020). The causes of sea-level rise since 1900. *Nature, 584*(7821), 393–397.

Gallacher, D. J., & Hill, J. P. (2013). Intensification of rangeland grazing in an oil-rich state: Causes, consequences and possible solutions. *Dubai Desert Conservation Reserve*, 27–38.

Ghazanfar, S. A. (2024). Biogeography and conservation in the Arabian Peninsula: A present perspective. *Plants, 13*(15), 2091.

Hamed, M. M., Al-Hasani, A. A. J., Nashwan, M. S., Sa'adi, Z., & Shahid, S. (2024). Assessing the growing threat of heat stress in the North Africa and Arabian Peninsula region connected to climate change. *Journal of Cleaner Production, 447*, Article 141639. https://doi.org/10.1016/j.jclepro.2023.141639

Han, W., Meehl, G. A., Rajagopalan, B., Fasullo, J. T., Hu, A., Lin, J., & Yeager, S. (2010). Patterns of Indian Ocean sea-level change in a warming climate. *Nature Geoscience, 3*(8), 546–550.

Hereher, M. E. (2020). Assessment of climate change impacts on sea surface temperatures and sea level rise—The Arabian Gulf. *Climate, 8*(4), 50.

Hijawi, U., Gastli, A., Hamila, R., Ellabban, O., & Unal, D. (2020, February). Qatar green schools initiative: Energy management system with cost-efficient and lightweight networked IoT. In *2020 IEEE International conference on informatics, IoT, and enabling technologies (ICIoT)* (pp. 415–421). IEEE.

Holmbukt, V. (2021). The UAE's "National Climate Change Plan 2017–2050" covered in Al Khaleej.

Huang, J., Li, Y., Fu, C., Chen, F., Fu, Q., Dai, A., & Wang, G. (2017). Dryland climate change: Recent progress and challenges. *Reviews of Geophysics, 55*(3), 719–778.

IPCC (2007) Climate change 2007: Impacts, adaptation and vulnerability. In M. L. Parry, O. F. Canziani, J. P. Palutikof, P. J. van der Linden, & C. E. Hanson (Eds.), *Contribution of working group II to the fourth assessment report of the intergovernmental panel on climate change*. Cambridge University Press.

IPCC (2021). *Climate change 2021—The physical science basis, Contribution of Working Group I to the sixth assessment report of the intergovernmental panel on climate change*. Cambridge University Press.

Jiang, K., Ashworth, P., Zhang, S., Liang, X., Sun, Y., & Angus, D. (2020). China's carbon capture, utilization and storage (CCUS) policy: A critical review. *Renewable and Sustainable Energy Reviews, 119*, Article 109601.

Korkmaz, M., Mangıt, F., Dumlupınar, İ, Çolak, M. A., Akpınar, M. B., Koru, M., & Jeppesen, E. (2023). Effects of climate change on the habitat suitability and distribution of endemic freshwater fish species in semi-arid central Anatolian ecoregion in Türkiye. *Water, 15*(8), 1619.

Kotwicki, V., & Al Sulaimani, Z. (2009). Climates of the Arabian Peninsula–past, present, future. *International Journal of Climate Change Strategies and Management, 1*(3), 297–310.

Kumar, R., & Gautam, H. R. (2014). Climate change and its impact on agricultural productivity in India. *Journal of Climatology & Weather Forecasting, 2*(1), 1–3.

Kumar, U., Singh, D. K., Panday, S. C., Bisht, J. K., & Kant, L. (2023). Spatio-temporal trend and change detection of rainfall for Kosi River basin, Uttarakhand using long-term (115 years) gridded data. *Arabian Journal of Geosciences, 16*(3), 173.

Kumar, A., Nagar, S., & Anand, S. (2021). Climate change and existential threats. In *Global climate change* (pp. 1–31). Elsevier.

Kuniyal, J. C., Kanwar, N., Bhoj, A. S., Rautela, K. S., Joshi, P., Kumar, K., & Singh, H. B. (2021). Climate change impacts on glacier-fed and non-glacier-fed ecosystems of the Indian Himalayan Region: People's perception and adaptive strategies. *Current Science, 120*(5), 888–899.

Madakam, S., & Bhawsar, P. (2021). NEOM smart city: The city of future (the urban oasis in Saudi desert). In *Handbook of smart cities* (pp. 1–23). Springer. https://doi.org/10.1007/978-3-030-47420-8_19-1

Makki, T., Mostafavi, H., Matkan, A. A., Valavi, R., Hughes, R. M., Shadloo, S., & Coad, B. W. (2023). Predicting climate heating impacts on riverine fish species diversity in a biodiversity hotspot region. *Scientific Reports, 13*(1), 14347.

Metwally, W. M. (2019). LEED and green building designs. In *GCEC 2017: Proceedings of the 1st global civil engineering conference 1* (pp. 307–321). Springer Singapore.

Milly, P. C., Dunne, K. A., & Vecchia, A. V. (2005). Global pattern of trends in streamflow and water availability in a changing climate. *Nature, 438*(7066), 347–350.

Mushtaq, F., Haq, A. U., Bharti, S., Alhems, L. M., & Farooq, M. (2025). Regional climate dynamics in the Arabian Peninsula: A study of temperature trends and precipitation uncertainty (No. EGU25-3014). Copernicus Meetings. https://doi.org/10.5194/egusphere-egu25-3014

Nasrallah, H. A., & Balling, R. C. (1995). Impact of desertification on temperature trends in the Middle East. *Environmental Monitoring and Assessment, 37*, 265–271.

Odhiambo, G. O. (2017a). Water scarcity in the Arabian Peninsula and socio-economic implications. *Applied Water Science, 7*(5), 2479–2492.

Odhiambo, G. O. (2017b). Water scarcity in the Arabian Peninsula and socio-economic implications. *Applied Water Science, 7*(5), 2479–2492. https://doi.org/10.1007/s13201-016-0440-1

Pal, J. S., & Eltahir, E. A. (2016). Future temperature in southwest Asia projected to exceed a threshold for human adaptability. *Nature Climate Change, 6*(2), 197–200.

Patlakas, P., Stathopoulos, C., Flocas, H., Kalogeri, C., & Kallos, G. (2019a). Regional climatic features of the Arabian Peninsula. *Atmosphere, 10*(4), 220.

Patlakas, P., Stathopoulos, C., Flocas, H., Kalogeri, C., & Kallos, G. (2019b). Regional climatic features of the Arabian Peninsula. *Atmosphere, 10*(4), 220. https://doi.org/10.3390/atmos10040220

Ragab, R., & Prudhomme, C. (2000, March). Climate change and water resources management in the southern Mediterranean and Middle East countries. In *The second world water forum* (pp. 17–22). The Hague.

Riegl, B. (2003). Climate change and coral reefs: Different effects in two high-latitude areas (Arabian Gulf, South Africa). *Coral Reefs, 22*, 433–446.

Rogelj, J., McCollum, D. L., O'Neill, B. C., & Riahi, K. (2013). 2020 emissions levels required to limit warming to below 2° C. *Nature Climate Change, 3*(4), 405–412.
Šakić Trogrlić, R., van den Homberg, M., Budimir, M., McQuistan, C., Sneddon, A., & Golding, B. (2022). Early warning systems and their role in disaster risk reduction. *Towards the "perfect" weather warning: Bridging disciplinary gaps through partnership and communication* (pp. 11–46). Springer International Publishing.
Salah, A. A., Shalby, M. M., & Basim Ismail, F. (2023). The status and potential of renewable energy development in Jordan: Exploring challenges and opportunities. *Sustainability: Science, Practice and Policy, 19*(1), 2212517.
Salim, A. M., & Abu Dabous, S. (2018). Quantitative analysis of sustainable housing energy systems based on Estidama pearl rating system. *International Journal of Green Energy, 15*(14–15), 905–917.
Sankaran, V., & Chopra, A. (2020, December). Creating global sustainable smart cities (A case study of Masdar City). In *Journal of Physics: Conference Series, 1706*(1), 012141. https://doi.org/10.1088/1742-6596/1706/1/012141
Saud, A. G., Mohamed, F. S., & Asif, U. Z. (2013). Saudi Arabia confronts with water scarcity: An insight. *International Journal of Water Resources & Arid Environments, 2*(4), 218–225.
Sheppard, C., Al-Husiani, M., Al-Jamali, F., Al-Yamani, F., Baldwin, R., Bishop, J., & Zainal, K. (2010). The Gulf: A young sea in decline. *Marine Pollution Bulletin, 60*(1), 13–38.
Shindell, D., Borgford-Parnell, N., Brauer, M., Haines, A., Kuylenstierna, J. C. I., Leonard, S. A., & Srivastava, L. (2017). A climate policy pathway for near- and long-term benefits. *Science, 356*(6337), 493–494. https://doi.org/10.1126/science.aak9521
Singla, C., Aggarwal, R., Kaur, S., & Sharma, R. (2023). Analysis of meteorological parameter changes using Mann-Kendall statistical tests in Indian Punjab. *Mausam, 74*(1), 207–213.
Smadi, M. M., & Zghoul, A. (2005). A sudden change in rainfall characteristics in Amman, Jordan during the mid 1950s.
Soultan, A., Wikelski, M., & Safi, K. (2019). Risk of biodiversity collapse under climate change in the Afro-Arabian region. *Scientific Reports, 9*(1), 955. https://doi.org/10.1038/s41598-018-37851-6
Stanković, S., Vasović, D., & Živković, N. (2019). Impacts of extreme hydrological events on sustainable water resources management and human well-being. *Safety Engineering, 9*(1), 37–42.
Talhouk, S. N., & Abboud, M. (2009). Ecosystems and biodiversity. In *Arab environment: Climate change* (pp. 101–120). Arab Forum for Environment and Development. https://www.afedonline.org/afedreport09/Full%20English%20Report.pdf
Vinodhkumar, B., Ullah, S., Kumar, T. L., & Al-Ghamdi, S. G. (2024). Amplification of temperature extremes in Arabian Peninsula under warmer worlds. *Scientific Reports, 14*(1), 16604. https://doi.org/10.1038/s41598-024-49107-6
Water, U. N. (2012). *Global Annual Assessment of Sanitation and drinking-water (GlAAS) 2012 report: The challenge of extending and sustaining services.* UN.
Wei, T., Dong, W., Yan, Q., Chou, J., Yang, Z., & Tian, D. (2016). Developed and developing world contributions to climate system change based on carbon dioxide, methane and nitrous oxide emissions. *Advances in Atmospheric Sciences, 33*, 632–643.
White, F., & Léonard, J. (1991). Phytogeographical links between Africa and southwest Asia.
Wikipedia Contributors. (2024, May 13). *Arabian Peninsula.* Wikipedia. https://en.wikipedia.org/wiki/Arabian_Peninsula
World Bank Publications. (2014). *Turn down the heat: Confronting the new climate normal.* World Bank Publications.
World Bank. (2019). *Sustainable land management and restoration in the Middle East and North Africa Region: Issues, challenges, and recommendations.* World Bank. https://documents.worldbank.org/en/publication/documents-reports/documentdetail/507581559050957140/

Chapter 2
Geographical and Climatic Characteristics of the Arabian Peninsula: Evolution of Landforms and Climate Regimes

Ruby Gorka and Rakesh Kumar

2.1 Strategic Location and Historical Context of the Arabian Peninsula

The Arabian Peninsula, commonly referred to as Arabia, is the largest peninsula in the world, situated at the northeastern edge of Africa. Located in strategic West Asia, it is bounded by the Arabian Sea and the Indian Ocean to the southeast, the Levant and Mesopotamia to the north, the Arabian Gulf to the northeast and the Red Sea to the west. This unique geographic position has historically shaped global geopolitics, trade routes and cultural exchanges. The discovery of significant oil reserves beginning in Iraq (1927), followed by Bahrain (1932) and Saudi Arabia (1938) transformed the region into a global economic hub (Kumar, 2013). This ushered in rapid urbanization and industrialization, leading to the rise of modern cities. The region now attracts tourists from around the world and hosts a large expatriate population, primarily youthful migrant workers from South Asia and Africa (Dahy et al., 2024; Mirkin, 2010).

Geopolitically, the Peninsula comprises seven principal countries: Bahrain, Kuwait, Oman, Qatar, Saudi Arabia, the United Arab Emirates (UAE) and Yemen. The northern fringes also extend into parts of Jordan and Iraq. Historically, the region has been divided into several subregions: the Central Plateau (Najd & Al-Yamama), Eastern Arabia (Al-Hassa or Al-Bahrain), South Arabia (Yemen, Hadhramaut and Oman) and the Hejaz including Tihamah along the Red Sea. Saudi Arabia, the largest country, occupies more than four-fifths of the Peninsula and plays a dominant role in the region's economic and geopolitical dynamics. Yemen, the second-largest by area, has held historic significance due to its proximity to the Bab-el-Mandeb Strait,

R. Gorka · R. Kumar (✉)
Department of Environmental Sciences, University of Jammu, Jammu, JKUT, India
e-mail: rakeshkumaratri@gmail.com

F. Mushtaq et al. (eds.), *Climate Change in the Arabian Peninsula*,
https://doi.org/10.1007/978-3-032-02481-7_2

a critical maritime passage linking the Red Sea to the Gulf of Aden (Khoshnaw, 2022). Oman, with coastal access to both the Arabian Sea and the Gulf of Oman, has also played a key role in maritime trade and regional diplomacy (Osman et al., 2022). The smaller Gulf states such as Kuwait, Qatar and Bahrain have developed thriving economies focused on oil, finance, tourism and technology.

The Peninsula's geographic location positioned between Africa, the Iranian plateau and the Levant is crucial for understanding early human migration. Flanked on three sides by the Arabian Gulf, Arabian Sea and Red Sea, the region presented natural barriers to movement. However, the northern expanse, consisting of open grasslands extending to the Mediterranean, facilitated relatively unrestricted population flows (Petraglia, 2003). Acting as a natural land bridge connecting Africa, Europe and Asia, the Peninsula was pivotal in ancient human dispersal. Two major migratory pathways are identified: one through the Nile Valley into the Levant and the other crossing the Red Sea from the Horn of Africa via the Bab al-Mandab Strait. Once established in the Levant and the Peninsula, early hominins likely expanded eastward across the Iranian Plateau and along the Zagros Mountains (Petraglia, 2003).

2.2 Physical Geography and Cultural Landscape of the Arabian Peninsula

Geographically, the Arabian Peninsula extends over 3.2 million square kilometers in West Asia. Owing to its unique location and complex geological history, the Peninsula's geographical features are shaped by tectonic activity, harsh climatic conditions and erosional processes (Holm, 1960). The region is culturally significant as the birthplace of Islam and the location of two of its most sacred cities, Mecca and Medina situated in Saudi Arabia (Nasr, 2015). The culture of the Peninsula is deeply rooted in Islamic traditions and enduring artistic heritage. It is marked by strong literary and poetic expressions, particularly the *qasida* (odes), a classical form with significant influence across South Asia. Folk music traditions have been shaped by pilgrims and the nomadic Bedouins who brought musical elements from across the world. Folk dances often termed Oriental dances like the "Ardah" sword dance are prominently showcased during festivals and ceremonial events. Ethnically, the diversity of the Arabian Peninsula stems from its historical role as a nexus for trade and migration, attracting communities from Africa, Asia and Europe (Al-Rasheed, 2005). The Arabs remain the predominant ethnic group, unified by shared language, history and culture. Among them, the Bedouins nomadic desert dwellers have maintained rich oral traditions and tribal affiliations. Though many Bedouins have transitioned to urban life, their cultural imprint endures (Cole, 2003).

The Hadhrami Arabs from Yemen's Hadhramaut region have historically spread Islamic traditions across the Indian Ocean, establishing diasporas in South Asia and East Africa. In Oman, the presence of Baloch and Swahili communities adds to the region's cultural pluralism (Jahani, 2014). The Baloch, originally from present-day

Pakistan and Iran, have been prominent in Oman's security forces. Meanwhile, the Swahili community reflects Oman's deep maritime connections with East Africa. The Gulf states also host large Persian and South Asian populations, drawn by economic opportunities, especially in construction and service sectors. Migrants from India, Pakistan and Bangladesh have played a pivotal role in the economic transformation of the region while contributing diverse cultural practices (Kemp, 2012). African-descended communities such as the Zanj in Oman and southern Saudi Arabia have further enriched the cultural scene through music, dance and performance arts (Boulos & Ayari, 2021). This ethnocultural diversity is visibly embedded in the Peninsula's architecture, cuisine and social norms. For instance, the mud-brick structures of Hadhramaut and coral-stone houses along the Gulf coast reflect centuries of architectural adaptation and intercultural exchange. Complementing this cultural richness is the region's remarkable geomorphological diversity, featuring expansive deserts, rugged mountain ranges, coastal plains and elevated plateaus.

2.3 Tectonic Framework and Mineral Resources of the Arabian Peninsula

The Arabian Peninsula lies atop the Arabian Plate, a minor lithospheric plate surrounded by the African, Eurasian and Indian Plates. The Arabian Plate originated approximately 25 million years ago (Ma) as a result of rifting in northeastern Africa, which led to the formation of the Gulf of Aden and the Red Sea separating it as a fragment of the African continent. It is currently moving northward, away from the African Plate and converging with the Eurasian Plate. This plate is one of the youngest and twelfth largest of Earth's lithospheric plates (Leroy et al., 2013). The Arabian Plate's crust comprises a Precambrian crystalline basement well exposed in the west and locally in the east overlain by a thick Phanerozoic sedimentary succession (up to 10 km thick), as well as Cenozoic basalt fields (*harrat*) covering both crystalline and sedimentary rocks in the western and northwestern regions. The convergence with the Eurasian Plate is responsible for significant tectonic activity along the Zagros Mountain belt in Iran (Heydari, 2008). This ongoing activity has produced fault zones, uplifted mountain ranges and sedimentary basins throughout the region. A significant geological feature of the Arabian Peninsula is the extensive sedimentary cover, which contains vast oil and gas reserves particularly along the Arabian Gulf. These formations, dating from the Paleozoic to the Cenozoic eras, consist mainly of limestones, sandstones and shales (Alsharhan & Kendall, 2021). The interaction of tectonic, volcanic and sedimentary processes has created a geologically distinct and resource-rich region (Hadley & Schmidt, 1980; Stern & Johnson, 2010).

The Arabian Plate is subdivided into two major geological regions: the Arabian Shield in the west and the Arabian Platform in the east. The Arabian Shield, running parallel to the Red Sea coast, is topographically higher and composed primarily of ancient crystalline rocks more than 540 million years old from the Precambrian era. These rocks mainly granites, gneisses and volcanic types are remnants of early Earth's crust. The Shield is rich in mineral resources such as gold, copper and zinc due to its complex geologic history and mineralization processes (Ahmed, 2022). Mineralization in the Shield is largely associated with hydrothermal processes affecting ancient volcanic and intrusive rocks. Fault zones and fractures have facilitated the movement of mineral-rich fluids, creating ore deposits. The Arabian Shield has been historically significant for gold mining, which has been active for over 3000 years. Even today, gold extraction continues in several regions. The Shield also hosts polymetallic massive sulfide deposits of volcanic origin and mineral belts where ore formation is constrained by tectonic and hydrothermal conditions (Nehlig et al., 1999). Additionally, phosphate and bauxite deposits enhance the Shield's economic potential (Duncan & Al-Amri, 2013; Groucutt, 2020). On the other hand, the Arabian Platform comprises sedimentary rocks such as limestone, sandstone and shale (Alsharhan & Kendall, 2021; Droste, 1990), deposited during the Phanerozoic Eon, which spans approximately 541 million years. These thick sedimentary layers increase in thickness toward the Arabian Gulf (Allison, 1997; Coleman, 1993). The Platform contains the world's largest oil and gas reserves, making it the economic backbone of the Arabian Peninsula.

2.4 Geological and Geomorphological Evolution

The Arabian Peninsula exhibits a complex geological and geomorphological history spanning millions of years, which has fundamentally shaped its ecological and geographical diversity. Its foundation is composed of ancient igneous and metamorphic rocks from the Precambrian era (4600–541 million years ago), prominently represented by the Arabian-Nubian Shield in the west and the Tuwaiq Mountains in central Saudi Arabia (Brook et al., 2006; Ghazanfar & Fisher, 1998; Johnson & Woldehaimanot, 2003). Over deep geological time, tectonic forces including the divergence of the Arabian and African Plates contributed to the formation of the Red Sea and Gulf of Aden. Volcanism has further influenced the region's morphology, evidenced by volcanic fields such as Harrat ar Rahat in western Saudi Arabia, underscoring ongoing geological activity (Bosworth et al., 2005; Camp & Roobol, 1989). Much of the Peninsula's surface is dominated by sedimentary rock formations. Vast basins like the Rub' al Khali (Empty Quarter) in the southeast and the Arabian Gulf basin in the northwest bear evidence of climatic and hydrological transitions over millennia (Glennie, 2010). These processes have produced a diverse landscape of mountains, plateaus, deserts and wadis. Mountain ranges such as the Asir, Hijaz and Sarawat extend along the Red Sea coast, with peaks reaching 3660 m such as Jabal Shuayb in Yemen. Oman's Hajar and Dhofar Mountains, marked by deep gorges and valleys carved by erosion and flash floods, are integral to regional hydrological systems and the sustenance of freshwater ecosystems (Parker, 2010).

Central Arabia is dominated by the Nejd Plateau, an elevated terrain with historic significance due to its oases that supported pastoral communities for millennia. Rich in archaeological remains, including petroglyphs and inscriptions, the plateau provides insights into the Peninsula's role as an ancient trade corridor (Parker, 2010). The deserts of the Peninsula, molded by paleoclimatic shifts, are among its most iconic features. The Rub' al Khali, spanning Saudi Arabia, Oman, the UAE and Yemen, is one of the largest continuous sand deserts in the world, with dunes over 250 m high interspersed with gravel plains and relic fluvial systems indicating wetter ancient conditions. Other major deserts such as An Nafud in north-central Saudi Arabia and Ad Dhana in the south act as ecological dividers that shape species distribution and human settlement. The An Nafud, with its dunes, rocky plateaus and dry riverbeds, has historically acted both as a natural barrier and a conduit for trade and migration.

The Peninsula's 9179 km-long coastline, stretching along the Red Sea, Arabian Gulf and Arabian Sea, supports a variety of marine and coastal ecosystems. These include coral reefs, mangrove forests and intertidal habitats. Avicennia marina mangroves flourish along the Red Sea coasts of Saudi Arabia, Yemen and Oman, offering habitat to marine biodiversity and aiding in coastal protection (Ghazanfar et al., 2019). Coral reefs, particularly in the Arabian Gulf, are noted for their resilience to extreme temperatures and are focal points in climate adaptation research. Notably, in-situ breeding experiments have shown coral survival rates of up to 84%, highlighting the importance of host-symbiont dynamics in warming seas. Another unique geomorphological feature is the *sabkha* coastal salt flats shaped by high evaporation and saline groundwater conditions. These environments host specialized microbial communities with valuable genetic traits relevant for ecological and biotechnological research (Barth et al., 2002; Chatziefthimiou et al., 2018).

2.5 Major Landforms and Surface Processes of the Arabian Peninsula

The Arabian Peninsula's mountain ranges, primarily located along its western and southern margins, have evolved due to tectonic uplift associated with the Red Sea Rift and the collision between the Arabian and Eurasian plates. The Hijaz Mountains extend from the Gulf of Aqaba southward into the Asir Mountains, which continue into Yemen. These ranges exceed altitudes of 3000 m and feature steep slopes, rugged terrains and occasional volcanic cones. The Hajar Mountains of Oman and eastern UAE were also formed through tectonic collisions, resulting in the uplift and folding of sedimentary layers. These mountains contain deep canyons, such as Wadi Nakhr, shaped by erosion and weathering. The Asir region experiences orographic rainfall from moist Red Sea air, fostering a relatively lush landscape with greater vegetation diversity compared to the surrounding arid zones. The Peninsula's coastal plains have developed through sedimentation and tectonic subsidence.

Two principal plains include:

1. The Tihama Plain, extending from the Gulf of Aqaba to the Bab-el-Mandeb Strait, is narrow and low-lying, bordered by the Hijaz and Asir ranges. Seasonal floods transport mountain sediments to the plain, where salt flats (sabkhas) frequently form under high evaporation.
2. The Arabian Gulf coastal plain is broader, characterized by widespread sabkhas with halophytic vegetation, shallow bays, mangrove forests, coral reefs and seagrass meadows.

Internally, the Peninsula is dominated by plateaus and sedimentary basins. The Najd Plateau and Rub' al Khali Basin occupy much of the central terrain. The Najd Plateau is marked by gently rolling topography, escarpments and fault lines (Bar et al., 2016). These features and broader lithological units are well illustrated in the geological map of the Peninsula (Fig. 2.1). This region includes sedimentary formations of sandstone and limestone, affected by the Najd Fault System, leading to escarpments, solitary hills and valleys shaped by erosion (Drechsler, 2012). The tectonic history that shaped these formations and fault systems is illustrated in Fig. 2.2, showing the Arabian Plate's evolution through distinct geological phases.

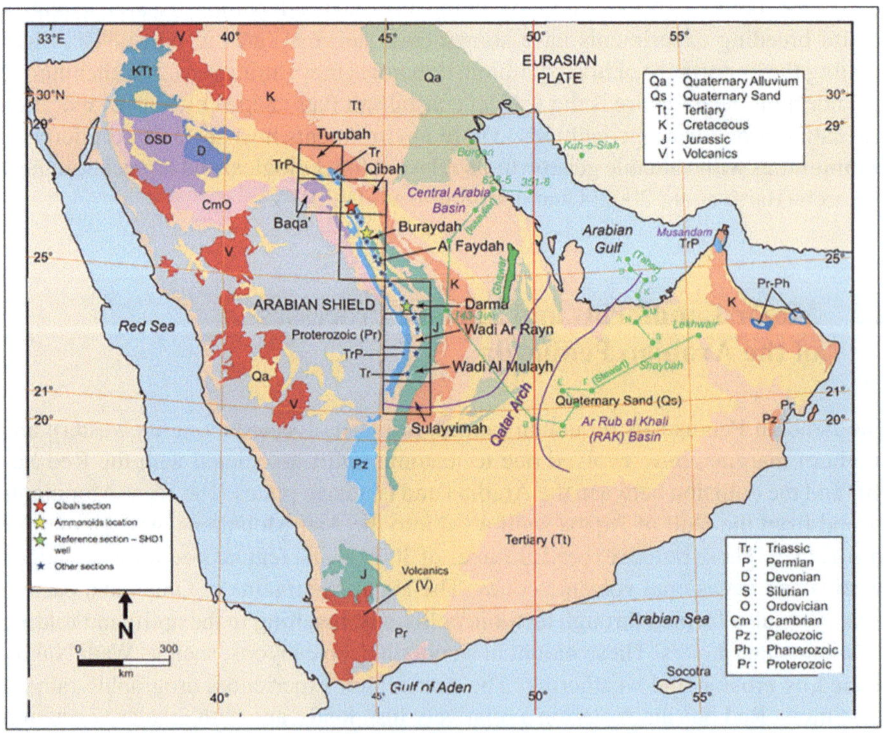

Fig. 2.1 Geological map of the Arabian Peninsula showing major formations adopted from Le Nindre et al. (2023). *Source* Le Nindre et al. (2023)

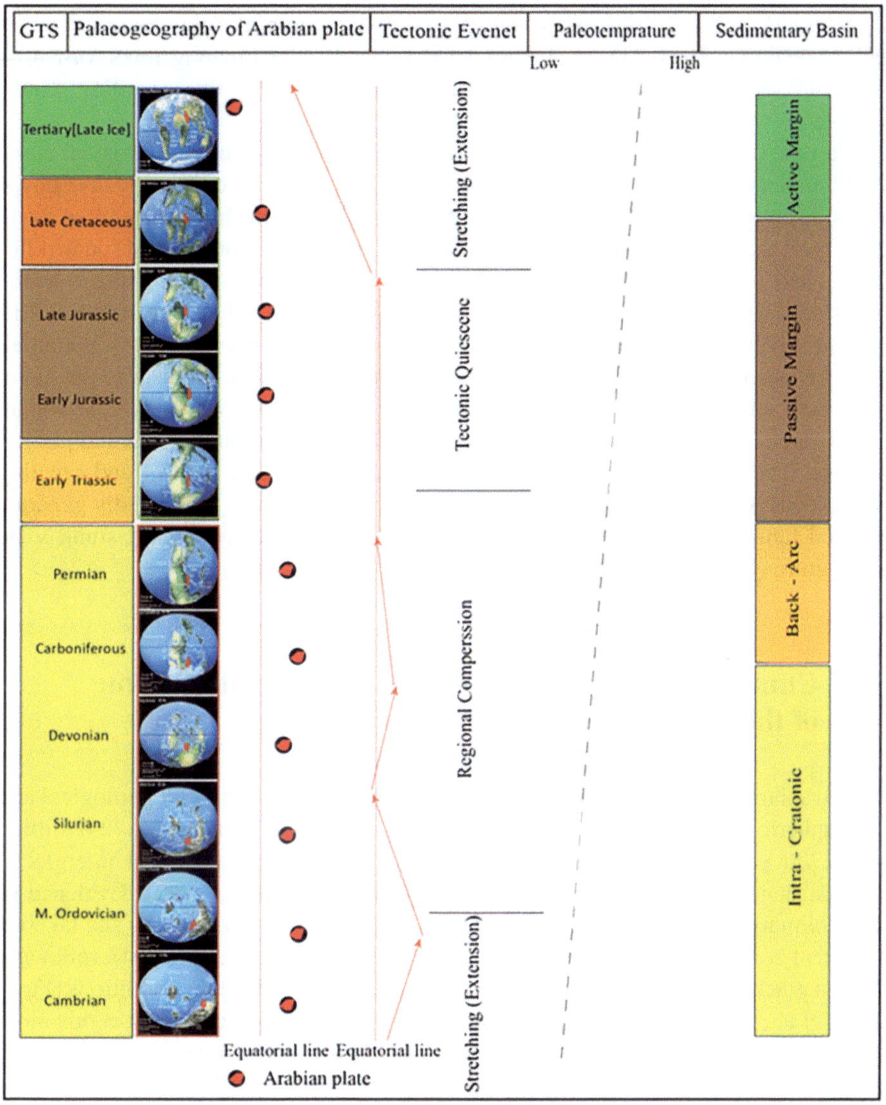

Fig. 2.2 Palaeogeographic and tectonic evolution of the Arabian Plate across geologic time, showing changes in plate position, tectonic activity, paleotemperature and sedimentary basin types adopted from Jaju et al. (2016). *Source* Jaju et al. (2016)

Deserts, shaped by aeolian processes, are defining features of the Peninsula. The Rub' al Khali (Empty Quarter) spans ~ 650,000 km^2 in the southeast and features dunes > 250 m high, interspersed with gravel plains and ancient fluvial traces. These sands overlie sedimentary basins with hydrocarbon significance. In northern Saudi Arabia, the An-Nafud Desert (~ 68,000 km^2) exhibits reddish barchan dunes formed by seasonal winds. Despite extreme aridity, occasional rainfall supports sparse vegetation and fauna. The Ad Dahna Desert, approximately 1000 km long, links An-Nafud and Rub' al Khali. It features longitudinal dunes aligned with prevailing winds, sculpted by continuous sedimentation (Brown et al., 1989).

Harrats, or volcanic fields, represent another prominent geomorphological class. Predominantly located in western Saudi Arabia, these consist of basaltic lava flows, volcanic cones and craters formed by tectonic rifting over millions of years. Among the largest are Harrat Rahat and Harrat Khaybar, which span thousands of square kilometers and comprise extensive basalt sheets and pyroclastic features. These volcanic rocks are largely basaltic and are linked to mantle plumes along the Red Sea Rift (Saif & Shah, 1988). Importantly, these regions have geothermal energy potential, vital for sustainable energy development (Al-Shaibani, 2020). Wadis, or ephemeral riverbeds, such as Wadi Ad Dawasir in Saudi Arabia and Wadi Bani Khalid in Oman, are shaped by episodic flash floods, particularly in mountainous and plateau regions. These floods carry sediments, forming gorges and valleys and creating micro-ecosystems that support unique flora and fauna. Over time, wadis generate alluvial fans, terraces and fertile deltas, which in some areas support small-scale agriculture (Shahin & Shahin, 2007).

2.6 Climatic Evolution and Present-Day Climate Regime of the Arabian Peninsula

The Arabian Peninsula's climate regime has evolved in tandem with geological and atmospheric changes spanning millennia. During the Pleistocene epoch (2.6 million to 11,700 years ago), the region experienced alternating glacial and interglacial periods that produced dry and humid conditions (Sanlaville, 1992). Evidence of these climatic fluctuations is preserved in ancient soils, fossilized fauna and lakebed sediment cores. The onset of the Holocene (11,700 years ago to present), following the last glaciation, brought about comparatively warmer and wetter conditions (Fleitmann et al., 2004). Archaeological traces from the early Holocene reflect this more favorable climate, with indications of early agriculture and permanent human settlement. Over time, however, the Peninsula became increasingly arid, a trend that shaped both vegetation dynamics and ecosystem structures. Xeric shrubs, succulents and relict taxa dominate the current flora, adapted to low precipitation and high evapotranspiration.

Today, the Arabian Peninsula is characterized by a predominantly hot desert climate, classified as BWh under the Köppen-Geiger system (Geiger, 1954; Patlakas et al., 2019). The interior deserts regularly experience summer highs approaching 45 °C, while winter temperatures in mountainous regions such as those in Yemen, Saudi Arabia and Oman can drop to 2 °C. The lower hills and plains typically record winter averages of about 22 °C between November and January. The north of the Peninsula is generally cooler than the south, while rainfall occurs mostly during spring and winter. Although the average annual rainfall is modest, isolated storm events can deliver substantial precipitation. For instance, weather systems moving eastward from the Red Sea during winter often produce orographic lifting and local downpours, particularly in mountainous regions. Flash flooding becomes a concern in such events, especially where vegetative cover is sparse. A study of rainfall at the Tabuk Gauging Station in northwestern Saudi Arabia found that a 6-h storm can deliver up to 60 mm of rainfall once every 50 years, with a potential peak intensity reaching 100 mm/hour (Almazroui et al., 2012a). Rainfall distribution across the Peninsula is highly variable. Comparative analysis of climate station data from 1994 to 2009 and long-term records from 1979 to 2009 reveal spatial and temporal variation. Notably, there has been an increase in rainfall along the southern Arabian coastlines, particularly near the Red Sea, whereas inland regions have not shown significant change (Almazroui et al., 2012b). These localized patterns underscore the complex interaction between topography and prevailing wind systems in modulating the Peninsula's rainfall regime.

2.7 Climatic Variability and Aridification History of the Arabian Peninsula

Building upon the climatic framework outlined earlier, this section explores the temporal variability, regional contrasts and progressive aridification that have shaped the Arabian Peninsula's environment and human history. Until the late Tertiary period, the Peninsula remained geologically connected to the African continent. The formation of the Red Sea Rift altered regional climate dynamics and initiated widespread aridification (Boone et al., 2021; Ghebreab, 1998). The Rub' al Khali desert, one of the largest sand deserts in the world, began forming over two million years ago. Since then, its climate has oscillated between temperate and hyper-arid conditions (Hasanean & Labban, 2022). Archaeological and paleoenvironmental evidence attests to episodic humid phases in the geological past. Excavations in present-day desert regions have uncovered remains of fauna such as hippopotamuses, buffaloes, zebras, wild cattle and hyenas indicative of wetter conditions. Numerous paleo-lakes formed during periods between 37,000–17,000 years ago and again between 10,000–5000 years ago (McClure, 1994). These fluctuations were primarily driven by shifts in Indian Ocean monsoons and westerly atmospheric circulation during the late Quaternary (Kotwicki & Al Sulaimani, 2009).

Human occupation of the Peninsula coincided with these humid intervals. Stone tools dating back to ~ 100,000 years ago have been found along the edges of the Rub' al Khali, while archaeological evidence suggests human expansion around 10,000 years ago. The transition from the Pleistocene to the Holocene marked a general shift toward drier conditions, which likely forced early populations to migrate toward the riverine environments of the Tigris and Euphrates. The adoption of irrigated agriculture in Mesopotamia supported surplus food production and early societal development (Kotwicki & Al Sulaimani, 2009). One major event the 4200 BP aridification episode caused significant cultural and socio-economic disruptions across the broader Middle East. Paleoclimatic reconstructions suggest that despite today's extreme dryness, the region may have experienced even more severe aridity in earlier times. For instance, during episodes between 135,000 and 75,000 years ago, Lake Malawi in East Africa experienced a > 95% reduction in water volume, indicating profound drought conditions (Willoughby, 1993). Some studies even propose that aridity during this period exceeded that of the Last Glacial Maximum (Scholtz et al., 2007).

Both scientific evidence and historical accounts point to a time when the Arabian Peninsula had a tropical climate that supported diverse megafauna, including Asian cheetahs, camels, leopards, wolves and hyenas. However, progressive drying over the past several thousand years has led to reduced surface water availability, desert expansion and cycles of human migration in search of more hospitable environments (Kotwicki & Al Sulaimani, 2009). In the present day, around 80% of the Peninsula experiences extremely hot and arid conditions (Abdullah & Al-Mazroui 1998; Ragab & Prudhomme, 2000). The Köppen-Geiger climate classification designates most of the region as Hot Desert (BWh), while southwestern highlands such as Asir and parts of Yemen receive enough convective precipitation to be classified under mild steppe climate zones (Almazroui et al., 2012a). Mean annual temperatures exceed 20 °C across 90% of the Peninsula and may reach 30 °C in parts of the interior (Kotwicki & Al Sulaimani, 2009). Summer (May–September) temperatures in lowland deserts frequently reach 40–45 °C, often surpassing 50 °C.

These high temperatures are attributed to the region's location beneath the subtropical high-pressure belt, characterized by descending air masses, minimal cloud cover and intense solar radiation. This results in substantial diurnal temperature variation, amplified by the high albedo of desert sands and minimal vegetation (Ghazanfar & Fisher, 1998). Rainfall is among the lowest in the world and varies drastically due to the interaction of complex atmospheric systems. Northern parts are influenced by Mediterranean synoptic systems, while southern regions are affected by the Indian summer monsoon. Geographic factors such as the vast Rub' al Khali and the western Sarawat Mountains further modulate precipitation (Alghamdi, 2023). Extreme climatic events particularly heatwaves and flash floods pose mounting risks. The Rub' al Khali receives < 50 mm/year and most deserts receive < 100 mm. Coastal areas fare slightly better, with 70–150 mm/year. Heatwaves, often linked to the Arabian Trough, are expected to increase in severity due to global warming. These events threaten infrastructure and public health and reflect the region's high vulnerability to climate change (Almazroui & Saeed, 2020; Singh & Chudasama, 2021; Patlakas et al., 2019).

Regional contrasts further shape the climate: southwestern mountains receive ~ 300 mm annually via orographic lifting, while coastal Yemen is influenced by the Red Sea Convergence Zone. In contrast, convective storms deliver occasional winter rainfall to the eastern Arabian Gulf (Hasanean & Almazroui, 2015). Winters in the northern Peninsula are markedly colder (Syed et al., 2024) and temperature moderation occurs along the Red Sea and Arabian Gulf coasts due to maritime influence (Patlakas et al., 2019; Raymond et al., 2024). Historical and geological records underscore the importance of understanding past climatic variability in informing projections of future risk. Fluctuating storm tracks and atmospheric circulation patterns, including El Niño–Southern Oscillation (ENSO) and North Atlantic Oscillation (NAO), have long governed rainfall distribution. Recognizing these drivers is essential to anticipating how climate change may affect water availability, settlement patterns and ecological stability in the Arabian Peninsula (Groucutt & Petraglia, 2012; Rose & Petraglia, 2010).

2.8 Conclusion

This chapter has provided a comprehensive overview of the geographical and climatic characteristics of the Arabian Peninsula. It has outlined how the region's landforms, tectonic activity and climate patterns have developed over millions of years and continue to shape the present-day landscape and environment. The Arabian Peninsula is characterized by diverse landforms, including mountain ranges, plateaus, deserts, coastal plains, volcanic fields and wadis. These features are largely the result of tectonic activity, particularly the movement of the Arabian Plate and its interaction with the African, Eurasian and Indian Plates. The Arabian Shield and the Arabian Platform represent the two main geological zones of the Peninsula. The Shield, made up of ancient crystalline rocks, is mineral-rich, while the Platform is composed mostly of sedimentary rocks and contains significant oil and gas reserves. Tectonic processes have led to the formation of various features, including the Red Sea Rift and the Zagros Mountain belt. The Peninsula is also marked by volcanic activity, particularly in the form of harrats (lava fields), which are found mainly in western Saudi Arabia. These features not only provide insights into the region's geological past but also have implications for natural resource exploration and potential geothermal energy use.

The geomorphological landscape of the Peninsula includes prominent features like the Rub' al Khali (Empty Quarter), the world's largest continuous sand desert and other deserts such as An Nafud and Ad Dhana. These desert landscapes have been shaped by wind-driven processes, sediment deposition and climatic changes over time. Plateaus like Najd and mountain ranges such as Hijaz, Asir and Hajar play important roles in defining the region's topography. Coastal plains along the Red Sea and the Arabian Gulf host ecosystems including mangroves, coral reefs and sabkhas. The climate of the Arabian Peninsula has varied significantly throughout its geological history. During the Pleistocene and Holocene epochs, the region experienced alternating wet and dry phases due to glacial-interglacial cycles and monsoonal influences. Evidence from paleoclimatic studies, such as lake sediments and archaeological findings, indicates that the region once supported more vegetation and diverse fauna, including species no longer found in the area.

At present, the Peninsula has a predominantly arid climate classified as "Hot Desert Climate" (BWh) under the Koppen-Geiger system. Summers are extremely hot, while winter temperatures in highland regions can fall to 2 °C. Rainfall is generally low and irregular, with the southwestern highlands receiving the most precipitation due to orographic effects. The climate also varies spatially. The northern regions tend to be cooler, while the southern areas are warmer. Coastal regions benefit from maritime influences, which moderate temperature extremes. Rainfall patterns are influenced by large-scale atmospheric systems such as the El Niño-Southern Oscillation (ENSO) and the North Atlantic Oscillation (NAO), contributing to interannual variability. Climatic variability is evident not only in the historical record but also in recent decades. Observations indicate a rise in mean temperatures and shifts in rainfall patterns, with some areas experiencing slight increases in precipitation and others seeing declines.

The region is vulnerable to climate change due to its natural aridity, limited freshwater resources and high exposure to extreme heat. Increasing frequency and intensity of heatwaves, droughts and flash floods pose risks to infrastructure, public health and ecosystems. The high albedo of desert sands and limited vegetation contribute to large temperature differences between day and night, further stressing environmental and human systems. Understanding the region's physical geography comprising its geology, geomorphology and climate is essential for developing effective policies to manage natural resources and adapt to future climate scenarios. Past climatic records, geological evolution and human adaptation strategies provide important context for contemporary challenges. Integrating this multidisciplinary knowledge can support sustainable planning, disaster risk reduction and long-term environmental management.

References

Abdullah, M. A., & Al-Mazroui, M. A. (1998). Climatological study of the southwestern region of Saudi Arabia. I. Rainfall analysis. *Climate Research, 9*(3), 213–223.

Ahmed, A. H. (2022). *Mineral deposits and occurrences in the Arabian–Nubian Shield*. Springer Nature.

Alghamdi, A. S. (2023). Climatology and changes in temperature seasonality in the Arabian Peninsula. *Atmosphere, 15*(1), 26.

Allison, R. J. (1997). Middle East and Arabia. In D. S. G. Thomas (Ed.), *Arid zone geomorphology: Process, form and change in drylands* (pp. 507–521). Wiley, Chichester.

Almazroui, M., & Saeed, S. (2020). Contribution of extreme daily precipitation to total rainfall over the Arabian Peninsula. *Atmospheric Research, 231*, Article 104672.

Almazroui, M., Islam, M. N., Jones, P. D., Athar, H., & Rahman, M. A. (2012a). Recent climate change in the Arabian Peninsula: Seasonal rainfall and temperature climatology of Saudi Arabia for 1979–2009. *Atmospheric Research, 111*, 29–45.

Almazroui, M., Nazrul Islam, M., Athar, H., Jones, P. D., & Rahman, M. A. (2012b). Recent climate change in the Arabian Peninsula: Annual rainfall and temperature analysis of Saudi Arabia for 1978–2009. *International Journal of Climatology, 32*(6), 953–966.

Al-Rasheed, M. (Ed.). (2005). *Transnational connections and the Arab Gulf* (p. 149). Routledge.

Al-Shaibani, A. M. (2020). Lava fields as potential groundwater sources in western Saudi Arabia. In *Hydrology and water resources* (pp. 103–120). CRC Press.

Alsharhan, A. S., & Kendall, C. G. S. C. (2021). Paleozoic sequence stratigraphy, depositional systems, and hydrocarbon habitats across the Arabian plate. *AAPG Bulletin, 105*(6), 1149–1198.

Bar, O., Zilberman, E., Feinstein, S., Calvo, R., & Gvirtzman, Z. (2016). The uplift history of the Arabian Plateau as inferred from geomorphologic analysis of its northwestern edge. *Tectonophysics, 671*, 9–23.

Barth, H. J., & Böer, B. (Eds.). (2002). *Sabkha ecosystems: Volume I: The Arabian Peninsula and adjacent countries* (Vol. 1). Springer Science & Business Media.

Boone, S. C., Balestrieri, M. L., & Kohn, B. (2021). Thermo-tectonic imaging of the Gulf of Aden-Red Sea rift systems and Afro-Arabian hinterland. *Earth-Science Reviews, 222*, Article 103824.

Bosworth, W., Huchon, P., & McClay, K. (2005). The red sea and gulf of Aden basins. *Journal of African Earth Sciences, 43*(1–3), 334–378.

Boulos, I., & Ayari, Y. (2021). The Art of the Ṭambūra in Qatar: African Identity Reimagined. *Music in Arabia: Perspectives on heritage, mobility, and nation*, 138–159.

Brook, M. C., Al Shoukri, S., Amer, K. M., Böer, B., & Krupp, F. (2006). Physical and environmental setting of the Arabian Peninsula and surrounding seas. In *Policy perspectives for ecosystem and water management in the Arabia Peninsula. UNESCO Doha and United Nations University, Hamilton, Ontario*, 1–16.

Brown, G. F., Schmidt, D. L., & Huffman, A. C., Jr. (1989). *Geology of the Arabian Peninsula; Shield area of western Saudi Arabia* (No. 560-A). US Geological Survey.

Camp, V. E., & Roobol, M. J. (1989). The Arabian continental alkali basalt province: Part I. Evolution of Harrat Rahat, Kingdom of Saudi Arabia. *Geological Society of America Bulletin, 101*(1), 71–95.

Chatziefthimiou, A. D., Deitch, E. J., Glover, W. B., Powell, J. T., Banack, S. A., Richer, R. A., & Metcalf, J. S. (2018). Analysis of neurotoxic amino acids from marine waters, microbial mats, and seafood destined for human consumption in the Arabian Gulf. *Neurotoxicity Research, 33*, 143–152.

Cole, D. P. (2003). Where have the Bedouin gone? *Anthropological Quarterly, 76*(2), 235–267.

Coleman, R. G. (1993). Geological evolution of the Red Sea. In *Oxford Monographs on geology and geophysics* (p. 186). Oxford University Press.

Dahy, B., Al-Memari, M., Al-Gergawi, A., & Burt, J. A. (2024). Remote sensing of 50 years of coastal urbanization and environmental change in the Arabian Gulf: A systematic review. *Frontiers in Remote Sensing, 5*, 1422910.

Drechsler, P. (2012). The Arabian Peninsula. In *A companion to the archaeology of the ancient Near East*, 485–500.

Droste, H. (1990). Depositional cycles and source rock development in an epeiric intra-platform basin: The Hanifa Formation of the Arabian Peninsula. *Sedimentary Geology, 69*(3–4), 281–296.

Duncan, R. A., & Al-Amri, A. M. (2013). Timing and composition of volcanic activity at Harrat Lunayyir, western Saudi Arabia. *Journal of Volcanology and Geothermal Research, 260*, 103–116.

Fleitmann, D., Burns, S. J., Neff, U., Mudelsee, M., Mangini, A., & Matter, A. (2004). Palaeoclimatic interpretation of high-resolution oxygen isotope profiles derived from annually laminated speleothems from Southern Oman. *Quaternary Science Reviews, 23*(7–8), 935–945.

Geiger, R. (1954). Klassifikation der Klimate nach W. Koppen [Classification of climates after W. Koppen]. In Landolt-Bornstein—Zahlenwerte und Funktionen aus Physik, Chemie, Astronomie, Geophysik und Technik, alte Serie (Vol. 3, pp. 603–607). Springer.

Ghazanfar, S. A., & Fisher, M. (Eds.). (1998). *Vegetation of the Arabian Peninsula* (No. 25). Springer Science & Business Media.

Ghazanfar, S. A., Böer, B., Al Khulaidi, A. W., El-Keblawy, A., & Alateeqi, S. (2019). Plants of Sabkha ecosystems of the Arabian Peninsula. *Sabkha Ecosystems: Volume VI: Asia/Pacific*, 55–80.

Ghebreab, W. (1998). Tectonics of the Red Sea region reassessed. *Earth-Science Reviews, 45*(1–2), 1–44.

Glennie, K. W. (2010). *Desert sedimentary environments* (Vol. 14). Elsevier.

Groucutt, H. S. (2020). Volcanism and human prehistory in Arabia. *Journal of Volcanology and Geothermal Research, 402*, Article 107003.

Groucutt, H. S., & Petraglia, M. D. (2012). The prehistory of the Arabian Peninsula: Deserts, dispersals, and demography. *Evolutionary Anthropology: Issues, News, and Reviews, 21*(3), 113–125.

Hadley, D. G., & Schmidt, D. L. (1980). Sedimentary rocks and basins of the Arabian Shield and their evolution. In P. G. Cooray & S. A. Tahoun (Eds.), *Evolution and mineralization of the Arabian-Nubian Shield*, 26–50.

Hasanean, H., & Almazroui, M. (2015). Rainfall: Features and variations over Saudi Arabia, a review. *Climate, 3*(3), 578–626.

Hasanean, H. M., & Labban, A. H. (2022). Features of climatic temperature over Saudi Arabia: A review. *Időjárás/quarterly Journal of the Hungarian Meteorological Service, 126*(4), 511–543.

Heydari, E. (2008). Tectonics versus eustatic control on supersequences of the Zagros Mountains of Iran. *Tectonophysics, 451*(1–4), 56–70.

Holm, D. A. (1960). Desert Geomorphology in the Arabian Peninsula: Distinctive land forms provide new clues to the Pleistocene and recent history of a desert region. *Science, 132*(3437), 1369–1379.

Jahani, C. (2014). The Baloch as an ethnic group in the Persian Gulf region. *The Persian Gulf in modern times: People, ports, and history* (pp. 267–297). Palgrave Macmillan US.

Jaju, M. M., Nader, F. H., Roure, F., & Matenco, L. (2016). Optimal aquifers and reservoirs for CCS and EOR in the Kingdom of Saudi Arabia: An overview. *Arabian Journal of Geosciences, 9*, 1–15.

Johnson, P. R., & Woldehaimanot, B. (2003). Development of the Arabian–Nubian Shield: Perspectives on accretion and deformation in the northern East African Orogen and the assembly of Gondwana.

Kemp, G. (2012). *The East moves West: India, China, and Asia's growing presence in the Middle East*. Brookings Institution Press.

Khoshnaw, Y. (2022). How Yemen became and has remained a territory of strategic importance.

Kotwicki, V., & Al Sulaimani, Z. (2009). Climates of the Arabian Peninsula–past, present, future. *International Journal of Climate Change Strategies and Management, 1*(3), 297–310.

Kumar, A. (2013). Natural hazards of the Arabian Peninsula: Their causes and possible remediation. In *Earth system processes and disaster management*, 155–180.

Le Nindre, Y. M., Vaslet, D., Vrielynck, B., Krystyn, L., Manivit, J., Memesh, A., & Davies, R. B. (2023). The Middle to Late Triassic of Central Saudi Arabia with emphasis on the Jilh Formation. Part I: Lithostratigraphy, facies and paleoenvironments, palaeontology and biostratigraphic age calibration from outcrop studies. *Comptes Rendus. Géoscience, 355*(S2), 67–97.

Leroy, S., Razin, P., Autin, J., Bache, F., d'Acremont, E., Watremez, L., & Lazki, A. A. (2013). From rifting to oceanic spreading in the Gulf of Aden: A synthesis. In *Lithosphere dynamics and sedimentary basins: The Arabian plate and analogues* (pp. 385–427). Springer.

McClure, H. A. (1994). A new Arabian stone tool assemblage and notes on the Aterian industry of North Africa. *Arabian Archaeology and Epigraphy, 5*(1), 1–16.

Mirkin, B. (2010). *Population levels, trends and policies in the Arab region: Challenges and opportunities*. United Nations Development Programme, Regional Bureau for Arab States.

Nasr, S. H. (2015). *Mecca the blessed and Medina the radiant (Bilingual): The holiest cities of Islam* (Bilingual Edition). Tuttle Publishing.

Nehlig, P., Salpeteur, I., Asfirane, F., Bouchot, V., Eberlé, J. M., & Genna, A. (1999, November). The mineral potential of the Arabian shield: A reassessment. In *Proceedings of the IUGS/UNESCO Meeting on the "Base and Precious Metal Deposits in the Arabian Shield", Jeddah, November* (pp. 12–19).

Osman, R. A. H., Gheraissa, M., & al-Zuhimi, S. (2022). History of economic relations between Oman and the Malay archipelago via maritime networks. *Journal of Positive School Psychology*, *6*(2), 301–312.

Parker, A. G. (2010). Pleistocene climate change in Arabia: Developing a framework for hominin dispersal over the last 350 ka. In *The evolution of human populations in Arabia: Paleoenvironments, prehistory and genetics*, 39–49.

Patlakas, P., Stathopoulos, C., Flocas, H., Kalogeri, C., & Kallos, G. (2019). Regional climatic features of the Arabian Peninsula. *Atmosphere, 10*(4), 220.

Petraglia, M. D. (2003). The Lower Paleolithic of the Arabian Peninsula: Occupations, adaptations, and dispersals. *Journal of World Prehistory, 17*(2), 141–179.

Ragab, R. A. G. A. B., & Prudhomme, C. H. R. I. S. T. E. L. (2000, March). Climate change and water resources management in the southern Mediterranean and Middle East countries. In *The second world water forum* (pp. 17–22).

Raymond, C., Matthews, T., & Tuholske, C. (2024). Evening humid-heat maxima near the southern Persian/Arabian Gulf. *Communications Earth and Environment, 5*(1), 591.

Rose, J. I., & Petraglia, M. D. (2010). Tracking the origin and evolution of human populations in Arabia. In: *The evolution of human populations in Arabia: Paleoenvironments, prehistory and genetics*, 1–12.

Saif, S. I., & Shah, S. M. A. (1988). Field and petrographic characteristics of Cenozoic basaltic rocks, Northwestern Saudi Arabia. *Journal of African Earth Sciences (and the Middle East), 7*(5–6), 805–809.

Sanlaville, P. (1992). Changements climatiques dans la péninsule arabique durant le Pléistocène supérieur et l'Holocène. *Paléorient*, 5–26.

Scholz, C. A., Johnson, T. C., Cohen, A. S., King, J. W., Peck, J. A., Overpeck, J. T., & Pierson, J. (2007). East African megadroughts between 135 and 75 thousand years ago and bearing on early-modern human origins. *Proceedings of the National Academy of Sciences, 104*(42), 16416–16421.

Shahin, M. (2007). *Water resources and hydrometeorology of the Arab region* (Vol. 59). Springer Science and Business Media.

Singh, P. K., & Chudasama, H. (2021). Pathways for climate change adaptations in arid and semi-arid regions. *Journal of Cleaner Production, 284*, Article 124744.

Stern, R. J., & Johnson, P. (2010). Continental lithosphere of the Arabian Plate: A geologic, petrologic, and geophysical synthesis. *Earth-Science Reviews, 101*(1–2), 29–67.

Syed, F. S., Al-Azemi, M. A., Zamreeq, A., Islam, M. N., & Ghulam, A. (2024). Observed heatwaves characteristics and variability over Saudi Arabia. *Meteorology and Atmospheric Physics, 136*(2), 13.

Willoughby, P. R. (1993). The middle stone age in East Africa and modern human origins. *African Archaeological Review, 11*(1), 3–20.

Chapter 3
Historical Climate Dynamics and Regional Variability in the Arabian Peninsula: Trends in Temperature and Precipitation Patterns

Reet Kamal Tiwari, Nitin Arora, Sartajvir Singh, and Vishakha Sood

3.1 Introduction

Human activities have changed the climate of the Earth since the Industrial Revolution in the late eighteenth century. This is primarily due to fossil fuel burning, deforestation and land-use change (Syed et al., 2019) (Fig. 3.1). As a result, carbon dioxide (CO_2) and methane (CH_4) have accumulated in the atmosphere, contributing to heat retention, reduced atmospheric stability and global warming (Madhukar et al., 2021). By the mid-twentieth century, climate change began to manifest across different spatial and temporal scales unlike anything previously recorded. The Fifth Assessment Report (AR5) (Collins et al., 2013) of the Intergovernmental Panel on Climate Change (IPCC) reported seven major indicators, including increases in global mean temperature, more frequent and severe extreme weather events (hurricanes, tornadoes, flooding), melting polar ice and mountain glaciers, shifting precipitation patterns, desertification and sea level rise. These changes have profound impacts on ecosystems, freshwater systems, agricultural productivity, human health and socio-economic stability.

In recent years, global warming has received considerable attention, largely attributed to an enhanced greenhouse effect. Scientists anticipate that continued

R. K. Tiwari (✉) · V. Sood
Department of Civil Engineering (Geomatics), Indian Institute of Technology, Ropar, Bara Phool, Punjab, India
e-mail: reetkamal@iitrpr.ac.in

N. Arora
Departement of Computer Science Engineering, Chandigarh University, Mohali, Punjab, India

S. Singh
Center of Excellence in Socio-Environmental Sustainability for River Sand Mining, Indian Institute of Technology, Ropar 140001, Punjab, India

F. Mushtaq et al. (eds.), *Climate Change in the Arabian Peninsula*,
https://doi.org/10.1007/978-3-032-02481-7_3

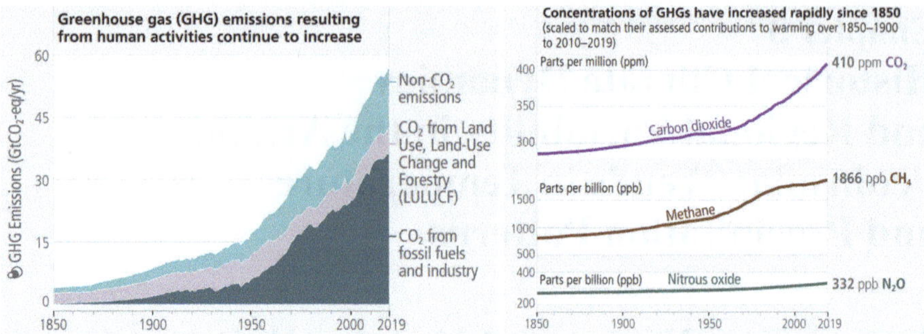

Fig. 3.1 Increased emission of greenhouse gas (GHGs) and increased concentration of GHGs in atmosphere as reported by Intergovernmental Panel on Climate Change (2023). *Source* IPCC (2023)

increases in long-lived greenhouse gas concentrations will further intensify these effects, leading to accelerated temperature rise, increased glacier melt, rising sea levels and more frequent droughts across various regions (Houghton et al., 1992). These anticipated changes carry substantial social and economic implications, calling for robust adaptation measures. To develop appropriate responses and improve understanding of projected climate outcomes, scientists have been extensively analyzing historical temperature and precipitation records. This includes assessing climate model performance and verifying predicted trends using long-term datasets (Folland et al., 1990; Wigley et al., 1990). Therefore, understanding the historical context of climate is crucial not only for defining and mitigating climate change-related impacts but also for informing climate modeling, guiding evidence-based policy-making and supporting adaptation and resilience-building efforts across ecosystems, economies and communities. The following section examines the climatic history of the Arabian Peninsula, the atmospheric circulation systems shaping its present-day climate, observed long-term variability and trends and insights from existing literature on regional climatic changes.

3.2 Climatic Evolution of the Arabian Peninsula

The Arabian Peninsula's climate history is marked by significant geological and climatic transformations. Between approximately 870 million years ago (Ma) and the end of the Precambrian period (542 Ma), the formation of the Arabian-Nubian Shield and the East African Orogen took place, involving major structural and tectonic events that fundamentally shaped the region (Stern et al., 2006). Throughout subsequent geological periods, particularly the Late Cretaceous, major tectonic activities continued to shape the Arabian Peninsula. Historically, the Arabian Peninsula has experienced oscillations between hyper-arid and temperate conditions. Fossil evidence, including remains of hippopotamuses, buffaloes, wild cows, zebras, hyenas, sheep and camels, supports periods of abundant rainfall and lush landscapes in regions now dominated by deserts. Climatic fluctuations during the Late Quaternary period (37–17 thousand years ago, ka) resulted in the formation of numerous lakes within what is now desert terrain. Human settlement evidence dates back over 100,000 years, with significant population expansions occurring around 10,000 years ago. Paleoclimatic studies, such as those by Jansen et al. (2007), utilize air trapped in polar ice caps to reconstruct atmospheric trace gases and aerosol time series dating back approximately 650,000 years. Such studies indicate that historical warm periods corresponded with higher atmospheric trace gas concentrations. McClure (1994) highlights critical climate shifts within the past 100,000 years, particularly noting climatic changes around 8000 BP, where a significant 200-year cooling event resulted in a 30% reduction in rainfall. This event triggered a shift from dry-land agriculture to irrigated farming along the Tigris and Euphrates rivers, significantly altering settlement patterns. Furthermore, the severe 4200-year-old drought, known as the 4.2 ka event, represents one of the most impactful climatic disruptions in human history, contributing to widespread cultural upheaval. Although modern climate conditions on the Arabian Peninsula are among the driest ever recorded, research indicates that even more severe arid phases existed. For instance, climatic studies during the Quaternary period have identified episodes significantly more arid than the last glacial maximum (Scholz et al., 2007). The peninsula's long-term trajectory has been characterized by gradual desiccation, marked by river depletion and the progressive encroachment of deserts onto previously fertile lands. This environmental shift, compounded by population growth, has profoundly impacted the region's agricultural productivity and human habitation patterns.

3.3 Atmospheric Circulation and Current Climatic Settings

At present, the Arabian Peninsula is 99% arid and hyper-arid (HA), as classified by UNESCO (1979) (Table 3.1). The region predominantly falls under the BWh (Hot Desert Climate) classification of the Köppen Climate System, characterized by extreme heat, minimal precipitation and high evaporation rates (Fig. 3.2). In addition to receiving some of the lowest amounts of rainfall on the planet, the region experiences unfavorable rainfall distribution, with sudden and erratic showers exhibiting high intra-annual and interannual variability. The combination of these climatic conditions results in widespread water scarcity, desertification and limited vegetation cover across the peninsula.

The Arabian Peninsula's climate is shaped by complex atmospheric circulation patterns, which govern temperature variations, precipitation trends and wind dynamics. Four major atmospheric systems influence the region:

1. Seasonally shifting global monsoon circulation, including the Indian Summer Monsoon (ISM) and the African Summer Monsoon (ASM), which influence precipitation over southern Arabia (Wang et al., 2017).

Table 3.1 Moisture regimes of the Arabian Peninsula

Moisture regime	Aridity index	%	km^2
Hyper-arid	< 0.03	32.02	1,020,107
Arid	0.03–0.2	66.72	2,25,415
Semi-arid	0.2–0.5	1.22	38,788
Sub-humid	> 0.5	0.04	1199

Source Kotwicki and Al Sulaimani (2009)

Fig. 3.2 Köppen climate classification system; a visual representation of global climate types, including arid (BWh, BWk) and semi-arid (BSh, BSk) climates, which dominate the Arabian Peninsula

2. Summer northwesterly Shamal winds from pressure gradients situated over northwestern Arabia cause recurrent dust storms (Yu et al., 2016).
3. Winter rain associated with mid-latitude cyclonic systems, which affect northern Arabia, the Levant and the Sinai Peninsula (Tyrlis et al., 2013).
4. The Red Sea Trough, a synoptic weather system that generates intense autumn rainfall in central Arabia (de Vries et al., 2013).

In winter, the Arabian Peninsula experiences polar continental air masses moving in from Central Asia, creating cold and dry air. In the boreal summer months, the Intertropical Convergence Zone (ITCZ) shifts northward, allowing more rain over the southern Arabian Peninsula as a result of the Indian and African summer monsoons (Wang et al., 2017). Northern Arabia, the Levant and parts of the Sinai Peninsula receive winter rainfall from mid-latitude cyclonic activity, while much of central Arabia can experience intense autumn rain from the Red Sea Trough (de Vries et al., 2013). The climate variability in the Arabian Peninsula is also affected by large-scale climatic oceanic and atmospheric phenomena, including the El Niño–Southern Oscillation (ENSO), the Pacific Decadal Oscillation (PDO), the Indian Ocean Oscillation (IOO) and the North Atlantic Oscillation (NAO). These climate drivers are influenced by the four (4) air masses responsible for the region's seasonal climatic characteristics.

There are complex atmospheric mechanisms that show these air masses interacting and one of those mechanisms is represented by the Subtropical Jet Stream (STJ), with its associated high- and low-pressure systems as well as residents circulations that account for most weather. The Iran Anticyclone (IA) and Arabian Anticyclone contribute to subsiding air that causes dry, hot conditions across much of

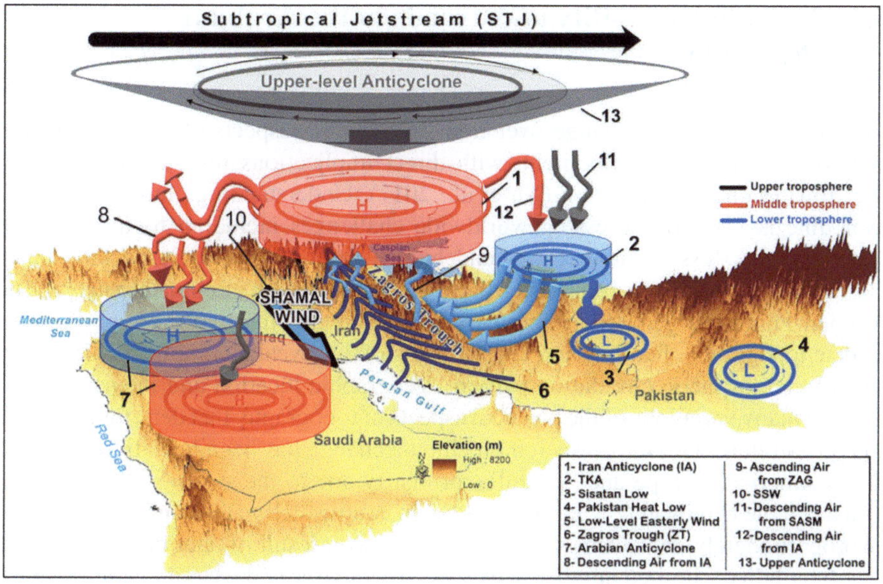

Fig. 3.3 Schematic depiction of atmospheric circulation processes over the Arabian Peninsula and surrounding regions and the role of both high- and low-pressure systems, the Subtropical Jet Stream (STJ), Shamal winds and the Zagros Trough (ZT) adopted from Mofidi and Zarrin (2022). It indicates the relationships among the upper, middle and lower troposphere and wind-pattern observations, areas of rising and descending air and an indication of regional atmospheric drivers (e.g., Iran Anticyclone (IA), Pakistan Heat Low and Sistan Low). *Source* Mofidi and Zarrin (2022)

the peninsula. In contrast, low-pressure systems such as the Pakistan Heat Low and the Sistan Low drive the influx of monsoonal moisture, particularly over Oman and Yemen (Fig. 3.3). The Shamal Wind, a northwesterly wind originating over Iraq, significantly impacts regional dust storm activity and influences surface temperatures. Additionally, the Zagros Trough (ZT) directs low-level easterly winds that interact with regional topography, further modifying local climatic conditions.

Today, the Arabian Peninsula exhibits a predominantly desert climate (BWh in the Köppen classification), characterized by extremely high daytime temperatures and notably cooler nighttime temperatures due to rapid radiative cooling. The region remains under the influence of a subtropical high-pressure system, which significantly impacts temperature variations, humidity levels and precipitation patterns (Lioubimtseva, 2004). Rainfall distribution is further dictated by the peninsula's diverse topography, with orographic effects playing a role in localized precipitation events. Understanding these atmospheric and climatic interactions is crucial for assessing regional climate trends, extreme weather risks and long-term adaptation strategies. The dominance of high-pressure systems and subtropical jet interactions underscores the aridity and hyper-aridity of the Arabian Peninsula, while interactions with low-pressure systems and monsoonal influences highlight occasional climatic variability.

3.4 Climate Variability and Trends in the Arabian Peninsula

Climate variability and change over time are inherent aspects of the environment and socio-economic development, with direct implications for the well-being of the growing global population (Xu et al., 2017). Understanding climate processes over time is vital to addressing worldwide challenges including food insecurity, water scarcity, loss of biodiversity and human health concerns, especially since the fundamental threads connecting these global issues lie in climate change. The Arabian Peninsula generally experiences arid climates through the world's most diverse topographic features and coastal areas often reveal a complicated interplay between atmospheric circulation patterns, geographic location and intensity, duration and many variations of the physical aspects of topography that determine rainfall and temperature. Previous research on climate change has been heavily focused on North America, Europe and Australia, primarily due to the superior observational data available (Freiwan & Kadioğlu, 2008). In contrast, climate change studies have disproportionately neglected Arabian Peninsula as a result of data limitations, data incompatibility, a limited number of weather stations due to the large swaths of desert landscape and lack of a reliable standard for climatic analyses (Kwarteng et al., 2009; Zhang et al., 2005).

The Arabian Peninsula is located in southwest Asia with physiographic zones that include sub-tropical conditions in the north (20°N and above) to tropical monsoonal flows in the south. Despite its predominantly dry climate, recent studies have made significant progress in mapping precipitation and temperature patterns across the Peninsula, utilizing global datasets to fill knowledge gaps. In one of the earliest and largest climate studies covering the Middle East, Zhang et al. (2005) examined climatic data for 52 stations across 15 countries and over the time period of 1950–2003. Their analysis indicated statistically significant and spatially consistent, warming trends and established an unequivocally rising trend in regional temperatures. In line with these results, Hulme (1996) and Jones and Reid (2001) found that increasing temperatures have led to decreasing precipitation-to-potential evapotranspiration (P/PE) ratios in many dryland areas throughout the globe. While Zhang et al. (2005) noted changes in precipitation across the Middle East were weak and without any clear spatial coherence, their research provides evidence to indicate a warming trend. There have been some additional studies pertinent to the Arabian Peninsula that have built upon these previous findings. For instance, Kwarteng et al. (2009) analysed rainfall in Oman from 1977 to 2003 and decided that average monthly and annual rainfall levels remained relatively stable (mean trends littered with noise, without a clear signal). Hamdi et al. (2009) examined climate datasets from six meteorological stations across Jordan and determined that there appears to be an increase in annual definite minimum temperatures and this was especially evident in the 1990s in comparison to earlier decades.

Although developed countries have made significant strides in moving the field of climate change science forward, because of the comparative resources committed and their access to vast datasets and advanced modelling technologies (Goddard et al., 2001), relatively less focus has been placed upon regions like the Arabian Peninsula. Globally, land areas are observed to be warming more than oceans at a rate of 0.27 °C/decade (Solomon, 2007). In the last two decades, there have been little attempts due to a lack of data; however, there has been a growing number of research in the area of climate variability in the Arabian Peninsula. The increasing research shows an alarming rate of warming across the Middle East, but the studies observe inconsistent results regarding the trend of rainfall. The IPCC emphasized the value of increased participation from Arab countries in climate change research and in shaping climate policy frameworks. Given the sensitivity of water resources, agriculture and ecosystems in arid and semi-arid regions, it is important for countries of the Arabian Peninsula to continue enhancing climate monitoring efforts, actively contribute to international climate dialogues and strengthen adaptation strategies that support sustainable development and long-term resilience. The next section identifies and discusses significant studies in this area and their key findings and provides context for current climatology in the Arabian Peninsula.

Recent Observed Climate Change in the Arabian Peninsula (1980–2008): A Study by AlSarmi and Washington (2011)

This study evaluated temperature and precipitation trends in the Arabian Peninsula using data from 21 stations from national meteorological services. This is the first comprehensive trend analysis (Seasonal, Monthly and Annual) for several Arabian Peninsula countries, including Oman, Qatar, the United Arab Emirates (UAE) and Yemen. It examines five climate variables across 21 meteorological stations spanning seven Arabian Peninsula countries. Monthly temperature and precipitation data were sourced from the National Meteorological Services of the Arabian Peninsula. The dataset comprises 21 stations: 17 from Oman, 4 from the UAE and others from Saudi Arabia, Yemen, Qatar, Bahrain and Kuwait. Selected stations had less than 5% missing values and passed quality control assessments. While most stations have data records spanning from 1980 to 2008, some possess records exceeding 50 years. Approximately 70% of the stations reported no missing values for temperature and precipitation variables.

Data quality was ensured through a two-stage quality control process. The first stage involved checking for erroneous values, while the second stage focused on identifying outliers by comparing them with data from neighboring stations. Erroneous values were set to missing after verification. Homogeneity checks were performed using RHtestV3 software to ensure that detected variations were due to climatic changes rather than inconsistencies in data collection. Significant change points, often associated with major climatic events such as the 1997/98 El Niño, were identified in multiple stations. Adjusted monthly time series were used for stations exhibiting significant change points. Linear trends for mean, maximum and minimum temperatures, diurnal temperature range (DTR) and precipitation were computed using a robust nonparametric approach, which is resistant to outliers and missing data. The Mann–Kendall test was employed to assess the statistical significance of trends at a 5% confidence level. Monthly, seasonal and annual trends were calculated across all 21 stations. To minimize biases from high mean value stations, regional average time series for the five climate variables were computed using a base reference period of 1985–2005. The regional standardized time series exhibited consistent trend patterns and statistical significance across the study region.

Temperature trend analysis revealed that 14 out of 21 stations exhibited significant warming in mean annual temperature. The highest warming trends were observed in Oman and the UAE, although some stations recorded localized cooling. Long-term analysis (1943–2008) confirmed significant warming in mean annual temperatures, with seasonal variations across different stations. Regional trends showed stronger warming in northern Arabian Peninsula compared to southern coastal areas. By categorizing the region into monsoonal and non-monsoonal subregions, the study revealed that warming was more pronounced in non-monsoonal areas. The annual mean temperature increased significantly across the Arabian Peninsula, with significant variations between subregions. The annual mean maximum temperature demonstrated substantial increases, particularly in non-monsoonal areas, with higher warming trends in the first half of the year. The annual mean minimum temperature trends were more pronounced than both mean and maximum temperatures, with significant warming observed across all seasons and peaks in specific months. A significant decrease in the annual mean DTR was observed across the Arabian Peninsula, particularly in the October–November (ON) season. Precipitation analysis indicated a general drying trend across the Arabian Peninsula, with annual total precipitation declining. While this trend was not statistically significant, it highlighted a potential shift in regional hydrological patterns. Sensitivity analysis was conducted by recalculating trends without data from Saudi Arabian stations, confirming that the temperature trends remained consistent, while precipitation trends were largely unaffected by station exclusions.

Key findings indicate that 14 out of 21 stations exhibited statistically significant warming in mean annual temperature at the 0.05 level, with most reaching 0.001 level. Only one station, Seeb, recorded a statistically significant cooling trend. The highest warming rates were observed in Oman (Sur: 1.03 °C per decade) and the UAE (Dubai: 0.81 °C per decade). The strongest seasonal warming was recorded in March–April, with Sur exhibiting the highest monthly mean temperature trend in May (1.47 °C per decade). Minimum temperature trends were statistically significant at 16 out of 21 stations, with the highest annual trends recorded in Dubai (1.24 °C per decade), Sohar (1.17 °C per decade) and Doha (1.13 °C per decade). The most pronounced warming for minimum temperatures was observed in October, with both Dubai and Kuwait reporting rates of 2.00 °C per decade. Maximum temperatures demonstrated significant increases at 12 out of 21 stations, with Bahrain reporting the highest monthly maximum temperature trend in March (2.27 °C per decade). The study also found a general decline in the diurnal temperature range, with six out of 21 stations exhibiting statistically significant negative trends. Only three stations recorded significant positive trends, suggesting a shift in regional temperature dynamics. Precipitation trends were less pronounced, with significant changes observed at only two stations, both of which reported decreasing trends. While the climatic conditions vary across the Arabian Peninsula, temperature trends showed significant spatial correlation within countries and subregions. A few stations had warming rates that were 1.5–3.5 °C times the global average.

The study underscores a consistent warming trend across the Arabian Peninsula, with 14 out of 21 stations indicating significant increases in mean annual temperatures. The increase in minimum temperatures was found to be greater than that of maximum temperatures, suggesting a trend toward enhanced nighttime warming. These findings highlight the importance of conducting regional climate change attribution studies to better understand the underlying drivers and associated impacts. The observed changes align with broader global concerns over climate variability and reinforce the need for informed adaptation and resilience strategies. The authors emphasize the importance of region-specific attribution analyses to further clarify the factors contributing to these climatic shifts. While global projections provide valuable guidance, the observed trends in this region suggest the potential for localized impacts that require closer analysis. Overall, the study encourages the advancement of climate monitoring and planning efforts across the Arabian Peninsula to support long-term sustainability and risk reduction.

Seasonal Trends in Rainfall and Temperature Across the Arabian Peninsula (1979–2009): Insights from Almazroui et al. (2012)

Almazroui et al. (2012) recently examined the seasonal climatology of the Arabian Peninsula by analysing regional- and station-scale climate variability for Saudi Arabia. Previous research has examined climate variability in Saudi Arabia using climate data from 1979 to 2009. They examined regional- and station-scale rainfall and temperature patterns during the wet season (November–April) and the dry season (June–September) using data from a combination of sources [surface observations, CPC Merged Analysis of Precipitation (CMAP), Climatic Research Unit (CRU) and the Tropical Rainfall Measuring Mission (TRMM)].

From the datasets it was shown that during the wet season Rub Al-Khali, the largest sand desert in the world, remained largely dry while belts of rainfall were observed to the north of 30 °N and south of 15 °N. It was also identified that during the dry season, the Arabian Peninsula was mostly arid to the north of 15 °N, with rainfall limited to regions south of this latitude. However, the location of heavy rainfall in the southwestern region of the peninsula was consistent, regardless of the dataset used. Temperature patterns showed that the highest wet season temperatures were recorded over the western region, while the hottest conditions in the dry season occurred across the central and northern parts. Analysis of surface observations highlighted a shift in rainfall trends over the study period. Between 1979 and 1993, rainfall exhibited an insignificant increase, followed by a significant decline from 1994 to 2009. The decrease was more pronounced during the wet season, with an average reduction of 35.1 mm per decade, compared to 5.5 mm per decade in the dry season. Meanwhile, temperature trends indicated a significant rise, with the dry season warming at a faster rate (0.72 °C per decade) compared to the wet season (0.51 °C per decade).

Further assessment of seasonal rainfall and temperature climatology, using both observed and gridded datasets, provided valuable insights into interannual and decadal climate trends over the Arabian Peninsula, particularly in Saudi Arabia. The spatial distribution of wet season rainfall (1979–2009 and 1998–2009) confirmed that the Rub Al-Khali experiences the least precipitation, while the southwest-to-northeast gradient in Saudi Arabia forms a relatively wetter zone (Fig. 3.4). In contrast, during the dry season, the northern Arabian Peninsula remains nearly rainless, with higher rainfall totals observed in the southern regions. Despite seasonal variations, the southwestern region of Saudi Arabia consistently receives significant rainfall, aligning with earlier studies. Temperature distribution patterns indicated that during the wet season, the highest temperatures were recorded in the western part of the Peninsula, while the lowest temperatures were observed in the northern Arabian Peninsula, particularly in the northwestern quadrant of Saudi Arabia and a small southwestern region extending into Yemen (Fig. 3.5). During the dry season, high temperatures prevailed across the most parts of Saudi Arabia, with fewer areas experiencing cooler conditions compared to the wet season.

Observational data from Saudi Arabia (1979–2009) showed that the average wet season rainfall was 66.8 mm, while the dry season received only 10.6 mm, contributing to an annual total of 93.5 mm. A significant decrease in rainfall was observed in recent decades (1994–2009), with reductions of 35.1 mm per decade during the wet season and 5.5 mm per decade during the dry season. Meanwhile, temperature trends demonstrated a clear warming pattern, with mean temperatures increasing at a rate of 0.51 °C per decade in the wet season and 0.72 °C per decade in the dry season. In addition, inland areas of the Arabian Peninsula experienced declining rainfall, an increase in precipitation was observed along the Red Sea coast and the southern Peninsula in recent decades compared to the 1980s, highlighting the spatial variability in climatic trends. This study serves as a critical reference for understanding past and present climatic variations across the Arabian Peninsula, providing a foundation for future projection and application-oriented research on regional climate dynamics. Apart from the studies discussed above, research on climate variability in the Arabian Peninsula has been carried out by various researchers over time. Several relevant studies highlighting key findings on current climate trends are presented in Table 3.2.

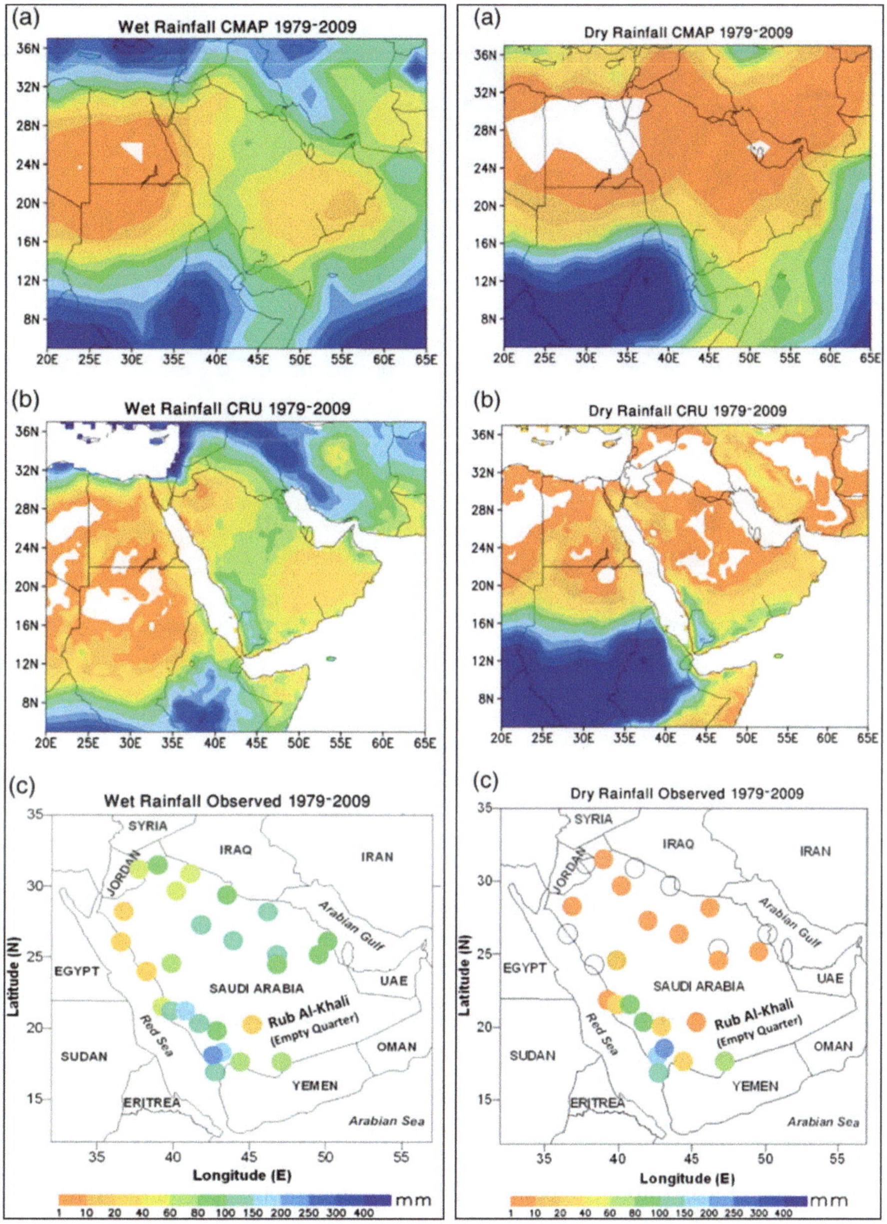

Fig. 3.4 Spatial distribution of rainfall during the wet and dry seasons (1979–2009) over the Arabian Peninsula, as analysed by Almazroui et al. (2012). *Source* Almazroui et al. (2012)

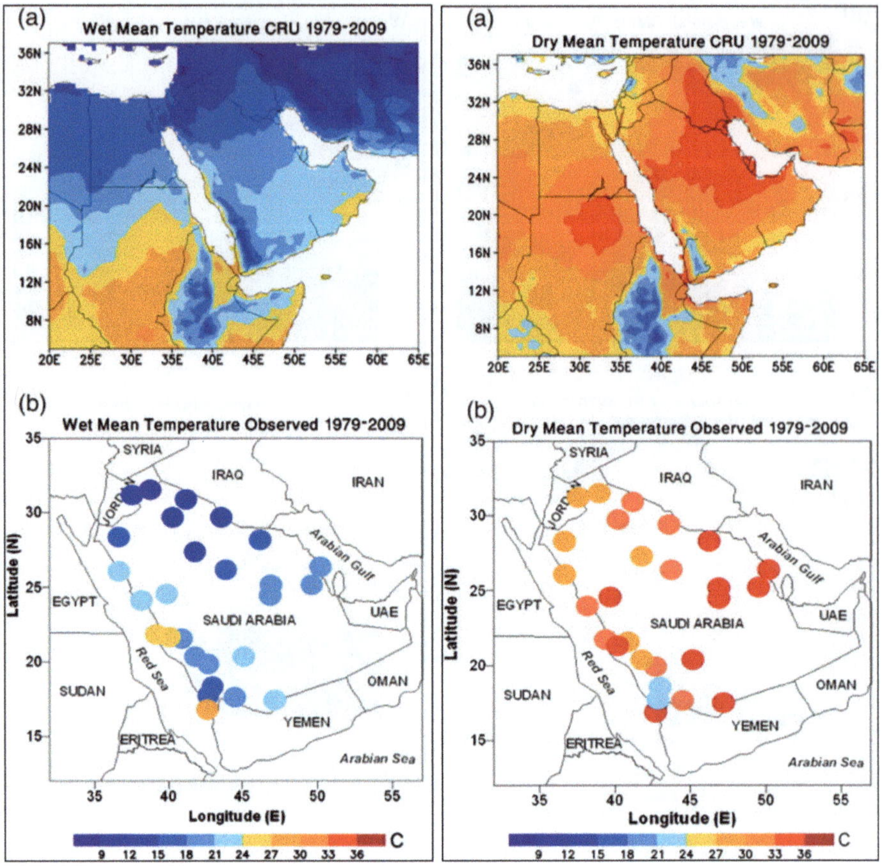

Fig. 3.5 Spatial distribution of mean temperature during the wet and dry seasons (1979–2009) over the Arabian Peninsula, as analysed by Almazroui et al. (2012). *Source* Almazroui et al. (2012)

3.5 Role of Historical Climate Trends in Understanding Climate Change

Studying historical climate trends plays a vital role in climate science. It helps scientists understand past climate patterns, which in turn deepens insights into current changes and guides future projections. This section explores several key reasons to focus on historical climate data. It includes validating climate models, guiding adaptive management strategies, supporting ecological balance, strengthening economic resilience, safeguarding social well-being, offering long-term perspectives and shaping effective policy decisions.

Table 3.2 Summary of key studies on climate variability in the Arabian Peninsula

Author(s) and year	Study period	Region covered	Data and methods used	Key findings
Mushtaq et al. (2025)	1960 to 2020	Arabian Peninsula	ERA5 reanalysis dataset from the European Centre for Medium-Range Weather Forecasts (ECMWF); Mann–Kendall test and Sen's slope estimator applied to high-resolution (0.5° × 0.5°) temperature and precipitation variables to assess the statistical significance and rate of long-term change	Warming trends are statistically significant across all countries, with minimum temperature (Tmin) increasing more rapidly than maximum temperature (Tmax). Iraq shows the highest rise in Tmax (+ 1.49 °C), while UAE records the largest Tmin increase (+ 2.91 °C). Saudi Arabia and Yemen also exhibit substantial Tmin warming. Precipitation trends are variable and mostly not significant, with declines in Iraq, Saudi Arabia, UAE and Yemen and slight increases in Bahrain, Kuwait, Qatar and Jordan. Seasonal warming is consistent for Tmax and Tmin ($p = 0.05$), except in Bahrain during DJF. Precipitation shows seasonal variability, with Kuwait's JJA rainfall being the only statistically significant increase

(continued)

Table 3.2 (continued)

Author(s) and year	Study period	Region covered	Data and methods used	Key findings
Alghamdi (2023)	1985–2022	Arabian Peninsula	Observational data from the Hadley Centre Integrated Surface Database (HadISD) and European Centre for Medium-Range Weather Forecasts (ECMWF) Reanalysis (ERA5-Land); hybrid hierarchical k-means clustering for seasonal classification; Mann–Kendall (MK) test and Sen's slope estimator for trend analysis	Summer is the longest season, lasting approximately 165 days, with its onset occurring in mid-April in the southern Arabian Peninsula and extending to late April in the northern regions. Winter is becoming shorter (92 days), while spring and autumn are the shortest (~ 52 days). Summer is starting earlier and becoming warmer, particularly at night. Winter is shortening, with increased daytime and nighttime temperatures. Spring is shifting earlier, with higher daytime warming. Autumn is starting later, but no significant temperature change was observed
Jaber et al. (2023)	1980–2018	Arab World (21 countries)	Climatologies at High Resolution for the Earth's Land Surface Areas (CHELSA) v2.1 high-resolution climate dataset; seasonal-Kendall trend analysis; Sen's slope estimator for temperature and precipitation trends; ArcGIS-based raster overlay analysis to identify vulnerable regions	Comprehensive assessment of climate change indicators over four decades. Nearly all areas exhibited significant warming trends. Maximum temperature increased by 0.027 to 0.714 °C/decade, with a mean of 0.318 °C/decade, while minimum temperature increased by 0.030 to 0.800 °C/decade, with a mean of 0.356 °C/decade. Precipitation trends were mixed, with most areas showing no significant change, while some regions experienced increases or decreases. Areas in Saudi Arabia, Somalia and Morocco showed the most significant rainfall reductions. Vulnerability hotspots were identified, with regions in Mauritania, Western Sahara, Libya, Jordan, Syria, Yemen and Saudi Arabia experiencing the highest climate risks

(continued)

Table 3.2 (continued)

Author(s) and year	Study period	Region covered	Data and methods used	Key findings
Elfeki et al. (2022)	1970–2012	Saudi Arabia	Monthly rainfall data from 28 meteorological stations across Saudi Arabia; statistical analysis including Kolmogorov–Smirnov (K–S) and Chi-Square tests for probability distribution; Empirical Orthogonal Function (EOF) analysis for dominant climate variability modes; Pearson correlation to assess teleconnections with large-scale atmospheric indices	The study examined the spatiotemporal variability of monthly rainfall and its relationship with global teleconnections. Rainfall variability is significantly influenced by El Niño-Southern Oscillation (ENSO), Pacific Decadal Oscillation (PDO) and Indian Ocean Dipole (IOD). Wet season (November–April) rainfall shows a strong correlation with ENSO and PDO, whereas dry season (June–September) rainfall is linked to IOD. Positive (warm) PDO and El Niño phases are associated with increased rainfall, while La Niña and negative (cold) PDO phases correspond to reduced rainfall
Syed et al. (2019)	1976–2100	Arabian Peninsula (Middle East and North Africa (MENA) & South Asia Domains)	Regional Climate Model (RCM) RCA4 downscaling of three Coupled Model Intercomparison Project Phase 5 (CMIP5) Global Climate Models (GCMs) [Centre National de Recherches Météorologiques–Climate Model 5 (CNRM-CM5), European Community Earth-System Model (EC-EARTH), Geophysical Fluid Dynamics Laboratory Earth System Model 2M (GFDL-ESM2M)]; observational validation using European Centre for Medium-Range Weather Forecasts (ECMWF) Reanalysis-Interim (ERA-Interim), Climatic Research Unit (CRU), Climate Prediction Center (CPC), Global Precipitation Climatology Centre (GPCC) and University of Delaware (UDEL) datasets; statistical trend analysis using Sen's slope estimator and Mann–Kendall (MK) test	Analyzed temperature and precipitation projections using the Rossby Centre Regional Climate Model (RCA4) under different Coordinated Regional Climate Downscaling Experiment (CORDEX) domains and lateral boundary conditions (LBCs). Temperature simulations were highly sensitive to domain selection (MENA vs. South Asia), while precipitation simulations were more influenced by LBCs. Future projections (2071–2100) under Representative Concentration Pathway 8.5 (RCP8.5) indicate up to 4.8 °C warming, with MENA simulations showing a generally warmer and drier climate than South Asia simulations. The study identified local biases, including a warm bias over eastern Saudi Arabia and a cold bias over the northwestern region

(continued)

Table 3.2 (continued)

Author(s) and year	Study period	Region covered	Data and methods used	Key findings
Subyani and Hajjar (2016)	1971–2012	Jeddah, Western Saudi Arabia	Daily rainfall data from six meteorological stations; statistical trend analysis using Mann–Kendall test; probability density function fitting using Kolmogorov–Smirnov test; spatial interpolation for missing data using inverse distance weighting (IDW)	Analysed rainfall variability, wet and dry spell durations and rainfall intensity in Jeddah. Identified three main rainy seasons: high-rain season (November–January), low-rain season (February–April, October) and dry season (May–September). The average dry spell duration is 80 days, while wet spells last an average of 1.39 days with an average rainfall intensity of 8.2 mm/day. Rainfall trends show increasing rainfall intensity in wet seasons and prolonged dry periods, aligning with global climate change patterns. Flood risk in Jeddah is increasing due to more intense rainfall events

(continued)

Table 3.2 (continued)

Author(s) and year	Study period	Region covered	Data and methods used	Key findings
Sherif et al. (2014)	1966–2010	United Arab Emirates (UAE)	Daily rainfall data from 42 stations (1966–2010) provided by the UAE Ministry of Environment and Water (MOEW); statistical techniques (Gumbel, Weibull and generalized extreme value distributions) for extreme rainfall analysis; drought severity index (DSI) estimation	The study analysed spatial and temporal rainfall variability across the UAE, dividing the region into four climate zones: East Coast, Mountains, Gravel Plains and Desert Foreland. Rainfall distribution showed high spatial variability, with the East Coast and Mountains receiving the highest rainfall, while the Desert Foreland had the lowest. Annual rainfall has significantly decreased since 1998, leading to prolonged drought conditions. The study estimated probable maximum precipitation (PMP) and developed intensity–duration–frequency (IDF) curves to support hydrological planning and flood risk assessment
Almazroui et al. (2012)	1979–2000	Arabian Peninsula (Northern & Southern Sub-Regions)	IPCC AR4 GCM simulations, Climate Prediction Center (CPC) Merged Analysis of Precipitation (CMAP), National Centers for Environmental Prediction/National Center for Atmospheric Research (NCEP/NCAR) sea level pressure data; statistical analysis including correlation tests, root mean square error (RMSE) and coefficient of variation (CV) to validate model performance	Interannual variability of rainfall was analyzed using 22 Global Climate Models (GCMs) from the Intergovernmental Panel on Climate Change (IPCC) Fourth Assessment Report (AR4) simulations. Only a subset of models accurately captured wet season rainfall in the northern sub-region and dry season rainfall in the southern sub-region. The Canadian Centre for Climate Modelling and Analysis (CCCMA) Coupled Global Climate Model (CGCM3.1-T47) performed best in simulating both seasons. The study confirmed that rainfall patterns are influenced by Mediterranean cyclones, the Indian monsoon and the Intertropical Convergence Zone (ITCZ), with the southwestern Arabian Peninsula receiving the most rainfall in both wet and dry seasons

(continued)

Table 3.2 (continued)

Author(s) and year	Study period	Region covered	Data and methods used	Key findings
Nasrallah and Balling (1996)	1891–1990	Arabian Peninsula	Historical climate data from Global Historical Climatology Network (GHCN); gridded temperature anomaly dataset; satellite-based lower tropospheric temperature data (1979–1991); long-term station-based monthly temperature and precipitation records; statistical trend analysis using standard deviation, z-score normalization and correlation tests	Analyzed long-term historical climate records to assess temperature and precipitation trends. The region experienced a linear temperature increase of 0.63 °C over 100 years, but this warming occurred primarily from 1911 to 1935. Over the most recent 50 years (1940–1990), the Arabian Peninsula has shown slight cooling rather than warming, despite the rapid buildup of greenhouse gases. Satellite-based lower tropospheric temperature measurements (1979–1991) did not show any significant warming. Unlike many regions experiencing decreasing diurnal temperature ranges, the Arabian Peninsula exhibited no long-term change in this parameter. Precipitation records showed a slight, statistically insignificant decrease over the past 40 years

Validating Climate Models

The validation of climate models is one reason to study historical climate trends. Climate models are complex tools that simulate the Earth's climate system and project future climatic conditions under various scenarios. These models rely on historical data for validation, as researchers assess model performance against observed records to evaluate their accuracy in replicating known climate behaviour. Since models are based on approximations and parameterizations, discrepancies between simulated and observed data often necessitate refinements in model structure or calibration of specific processes. This iterative process enhances the models' predictive capabilities and the robustness of future climate projections. Validating climate models is essential for planning and informed decision-making that supports societal resilience.

Informing Adaptive and Mitigation Strategies

Historical climate data are useful for developing informed adaptive and mitigation strategies. By understanding past climate changes, researchers can discover patterns and trends that shape responses to climate change today and into the future. This information is imperative to the development of strategies that promote a good ecological balance, strong economic base and a healthy society. For example, historical data can inform about specific regions that may be more susceptible to climate impacts and can promote informed targeted actions and resource allocation. The understanding of climate variability is also meaningful in the context of anticipating and preparing for climate changes that society may face in the future.

Maintaining Ecological Balance

Historical climate trends provide historical contexts in regard to understanding of how ecosystems have responded to past climate change. This is an important consideration in understanding the resilience of certain species and ecosystems in relation to climate change stressors. Examining past climate events, researchers are much better able to predict how ecosystems may respond to climate change in the recent and near futures, which is essential for conservation and biodiversity maintenance. Historical climate data can inform about changes in the distribution of certain species, changes in plant and animal phenology, changes in ecosystem structure and function, etc., which can all impact the ecological balance, which is useful when investigating and preserving the integrity of a system.

Ensuring Economic Stability

Climate change has significant economic impacts, affecting agriculture, infrastructure and overall economic health. Historical records of climate information can help identify trends related to agricultural productivity, such as shifts in precipitation and temperature extremes. This information is essential for farmers and policymakers in their decisions about crop selection, irrigation and disaster preparedness. Understanding climate variability historically allows businesses and productivity to collectively develop ways to minimise economic risks and learn to cope with climate change's impacts.

Promoting Societal Well-Being

Climate change has major implications for threats to human health and well-being. Historical climate studies can provide insight into how changes in climate have changed how diseases spread in the past or how illness associated with heat can lead to adverse health outcomes. Knowing the connections can help health professionals to mitigate the health impacts associated with climate change to benefit communities and public health. Historical climate records can also inform infrastructure plans and urban design based on climate hazards such as floods and heat waves.

Offering a Long-Term Perspective

By examining historical trends in climate, a long-term perspective on climate variability and change can be gained. Historical context is essential for comprehending natural climate cycles and the anthropogenic impacts that have emerged over recent decades. Linking present-day climate changes to their historical background allows for the assessment of the magnitude and significance of current trends, as well as the urgency associated with addressing climate change. Historical climate data also facilitates the evaluation of natural variability patterns, such as the El Niño–Southern Oscillation, in distinguishing human-induced influences from natural climate fluctuations.

Guiding Policy Development

Historical climate data is vital for policymakers in the science base for climate legislation and laws. By understanding climate change based on the margins of temperature and shifts moving forward, policymakers can make informed decisions based on historic data on greenhouse gas emissions, resilience and sustainable practices. Using climate trends from the historical context is particularly important for building public support for climate action and for the delivery of climate policies. Historical climate data can also support the contributions from individual nations in relation to global climate change.

3.6 Conclusion and Recommendations

The Arabian Peninsula has experienced considerable climate change in the last few decades due to a mixture of natural variation and human-induced change. The data presented in this chapter indicates a warming trend, especially in the driest season, with some areas of the region warming > 0.7 °C per decade. This warming trend has implications for regional climate stability, water resources and socio-economic sectors. Although the warming trends have been consistent across the region, precipitation has variable forms spatially and temporally. Some inland areas are experiencing reduced precipitation, whilst localised increases in precipitation are being seen along the Red Sea Coast and southwestern peninsula. The marked reduction or decline in wet season precipitation of 35.1 mm per decade and in dry season precipitation of 5.5 mm per decade infers a general trend to increased aridity in the region, with associated consequences for water security, agriculture and sustainability. The broader influence of global and regional drivers of climate upon variability in precipitation is seen. In particular, climatic events like ENSO (El Niño-Southern Oscillation), PDO (Pacific Decadal Oscillation), IOD (Indian Ocean Dipole) and NAO (North Atlantic Oscillation) all influence precipitation patterns, storm activity and drought cycles within the region. Apart from these climate events, Shamal and monsoonal influences also yield changes on a more localised scale and all impact long-term prediction uncertainties.

From a paleoclimatic viewpoint, past climate records suggest the Arabian Peninsula was considerably wetter in the past when there were expansive lakes and vegetation in most of the land currently characterised by arid regions. Over time, more progressive drying is acquired by greenhouse gas emissions, land-use changes and changes in atmospheric circulation that focus on enhancing climate-related stressors with human-induced desertification and water scarcity. Although progress has been made in climate science, some limitations still exist. One of the challenges is a scarcity of long-term, high-resolution climate data, especially for remote desert interiors. The limited availability of high-resolution data presents challenges in accurately reconstructing past climate trends across spatial and temporal scales. Additionally, regional climate models often face difficulties in capturing localized climatic variability, particularly in relation to extreme precipitation events and hyper-arid conditions. These limitations can contribute to uncertainties in localized future projections. Enhancing observational networks and refining climate models would greatly support the development of more reliable and context-specific climate projections.

To mitigate the challenges posed by rising temperatures and shifting precipitation patterns, the chapter underscores the necessity of proactive adaptation strategies. Strengthening climate monitoring systems through expanded meteorological networks and satellite observations will enhance climate data reliability. Urban planning should incorporate heat-resilient infrastructure, while sustainable water resource management must prioritize efficient groundwater extraction, desalination and conservation strategies. In the agricultural sector, adopting climate-smart practices, drought-resistant crops and advanced irrigation techniques will be essential for ensuring food security under increasingly arid conditions.

The Arabian Peninsula is experiencing growing climate-related challenges, highlighting the need for a comprehensive approach that integrates both adaptation and mitigation strategies. Future research efforts should prioritize high-resolution climate modeling, the formulation of transnational climate policies and the development of innovative water conservation strategies. As climate risks intensify, emphasis must be placed on strengthening climate resilience across the region by translating scientific insights into actionable policy. In summary, the region's climate change marked by rising temperatures, shifting precipitation patterns and heightened aridity poses growing risks to ecosystems, economies and human well-being. While the challenges are considerable, proactive adaptation strategies, sustainable management of natural resources and decisive political action can help avert the most severe impacts and pave the way for a climate-resilient future.

References

Alghamdi, A. S. (2023). Climatology and changes in temperature seasonality in the Arabian Peninsula. *Atmosphere, 15*(1), 26.

Almazroui, M., Islam, M. N., Jones, P. D., Athar, H., & Rahman, M. A. (2012). Recent climate change in the Arabian Peninsula: Seasonal rainfall and temperature climatology of Saudi Arabia for 1979–2009. *Atmospheric Research, 111*, 29–45.

AlSarmi, S., & Washington, R. (2011). Recent observed climate change over the Arabian Peninsula. *Journal of Geophysical Research: Atmospheres, 116*(D11).

Collins, M., Knutti, R., Arblaster, J., Dufresne, J.-L., Fichefet, T., Friedlingstein, P., Gao, X., Gutowski, W. J., Johns, T., Krinner, G., Shongwe, M., Tebaldi, C., Weaver, A. J., & Wehner, M. (2013). Long-term climate change: Projections, commitments and irreversibility. In T. F. Stocker, D. Qin, G.-K. Plattner, M. Tignor, S. K. Allen, J. Boschung, A. Nauels, Y. Xia, V. Bex, & P. M. Midgley (Eds.), *Climate change 2013: The physical science basis. Contribution of Working Group I to the fifth assessment report of the intergovernmental panel on climate change* (pp. 1029–1136). Cambridge University Press.

de Vries, A. J., Tyrlis, E., Edry, D., Krichak, S. O., Steil, B., & Lelieveld, J. (2013). Extreme precipitation events in the Middle East: Dynamics of the Active Red Sea Trough. *Journal of Geophysical Research: Atmospheres, 118*(13), 7087–7108.

Elfeki, A., Bahrawi, J., Latif, M., & Hannachi, A. (2022). Spatiotemporal analysis of monthly rainfall over Saudi Arabia and global teleconnections. *Geomatics, Natural Hazards and Risk, 13*(1), 2618–2648.

Folland, C. K., Karl, T. R., & Vinnikov, K. Y. (1990). Observed climate variations and change. *Climate Change: The IPCC Scientific Assessment, 195*, 238.

Freiwan, M., & Kadioğlu, M. (2008). Climate variability in Jordan. *International Journal of Climatology: A Journal of the Royal Meteorological Society, 28*(1), 69–89.

Goddard, L., Mason, S. J., Zebiak, S. E., Ropelewski, C. F., Basher, R., & Cane, M. A. (2001). Current approaches to seasonal to interannual climate predictions. *International Journal of Climatology: A Journal of the Royal Meteorological Society, 21*(9), 1111–1152.

Hamdi, M. R., Abu-Allaban, M., Elshaieb, A., Jaber, M., & Momani, N. M. (2009). Climate change in Jordan: A comprehensive examination approach.

Houghton, J. T., Callander, B. A., & Varney, S. K. (Eds.). (1992). *Climate change 1992.* Cambridge University Press.

Hulme, M. (1996). Recent climatic change in the world's drylands. *Geophysical Research Letters, 23*(1), 61–64.

IPCC. (2023). Summary for policymakers. In *Climate change 2023: Synthesis report. Contribution of Working Groups I, II and III to the sixth assessment report of the intergovernmental panel on climate change* [Core Writing Team, H. Lee, & J. Romero (Eds.)] (pp. 1–34). IPCC. https://doi.org/10.59327/IPCC/AR6-9789291691647.001

Jaber, S. M., Abu-Allaban, M. M., & Sengupta, R. (2023). Spatial and temporal patterns of indicators of climate change and variability in the Arab world in the past four decades. *Scientific Reports, 13*(1), 15145.

Jansen, E., Overpeck, J., Briffa, K. R., Duplessy, J.-C., Joos, F., Masson-Delmotte, V., Olago, D., Otto-Bliesner, B., Peltier, W. R., Rahmstorf, S., Ramesh, R., Raynaud, D., Rind, D., Solomina, D., Villalba, R., & Zhang, D. (2007). Palaeoclimate. In S. Solomon, D. Qin, M. Manning, Z. Chen, M. Marquis, K. B. Averyt, M. Tignor, & H. L. Miller (Eds.), *Climate change 2007: The physical science basis. Contribution of Working Group I to the fourth assessment report of the intergovernmental panel on climate change* (pp. 433–497). Cambridge University Press.

Jones, P. D., & Reid, P. A. (2001). Temperature trends in regions affected by increasing aridity/humidity. *Geophysical Research Letters, 28*(20), 3919–3922.

Kotwicki, V., & Al Sulaimani, Z. (2009). Climates of the Arabian Peninsula–past, present, future. *International Journal of Climate Change Strategies and Management, 1*(3), 297–310.

Kwarteng, A. Y., Dorvlo, A. S., & Vijaya Kumar, G. T. (2009). Analysis of a 27-year rainfall data (1977–2003) in the Sultanate of Oman. *International Journal of Climatology, 29*(4), 605.

Lioubimtseva, E. (2004). Climate change in arid environments: Revisiting the past to understand the future. *Progress in Physical Geography, 28*(4), 502–530.

Madhukar, A., Dashora, K., & Kumar, V. (2021). Spatial analysis of yield trends and impact of temperature for wheat crop across Indian districts. *International Journal of Plant Production, 15*(2), 325–335.

McClure, H. A. (1994). A new Arabian stone tool assemblage and notes on the Aterian industry of North Africa. *Arabian Archaeology and Epigraphy, 5*(1), 1–16.

Mofidi, A., & Zarrin, A. (2022). On the existence of summer Shamal wind induced by the Zagros Mountains in the Middle East. *Geophysical Research Letters, 49*(18), e2022GL100151.

Mushtaq, F., Haq, A. U., Bharti, S., Alhems, L. M., & Farooq, M. (2025). *Regional climate dynamics in the Arabian Peninsula: A study of temperature trends and precipitation uncertainty* (No. EGU25-3014). Copernicus Meetings.

Nasrallah, H. A., & Balling, R. C. (1996). Analysis of recent climatic changes in the Arabian Peninsula region. *Theoretical and Applied Climatology, 53*, 245–252.

Scholz, C. A., Johnson, T. C., Cohen, A. S., King, J. W., Peck, J. A., Overpeck, J. T., Talbot, M. R., Brown, E. T., Kalindekafe, L., Amoako, P. Y. O., Lyons, R. P., Shanahan, T. M., Castañeda, I. S., Heil, C. W., Forman, S. L., McHargue, L. R., Beuning, K. R., Gomez, J., & Pierson, J. (2007). East African megadroughts between 135 and 75 thousand years ago and bearing on early-modern human origins. *Proceedings of the National Academy of Sciences, 104*(42), 16416–16421.

Sherif, M., Almulla, M., Shetty, A., & Chowdhury, R. K. (2014). Analysis of rainfall, PMP and drought in the United Arab Emirates. *International Journal of Climatology, 34*(4).

Solomon, S. (Ed.). (2007). *Climate change 2007—The physical science basis: Working group I contribution to the fourth assessment report of the IPCC* (Vol. 4). Cambridge University Press.

Stern, R. J., Avigad, D., Miller, N. R., & Beyth, M. (2006). Evidence for the snowball Earth hypothesis in the Arabian-Nubian Shield and the east African Orogen. *Journal of African Earth Sciences, 44*(1), 1–20.

Subyani, A. M., & Hajjar, A. F. (2016). Rainfall analysis in the contest of climate change for Jeddah area, Western Saudi Arabia. *Arabian Journal of Geosciences, 9*, 1–15.

Syed, F. S., Latif, M., Al-Maashi, A., & Ghulam, A. (2019). Regional climate model RCA4 simulations of temperature and precipitation over the Arabian Peninsula: Sensitivity to CORDEX domain and lateral boundary conditions. *Climate Dynamics, 53*(11), 7045–7064.

Tyrlis, E., Lelieveld, J., & Steil, B. (2013). The summer circulation over the eastern Mediterranean and the Middle East: Influence of the South Asian monsoon. *Climate Dynamics, 40*, 1103–1123.

Wang, P. X., Wang, B., Cheng, H., Fasullo, J., Guo, Z., Kiefer, T., & Liu, Z. (2017). The global monsoon across time scales: Mechanisms and outstanding issues. *Earth-Science Reviews, 174*, 84–121.

Wigley, T. M. L., Barnett, T. P., & Golitsyn, G. S. (1990). Detection of greenhouse effect in the observations. In *Climate change: The IPCC scientific assessment* (pp. 239–256). Cambridge University Press.

Xu, Z., Tang, Y., Connor, T., Li, D., Li, Y., & Liu, J. (2017). Climate variability and trends at a national scale. *Scientific Reports, 7*(1), 3258.

Yu, Y., Notaro, M., Kalashnikova, O. V., & Garay, M. J. (2016). Climatology of summer Shamal wind in the Middle East. *Journal of Geophysical Research: Atmospheres, 121*(1), 289–305.

Zhang, X., Aguilar, E., Sensoy, S., Melkonyan, H., Tagiyeva, U., Ahmed, N., Kutaladze, N., Rahimzadeh, F., Taghipour, A., Hantosh, T. H., Albert, P., & Wallis, T. (2005). Trends in Middle East climate extreme indices from 1950 to 2003. *Journal of Geophysical Research: Atmospheres, 110*(D22).

Chapter 4
Future Climate Projections Over the Arabian Peninsula: Insights from CMIP6 Models

Fayma Mushtaq, Gowhar Meraj, and Mansour Almazroui

4.1 Introduction

Climate projections are simulations that estimate Earth's future climate over the coming decades, typically extending until 2100 (Lemos & Rood, 2010). These projections are based on different scenarios, which involve assumptions about future concentrations of greenhouse gases, aerosols and other atmospheric constituents that influence the planet's radiative balance (Roeckner, 1992). To generate these projections, numerical models are used to simulate the interactions between various components of the Earth's climate system. These models, called Global Climate Models (GCMs) or Regional Climate Models (RCMs), may cover the entire globe or focus on specific regions, such as Europe (Rummukainen, 2010). At the core of climate projections is the use of a GCM, which is a three-dimensional computer representation of the Earth's climate system (Hansen et al., 1988). These models simulate the fundamental processes driving weather and climate by solving mathematical equations that represent the physical laws governing the atmosphere, ocean, land and ice. Climate models are highly complex and require powerful computing resources to perform the necessary calculations. In a GCM, the globe is divided

F. Mushtaq (✉)
Applied Research Centre for Environment and Marine Studies, King Fahd University of Petroleum and Minerals, Dhahran, Saudi Arabia
e-mail: fayma.mushtaq@kfupm.edu.sa

G. Meraj
Department of Ecosystem Studies, Graduate School of Agricultural and Life Sciences, The University of Tokyo, Tokyo, Japan

M. Almazroui
Department of Meteorology, Center of Excellence for Climate Change Research, King Abdulaziz University, Jeddah, Saudi Arabia

Climatic Research Unit, School of Environmental Sciences, University of East Anglia, Norwich, UK

F. Mushtaq et al. (eds.), *Climate Change in the Arabian Peninsula*,
https://doi.org/10.1007/978-3-032-02481-7_4

into a grid of smaller boxes, typically ranging from 100 to 400 km on each side, with about 40 vertical layers that span from the surface to the top of the atmosphere (Skelly & Henderson-Sellers, 1996). The model calculates climate variables, such as temperature, rainfall and wind, within each grid box. These calculations are based on fundamental physical principles, including the conservation of mass, energy and momentum. The processes that are modeled in each box include solar radiation, ocean currents and other interactions that shape the Earth's climate system.

In addition to GCMs, RCMs provide more localized climate projections. RCMs focus on specific regions, offering higher resolution simulations of climate processes compared to GCMs (Giorgi, 2019). These models are based on the outputs of GCMs and downscale global projections to capture regional variations more accurately. RCMs typically operate at a spatial resolution of 10–50 km significantly finer than the 100–250 km grid spacing of GCMs which allows them to better represent complex terrain, coastlines and land–atmosphere interactions (Gutowski et al., 2020). RCMs are particularly useful for assessing climate impacts on smaller scales, such as regional weather patterns, local ecosystems and specific sectors like agriculture, water resources and urban development. RCMs provide enhanced detail on temperature, precipitation and wind patterns in a given region, which is essential for more targeted adaptation strategies. By running these models under different greenhouse gas emission scenarios, scientists can project future climate conditions and provide insights into potential changes in temperature, precipitation and other climate factors.

Projections of climate change from GCMs form a crucial foundation for planning mitigation, adaptation and risk management strategies across various sectors of society and ecosystems worldwide. These projections offer essential insights into the potential future climate conditions, allowing policymakers, industries and communities to make informed decisions about how to address and prepare for the impacts of climate change. Depending on the specific needs of a project or region, scientists can choose from a variety of climate modeling projection sets, each with distinguishing factors such as spatial resolution, time frame, emission scenarios and regional focus. This diversity allows for tailored projections that best suit the context of the application, whether it is related to infrastructure planning, agriculture, water management, public health, or disaster risk reduction. Climate projections are especially important for mitigation, which involves reducing the sources of greenhouse gas emissions to limit future warming (Kumar, 2018). By understanding the potential range of future climate scenarios, decision-makers can identify which emission pathways will lead to acceptable outcomes, determine the urgency of emission reduction efforts and prioritize actions that can prevent the most severe climate impacts. For example, if projections indicate an increase in extreme weather events, mitigation strategies can be focused on reducing carbon footprints and investing in renewable energy sources to limit the extent of global warming.

Similarly, adaptation strategies rely heavily on accurate climate projections (Dessai & Hulme, 2004). Since climate change impacts are already being felt, adapting to these changes is necessary for ensuring resilience in vulnerable communities, ecosystems and economies (Asmamaw et al., 2015). Projections help identify specific areas that will be most affected by climate change, such as rising sea levels, more frequent and severe droughts, or shifts in agricultural growing seasons. This information enables the development of region-specific adaptation plans, such as strengthening coastal infrastructure, improving water management practices, or adjusting agricultural techniques to cope with changing weather patterns. In addition to mitigation and adaptation, climate projections are essential for risk management (Deser, 2020). Understanding the potential risks posed by climate change allows governments and industries to prepare for and reduce vulnerabilities. By predicting future climate extremes, projections help prioritize investments in climate resilience, emergency response systems and long-term infrastructure planning.

While the future of climate change will likely be shaped by a combination of mitigation, adaptation and risk management efforts, the distinctions between these areas influence the choice of which climate projection set to use. The right set of projections can significantly impact the effectiveness of climate action, guiding the allocation of resources and shaping policies that will enable societies to navigate the challenges of a changing climate. Therefore, this chapter is presents future climate projections for the Arabian Peninsula based on Sixth Phase of the Coupled Model Intercomparison Project (CMIP6) data, at spatial resolution of 0.25° × 0.25° (approximately 25 km × 25 km) for temperature and precipitation.

4.2 Dataset Used and Methods Adopted

This study integrates historical climate observations and future projections to assess changing temperature and precipitation trends across the Arabian Peninsula. Historical climate data were primarily obtained from the Climatic Research Unit Time Series (CRU TS) version 4.05, developed by the University of East Anglia. This gridded dataset provides monthly values of surface air temperature and precipitation at a spatial resolution of 0.5° × 0.5°, covering the period from 1950 to 2014. To complement these observations and better capture atmospheric dynamics, the ERA5 reanalysis dataset from the European Centre for Medium-Range Weather Forecasts (ECMWF) was also utilized. ERA5 offers hourly climate variables at a finer resolution of 0.25° × 0.25°, derived from both observational inputs and numerical weather prediction models and was instrumental in characterizing variability and extremes during the historical baseline period.

For projecting future climatic conditions, this study relied on simulations from the CMIP6. A multi-model ensemble of GCMs was selected to represent a range of possible climate futures under different Shared Socioeconomic Pathways (SSPs), including SSP1-1.9, SSP1-2.6, SSP2-4.5, SSP3-7.0 and SSP5-8.5 for maximum surface air temperature, minimum surface air temperature and total precipitation. Additional data and graphical outputs used for scenario comparison and regional analysis were sourced from the World Bank Climate Change Knowledge Portal (CCKP, 2023), which provided relevant baseline data, future projections and visualization tools supporting this assessment. Climate metrics were evaluated with a specific focus on the mid-century (2041–2060) and end-century (2081–2100) periods and compared against the historical baseline (1995–2014), to assess the magnitude of projected changes under different emissions scenarios (UNESCWA, 2022). The combination of robust data sources, ensemble-based methodology, ensures a scientifically rigorous basis for interpreting climate risks and informing adaptation planning in the region.

4.3 Overview of the Global Climate Models (GCMs) in Arabian Peninsula

GCMs serve as fundamental tools for simulating climate dynamics, helping researchers understand regional climate patterns and future projections (Lau, 1992). The Arabian Peninsula, with its complex climatic variability, has been extensively studied using GCMs, particularly those from the Coupled Model Intercomparison Project (CMIP) (Barfus & Bernhofer, 2014). Recent advancements in climate modeling, especially in CMIP6, have improved the accuracy of climate simulations by incorporating enhanced parameterization and finer resolution (Almazroui et al., 2020; Zhu et al., 2020). Despite these improvements, the coarse spatial resolution of GCMs presents challenges in accurately representing the local climate processes of the Arabian Peninsula. To address this limitation, RCMs such as RegCM4 and WRF have been used to downscale GCM projections, offering more localized and refined climate data (Ajay et al., 2019; Almazroui et al., 2016). These models significantly enhance the representation of rainfall patterns, temperature variations and extreme climate events by integrating local topographical and atmospheric influences.

Several GCMs have been assessed for their reliability in modeling the Arabian Peninsula's climate (Table 4.1). Among these, the CCCMA-CGCM3 (T47) model has been identified for its strong ability to simulate rainfall distribution across both wet and dry seasons, making it particularly useful for hydrological studies (Almazroui et al., 2013). The CORDEX framework, which incorporates models such as CNRM-CM5, EC-EARTH and GFDL-ESM2M, has been instrumental in regional-scale climate modeling, providing insights into how domain selection and lateral boundary conditions influence climate projections (Syed et al., 2019a, 2019b). Comparative analyses between CMIP5 and CMIP6 models reveal that CMIP6 exhibits improved climate sensitivity, capturing more pronounced changes in temperature, precipitation and potential evapotranspiration (PET), which are crucial for understanding drought risks and water availability (Saharwardi et al., 2024). Additionally, CMIP5's 30-member ensemble simulations have demonstrated their reliability in projecting long-term climate trends under different emission scenarios (Almazroui, 2018). Downscaling techniques especially those employing RegCM4 driven by CMIP5 model outputs have significantly improved the representation of temperature and precipitation extremes, thereby enhancing assessments of climate variability across the Arabian Peninsula (Almazroui, 2019). However, despite these advancements, model biases remain a concern. Many GCMs and RCMs struggle to accurately simulate

Table 4.1 Examples of GCMs used in the Arabian Peninsula

Model name	Description and usage	References
CMIP6 models	Utilized to assess potential evapotranspiration changes and their implications for water resources	Ajjur and Al-Ghamdi (2022)
RegCM4	Regional climate model driven by CMIP5 outputs for the assessment of temperature and precipitation extremes	Almazroui (2019)
CNRM-CM5	Used in the Coordinated Regional Climate Downscaling Experiment (CORDEX) framework for downscaling to study temperature and precipitation patterns	Syed et al. (2019a, 2019b)
EC-EARTH	This model also used CORDEX framework, focusing on the effects of domain and lateral boundary conditions on climate simulations	Syed et al. (2019a, 2019b)
GFDL-ESM2M	Part of the CORDEX framework, used for downscaling to assess regional climate impacts	Syed et al. (2019a, 2019b)
CMIP5 models	A 30-member ensemble used to project surface air temperature changes under RCP4.5 and RCP8.5 scenarios	Almazroui (2018)
CCCMA-CGCM3 (T47)	Found to be relatively better in simulating rainfall climatology for both wet and dry seasons across the northern and southern sub-regions of the Arabian Peninsula	Almazroui et al. (2013)

historical precipitation trends, leading to uncertainties in future climate projections (Almazroui et al., 2013; Saharwardi et al., 2024). The selection of appropriate models, ensemble-based approaches and continuous evaluation using observational data remains crucial in improving the reliability of climate projections for the region. While CMIP6 models provide more refined climate projections compared to their predecessors, multi-model ensembles and regional downscaling remain essential for capturing the Arabian Peninsula's complex climate variability. The use of CCCMA-CGCM3, CORDEX models and high-resolution downscaled simulations plays a critical role in enhancing the accuracy of climate assessments, supporting climate adaptation and resource management strategies in the region.

4.4 Overview of Regional Climate Models (RCM) in the Arabian Peninsula

RCMs are essential tools for understanding and predicting climate dynamics in the Arabian Peninsula, a region characterized by extreme temperatures, aridity and significant climatic variability (Mushtaq et al., 2025; Patlakas et al., 2021). These models play a critical role in downscaling GCMs to provide high-resolution climate projections, which are essential for assessing localized climate impacts such as extreme heat events, precipitation variability and dust storms (Guo & Wang, 2016). Given the region's complex topography and evolving climate challenges, RCMs have become indispensable for climate impact assessments, adaptation planning and risk management (Ajjur & Al-Ghamdi, 2023). Several RCMs have been applied across the Arabian Peninsula to analyze different aspects of climate variability and change (Table 4.2). Among them, the Weather Research and Forecasting (WRF) model, particularly when enhanced with spectral nudging, has been instrumental in improving simulations of the Arabian Peninsula's summer climate (Attada et al., 2021). It helps reduce cold biases commonly observed in climate models and enhances the representation of synoptic features such as heatwaves, which are critical for understanding extreme summer conditions in the region. Additionally, the RCA4 model, used within the Coordinated Regional Climate Downscaling Experiment (CORDEX) framework, has highlighted the sensitivity of temperature simulations to the selected domain. Studies show that the Middle East and North Africa (MENA) domain projects a warmer and drier future compared to the South Asia (SA) domain, underscoring the significance of lateral boundary conditions (LBCs) in shaping precipitation projections (Syed et al., 2019a, 2019b).

A high-resolution atmospheric model has also been applied to study regional climatology, effectively capturing the spatial and temporal variations in temperature, wind and humidity across the Arabian Peninsula. It indicates positive trends in temperature extremes, particularly in the northwestern regions, suggesting that these areas may experience intensified heat stress in the future (Patlakas et al.,

Table 4.2 Example application of regional climate models in the Arabian Peninsula

Application area	Model used	Key findings	References
Summer climate simulation	WRF model with spectral nudging	Enhanced simulation of summer climate, reduced cold biases and better representation of synoptic features like heat waves	Attada et al. (2021)
Regional climatology	High-resolution atmospheric model	Good agreement with in situ measurements; positive trends in temperature extremes, especially in northwestern regions	Patlakas et al. (2019)
Temperature and precipitation sensitivity	RCA4 model	Sensitivity to CORDEX domain and lateral boundary conditions (LBCs); the Middle East and North Africa (MENA) domain projects warmer and drier conditions compared to the South Asia (SA) domain	Syed et al. (2019a, 2019b)
Future wind speed changes	WRF model	Predicted weakening of regional pressure gradient, leading to reduced wind speeds and fewer dust storms	Alsarraf and Broeke (2015)

2019). The RegCM model, another widely used RCM, has been particularly effective in simulating dust storms, a major environmental and health concern in the region. The model has been validated against satellite observations and other climate models, demonstrating its reliability in replicating dust storm dynamics (Mobarak et al., 2023). In terms of long-term climate projections, the WRF model suggests a substantial increase in summer surface temperatures by mid-century, further intensifying extreme heat events across the region. Additionally, simulations indicate a weakening of the regional pressure gradient, which could lead to a reduction in average wind speeds and a potential decline in the frequency of dust storms. These changes carry significant implications for human health, transportation and air quality management.

The COSMO-CLM model, known for its high-resolution capabilities, is frequently used for studying land surface interactions and urban climate dynamics, both of which are crucial for understanding climate change impacts on rapidly urbanizing areas in the Arabian Peninsula (Russo et al., 2020). While specific applications of COSMO-CLM in this region are still developing, it remains a valuable tool for ensemble climate projections and paleoclimate studies. The selection of a specific RCM depends on the research objectives, computational resources and the desired resolution of climate projections. Despite continuous advancements in regional climate modeling, uncertainties remain due to factors such as model parameterization, selection of driving GCMs and the accuracy of observational datasets used for validation. Therefore, continuous model improvements, validation efforts and the integration of ensemble-based approaches are essential for enhancing the reliability of climate projections in the Arabian Peninsula. Overall, RCMs provide valuable insights into the evolving climate of the Arabian Peninsula, aiding in climate-resilient planning and adaptation strategies. As climate modeling capabilities continue to advance, these models will play an increasingly vital role in supporting evidence-based policymaking, ensuring sustainable development in a region highly vulnerable to climate change impacts.

4.5 Projected Temperature Changes and Precipitation Patterns

Understanding future climate challenges is crucial for strategic planning, policymaking and disaster risk reduction across Saudi Arabia and the broader Arabian Peninsula. The region's arid and semi-arid climatic conditions make it particularly vulnerable to increasing temperatures, altered precipitation patterns and more frequent extreme climate events. Previous studies have highlighted that the Arabian Peninsula is warming at a rate approximately twice the global average, with some areas already experiencing a 1.5 °C rise since the pre-industrial era (Vinodhkumar et al., 2024). This trend is projected to continue, necessitating robust climate projections to inform adaptation and mitigation strategies. Recent advances in climate modeling have enabled a more refined understanding of future climatic scenarios through RCMs, CMIP5 and CMIP6 simulations. These high-resolution climate models, including Providing Regional Climates for Impacts Studies (PRECIS), CORDEX initiative for the Arabian Peninsula (CORDEX-AP) and downscaled CMIP6 projections, offer localized insights into future temperature and precipitation trends. For Example, high-resolution regional simulations driven by CMIP5 models, projections under the RCP8.5 scenario suggest that parts of the Arabian Peninsula particularly over central and eastern desert regions could experience temperature increases of up to 7.6 °C by the end of the twenty-first century (Malik et al., 2024).

The Arabian Peninsula, with its predominantly arid climate and socio-economic sensitivities, is increasingly susceptible to evolving climate challenges driven by anthropogenic influences. This study employs CMIP6 multi-model ensemble projections across five SSPs viz to explore future climate trajectories in terms of temperature and precipitation patterns. Country-wise projected trends are discussed in the following sections with focus on mid and end century projections under SSP2-4.5 and SSP5-8.5 scenarios These scenarios offer a spectrum of possible futures ranging from very low to very high greenhouse gas emissions, providing a comprehensive basis for assessing climate outlooks across the region.

Saudi Arabia: Multi-model climate projections for Saudi Arabia point to a robust and progressive warming trend across all emissions scenarios throughout the twenty-first century (Fig. 4.1). The baseline climatology (1995–2014) indicates an average annual maximum temperature of 32.3 °C, mean temperature of 25.5 °C and minimum temperature of 18.7 °C, while the annual precipitation stands at 58.3 mm/year. Under the SSP2-4.5 scenario, which represents a moderate mitigation pathway, Saudi Arabia is projected to experience an increase of 1.7 °C in maximum temperature (reaching 34.0 °C), 1.8 °C in mean temperature (27.3 °C) and 1.8 °C in minimum temperature (20.5 °C) relative to the baseline. Annual precipitation is expected to rise moderately to 64.7 mm/year, an increase of 6.4 mm, indicating potential for slight wetting tendencies in parts of the region. In the SSP5-8.5 high-emissions scenario, warming is more pronounced; maximum temperatures are projected to reach 34.7 °C (+ 2.4 °C), mean temperatures 28.0 °C (+ 2.5 °C) and minimum temperatures 21.3 °C (+ 2.6 °C). Precipitation levels are expected to increase to 66.7 mm/year, representing a gain of 8.4 mm from the baseline, possibly indicating increased convective rainfall episodes under warmer atmospheric conditions. By the end of the century, differences between emission trajectories become markedly clearer. Under SSP2-4.5, the projected maximum temperature increases to 35.1 °C (+ 2.8 °C), mean temperature to 28.3 °C (+ 2.8 °C) and minimum temperature to 21.6 °C (+ 2.9 °C). Annual precipitation is estimated at 66.4 mm/year, up 8.1 mm from baseline levels. In contrast, SSP5-8.5 projects substantial warming, with maximum temperatures reaching 37.8 °C, mean temperatures 31.2 °C and minimum temperatures 24.6 °C, reflecting respective increases of 5.5, 5.7 and 5.9 °C above the historical reference. Precipitation is projected to rise to 74.3 mm/year, indicating a cumulative increase of 16.0 mm.

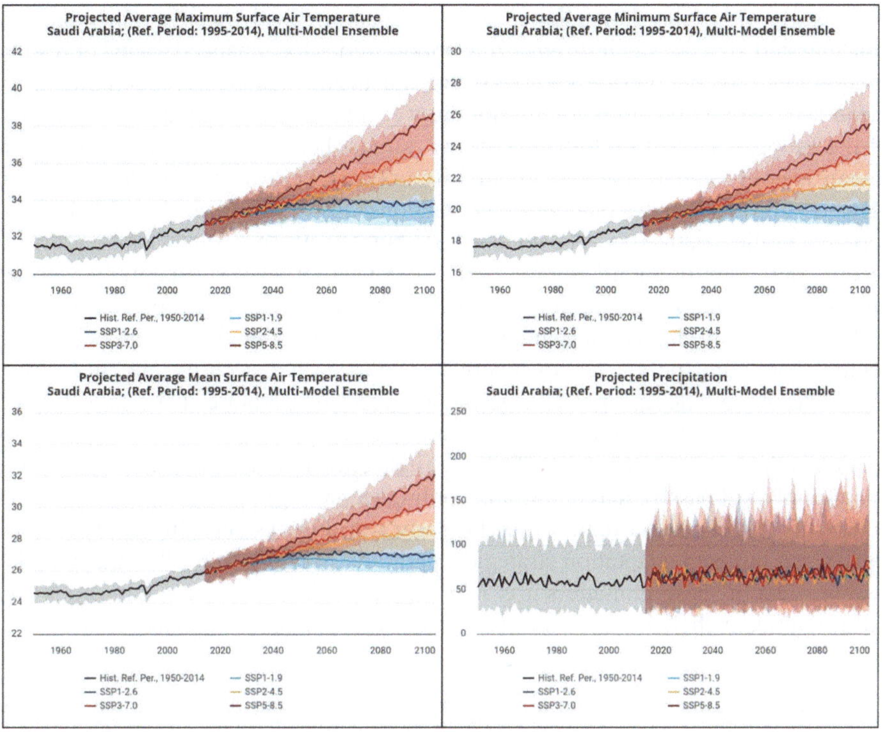

Fig. 4.1 Projected changes in maximum, minimum, mean surface air temperatures and precipitation for Saudi Arabia under different SSP scenarios, based on a multi-model ensemble relative to the 1995–2014 reference period

These results suggest that Saudi Arabia is likely to face intensifying heat stress particularly in summer and nighttime temperatures across all emission trajectories. Although precipitation shows a moderate increasing trend, especially under SSP5-8.5, the implications for hydrological stability remain uncertain due to possible increases in interannual variability and extreme precipitation events. These climatic shifts underscore the necessity for proactive adaptation strategies centered on sustainable water use, urban heat resilience and climate-informed infrastructure planning tailored to the region's arid and highly sensitive environment. Supporting these findings, Sharif (2015) projected widespread positive temperature anomalies across Saudi Arabia, with the most intense warming focused over the central interior regions. Their study revealed a clear progression toward a substantially hotter climate from the mid-century to end-century period, particularly under high-emission scenarios. Location-specific analyses at Jeddah, Riyadh, Dammam and Jazan indicated that Riyadh, situated on the central plateau, is likely to experience the greatest increase in temperatures, identifying it as a key hotspot for future thermal stress across the Kingdom. Komurcu et al. (2020) further reported a projected intensification in both the frequency and magnitude of precipitation events during August by mid-century, particularly along the mountainous western coast of Saudi Arabia. This spatially concentrated increase suggests emerging opportunities for enhanced water harvesting in those regions. In contrast, their analysis indicates a significant reduction in both mean and extreme precipitation during the wet season across the northern margins of the Empty Quarter. Additionally, rising August heat index values are expected to considerably impair regional habitability in coastal cities such as Jeddah, where elevated humidity and temperature interactions may exacerbate thermal discomfort and public health risks. The concurrence between the present ensemble-derived results and prior literature reinforces the expectation of a substantially warmer and more hydrologically variable climate in Saudi Arabia. The projected thermal and hydrological changes are anticipated to impose significant stress on critical sectors, including public health, energy systems, water security and infrastructure. The magnitude and persistence of projected temperature increases, coupled with the growing uncertainty in precipitation regimes, underscore the urgency of implementing adaptive measures. These include climate-responsive urban planning, enhancement of passive and renewable cooling technologies, investment in resilient water infrastructure and region-specific early warning systems. In the absence of timely and targeted adaptation strategies, particularly under SSP5-8.5, the cumulative impacts of these climatic shifts may surpass existing adaptive capacities and pose significant challenges to sustainable development across the Kingdom.

United Arab Emirates (UAE): Multi-model ensemble projections indicate that the UAE will experience substantial and consistent warming across all temperature indices throughout the twenty-first century, with the most pronounced increases occurring under the high-emissions SSP5-8.5 scenario (Fig. 4.2). During the baseline period (1995–2014), the annual average maximum temperature was 34.1 °C, mean temperature 27.8 °C and minimum temperature 21.5 °C, while the annual precipitation averaged 42.5 mm/year. By the mid-century period (2041–2060) under SSP2-4.5, the UAE is projected to experience an increase of 1.6 °C in both maximum and mean temperatures, reaching 35.7 °C and 29.5 °C, respectively. The minimum temperature is expected to rise by 1.7 °C to 23.3 °C. Annual precipitation is projected to increase moderately to 46.3 mm/year, reflecting a gain of 3.8 mm. Under SSP5-8.5 for the same period, warming intensifies, with maximum, mean and minimum temperatures reaching 36.3 °C, 30.1 °C and 23.9 °C, respectively each showing increases of approximately 2.2–2.4 °C. Precipitation under this scenario is slightly higher at 47.4 mm/year, an increase of 4.9 mm above the baseline. By the end of the century (2081–2100), the disparity between emissions pathways becomes more pronounced. Under SSP2-4.5, maximum temperature is projected to rise to 36.6 °C (+ 2.5 °C), mean temperature to 30.5 °C (+ 2.6 °C) and minimum temperature to 24.3 °C (+ 2.7 °C). Precipitation under this scenario decreases slightly to 44.8 mm/year, though it remains 2.3 mm above the baseline. In contrast, the SSP5-8.5 scenario projects a more substantial warming trend, with end-century maximum, mean and minimum temperatures reaching 39.2 °C, 33.1 °C and 27.0 °C, respectively. These values represent increases of 5.1–5.5 °C above the baseline. Annual precipitation is also expected to rise further to 52.9 mm/year, showing a total increase of 10.5 mm.

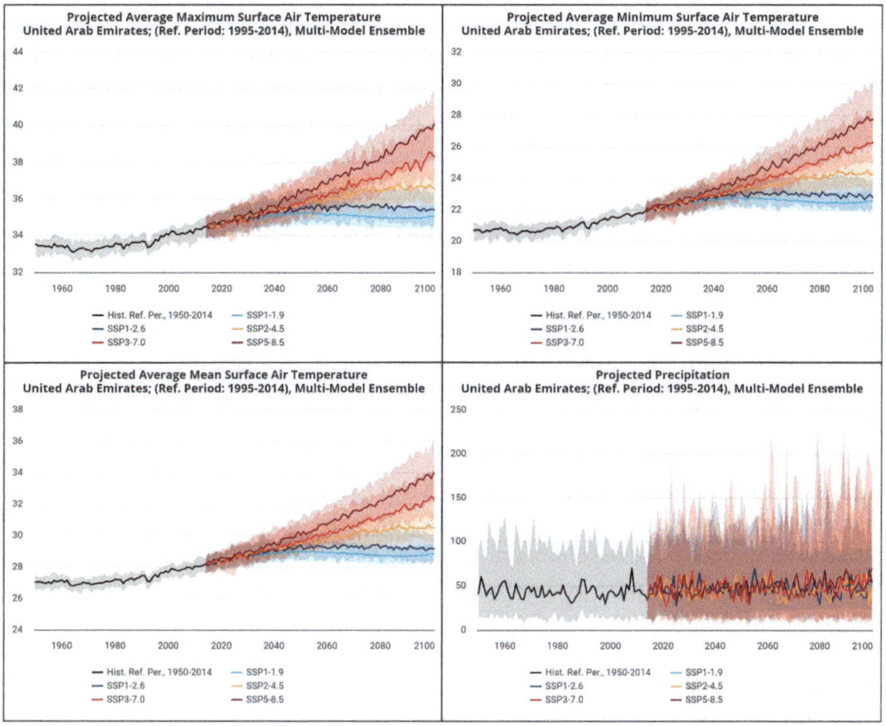

Fig. 4.2 Projected changes in maximum, minimum, mean surface air temperatures and precipitation for UAE under different SSP scenarios, based on a multi-model ensemble relative to the 1995–2014 reference period

These projected changes especially the sharp rise in minimum temperatures and moderate increases in precipitation necessitate detailed assessment given their broader implications. The projected increase in minimum temperatures, reaching up to 5.5 °C above baseline levels under SSP5-8.5, not only reduces nocturnal cooling but also carries significant consequences for human health, energy consumption and ecosystem functioning (Vinodhkumar et al., 2024). As higher nighttime temperatures become more common, thermal stress and cooling energy demands are expected to intensify, particularly in urban environments. In terms of precipitation, while the UAE is projected to experience moderate increase in annual totals, the more pressing concern lies in the potential for greater interannual variability and episodic extremes. Although ensemble median values suggest relatively stable or slightly wetter conditions, the likelihood of short-duration high-intensity rainfall events and extended dry periods remains elevated, particularly under SSP5-8.5. This evolving hydroclimatic regime may pose increasing challenges for water resource management, urban drainage systems and infrastructure resilience in the UAE. These projections align with the findings of Rao et al. (2024), who suggest that annual precipitation in the UAE could increase by as much as 30% from the mid-century onward across multiple SSPs. While the absolute increase in rainfall is moderate, even small gains may significantly enhance groundwater recharge potential, support vegetation productivity and mitigate drought risk in hyper-arid conditions. Understanding these precipitation trends and their variability is therefore essential for designing robust adaptation strategies that promote water sustainability and food-system resilience in the face of climatic uncertainty.

Qatar: Climate projections for Qatar based on CMIP6 multi-model ensemble outputs reveal a clear trajectory of progressive warming and modest increases in annual precipitation throughout the twenty-first century (Fig. 4.3). Using the baseline period of 1995–2014, the average annual maximum temperature in Qatar was 32.4 °C, with mean and minimum temperatures of 27.4 °C and 22.3 °C, respectively. Annual precipitation during this reference period was 52.0 mm/year. Under the SSP2-4.5 scenario, the mid-century period (2041–2060) is projected to experience increases of + 1.6 °C in both maximum and mean temperatures, reaching 34.1 °C and 29.0 °C, respectively. Minimum temperature is expected to rise to 24.0 °C, an increase of + 1.7 °C. Precipitation is projected to increase to 58.8 mm/year, representing a modest gain of 6.8 mm compared to the baseline. For the same period under the SSP5-8.5 scenario, projected warming is more pronounced. Maximum, mean and minimum temperatures are expected to reach 34.7 °C, 29.7 °C and 24.6 °C, respectively each showing increases of approximately + 2.3 °C. Annual precipitation is projected at 58.3 mm/year, an increase of 6.3 mm from historical levels. By the late century (2081–2100), under SSP2-4.5, Qatar is projected to experience a further rise in maximum temperature to 35.0 °C (+ 2.6 °C), mean temperature to 30.0 °C (+ 2.6 °C) and minimum temperature to 24.9 °C (+ 2.7 °C). Precipitation under this scenario will likely stabilize around 58.5 mm/year, marking a total increase of 6.5 mm from the baseline. In the high-emissions SSP5-8.5 scenario, end-of-century warming becomes significantly more intense. Maximum, mean and minimum temperatures are projected to reach 37.5 °C, 32.6 °C and 27.6 °C, respectively reflecting increases of + 5.1 °C, + 5.2 °C and + 5.3 °C. Annual precipitation is projected to rise to 61.3 mm/year, an increase of 9.3 mm compared to the baseline.

These projected shifts in temperature and precipitation are not merely numerical trends they carry substantial socio-environmental and infrastructural implications for Qatar, particularly in terms of energy demand, CO_2 emissions and the broader energy–water–food (EWF) nexus. Gurriaran et al. (2022) highlight a critical feedback loop between climate warming and electricity demand in Qatar, showing that each 1 °C increase in temperature may raise energy consumption by 1.7 GWh and CO_2 emissions by 0.4 $MtCO_2$. With CMIP6 projections indicating a potential rise in maximum temperatures of up to 5.1 °C under SSP5-8.5 by the end of the century, electricity demand could surge by 20–35% in the absence of decarbonization. These projections, paired with only moderate increases in precipitation, point to a future of intensified reliance on energy-intensive cooling systems and heightened stress on energy infrastructure. The study emphasized the necessity of incorporating warming-induced emissions into long-term planning, especially under high-emission scenarios where the frequency of extreme heat events is expected to

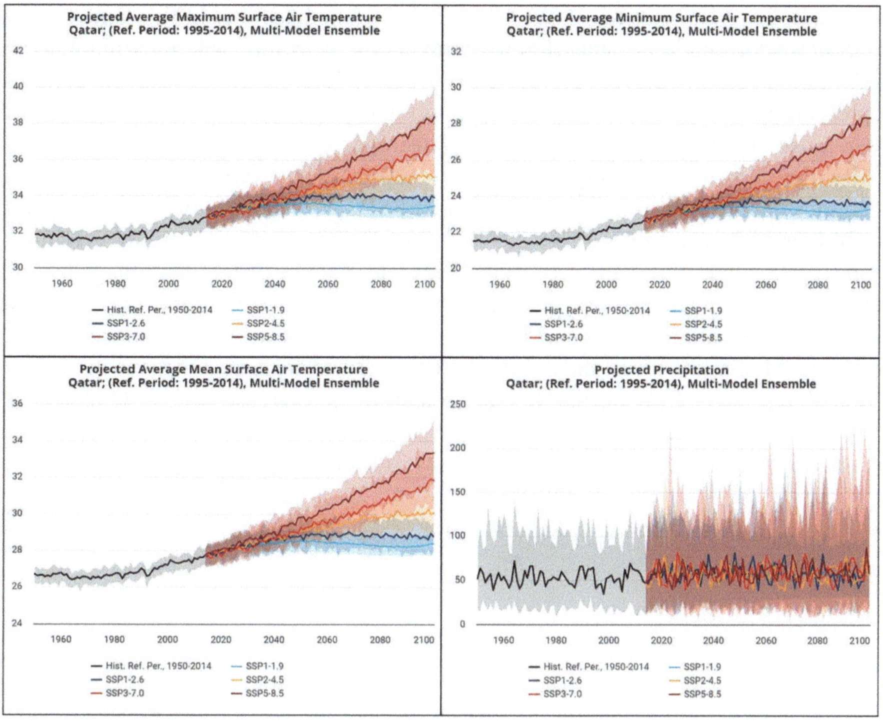

Fig. 4.3 Projected changes in maximum, minimum, mean surface air temperatures and precipitation for Qatar under different SSP scenarios, based on a multi-model ensemble relative to the 1995–2014 reference period

escalate. Complementing this perspective, Tareq et al. (2021) draw attention to the broader implications of climatic variability for the EWF nexus in arid regions like Qatar. Their findings suggest that shifts in temperature and precipitation not only elevate energy demand but also disrupt water availability and agricultural productivity. The projected climatic changes under SSP5-8.5 marked by significant warming and marginal rainfall increases are likely to exacerbate existing EWF vulnerabilities. Together, these studies reinforce the urgency of implementing integrated adaptation strategies that address the compounded challenges of thermal stress, carbon emissions and resource interdependencies in Qatar's climate future.

Bahrain: The climate projections for Bahrain indicate a clear and consistent warming trend across all future scenarios, with the intensity of change increasing under higher emissions pathways (Fig. 4.4). During the baseline period (1995–2014), the mean annual maximum, mean and minimum temperatures were 27.3 °C, 26.3 °C and 25.3 °C, respectively. Annual precipitation during this period averaged 54.0 mm/year. Under the SSP2-4.5 scenario by mid-century (2041–2060), maximum temperature is projected to increase by 1.6–29.0 °C, mean temperature by 1.6–27.9 °C and minimum temperature by 1.7–26.9 °C. Annual precipitation is projected to rise to 59.5 mm/

year, indicating an increase of 5.5 mm relative to the baseline. In the same period, the SSP5-8.5 scenario projects slightly more intense warming, with maximum, mean and minimum temperatures reaching 29.6 °C, 28.6 °C and 27.6 °C, respectively each representing an increase of 2.3 °C. Precipitation is projected to increase to 61.3 mm/year, a rise of 7.3 mm above the historical average.

By the end of the twenty-first century (2081–2100), the divergence between scenarios becomes more pronounced. Under SSP2-4.5, temperature increases from the baseline are projected to reach 2.6 °C for maximum, mean and minimum temperatures, with respective values of 30.0, 28.9 and 27.9 °C. Precipitation under this scenario is projected at 60.7 mm/year, or 6.7 mm above baseline levels. In the high-emissions SSP5-8.5 scenario, the late-century projections indicate substantial warming. Maximum, mean and minimum temperatures are expected to reach 32.4 °C, 31.4 °C and 30.4 °C, respectively each reflecting an increase of 5.1 °C above the 1995–2014 baseline. Precipitation is projected to rise further to 63.4 mm/year, an increase of 9.4 mm. The projected increase in annual precipitation for Bahrain,

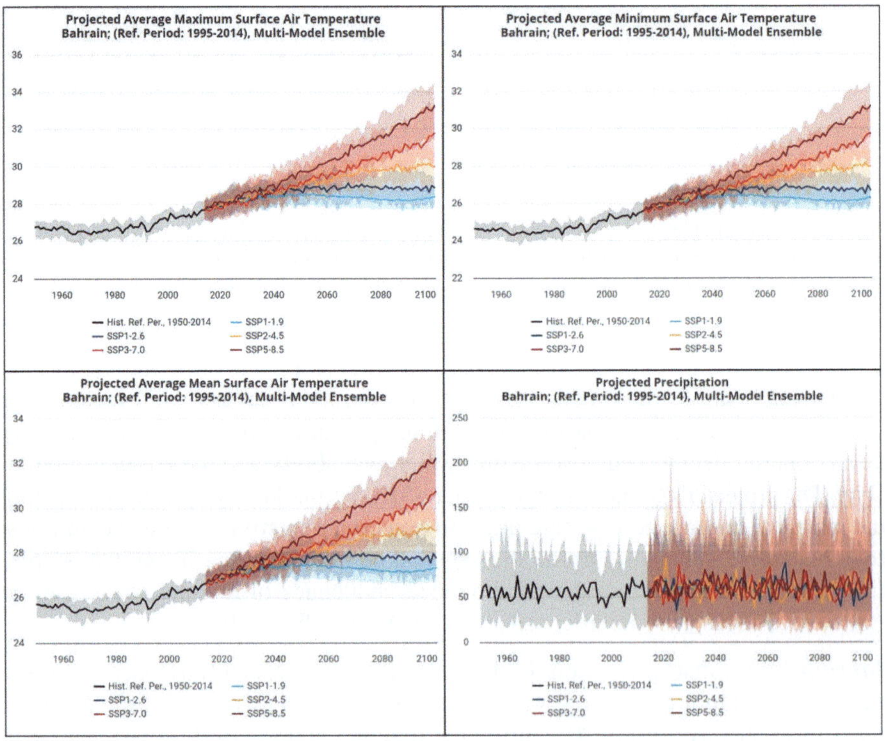

Fig. 4.4 Projected changes in maximum, minimum, mean surface air temperatures and precipitation for Bahrain under different SSP scenarios, based on a multi-model ensemble relative to the 1995–2014 reference period

though moderate in absolute terms, suggests a potential shift in rainfall characteristics rather than a substantial improvement in overall water availability. This projected intensification combined with significant warming of up to 5.1 °C under SSP5-8.5 implies an increased likelihood of irregular precipitation behavior, including both short-duration extreme rainfall events and prolonged dry periods. Given Bahrain's already limited freshwater resources, such variability could intensify existing water management challenges and underscores the need for proactive adaptation strategies that address both hydrological uncertainty and rising thermal stress (Alrowaili et al., 2024).

Kuwait: The CMIP6 multi-model ensemble indicates a robust warming trend over Kuwait throughout the twenty-first century, with the magnitude of projected change closely tied to the emissions scenario (Fig. 4.5). During the baseline period (1995–2014), the climatological means for maximum, mean and minimum temperatures were 32.4 °C, 26.2 °C and 19.9 °C, respectively. The corresponding annual precipitation averaged 114.9 mm/year. Under the SSP2-4.5 scenario for the mid-twenty-first century (2041–2060), projected temperature increases relative to the baseline are + 1.7 °C for maximum temperature (34.1 °C), + 1.8 °C for mean temperature (28.0 °C) and + 1.8 °C for minimum temperature (21.8 °C). Precipitation is projected to increase to 125.9 mm/year, representing a net gain of 11.0 mm. In contrast, under the SSP5-8.5 scenario for the same mid-century period, the projected warming is more pronounced: maximum, mean and minimum temperatures are expected to rise to 34.9 °C, 28.7 °C and 22.5 °C, indicating respective anomalies of + 2.5 °C, + 2.5 °C and + 2.5 °C. Annual precipitation is projected to increase to 125.4 mm/year, a change of + 10.6 mm relative to the baseline.

By the late twenty-first century (2081–2100), the divergence between scenarios becomes more evident. Under SSP2-4.5, maximum temperature is projected to increase by 2.8 °C (to 35.3 °C), mean temperature by 2.8 °C (to 29.0 °C) and minimum temperature by 2.9 °C (to 22.8 °C). Annual precipitation is projected to rise to 122.6 mm/year, which is 7.8 mm above the historical mean. Under the SSP5-8.5 scenario, late-century projections indicate a substantial intensification of warming, with maximum, mean and minimum temperatures reaching 38.0 °C, 31.9 °C and 25.7 °C, corresponding to increases of 5.6 °C, 5.7 °C and 5.8 °C, respectively. Annual precipitation is projected to reach 131.2 mm/year, representing an anomaly of + 16.4 mm. By the end of the century, Kuwait's maximum temperatures may exceed 38 °C under SSP5-8.5 and while annual precipitation is projected to increase moderately, the overall water balance is likely to remain constrained due to high evapotranspiration. In Kuwait's case, where the combination of aridity, extreme summer temperatures and dependence on desalinated water already strain energy and infrastructure systems, these projections carry serious policy implications (Aleisa, 2024). Rising temperatures will increase peak electricity loads, while erratic rainfall can disrupt already scarce freshwater planning. A strategic focus on heat-adaptive urban design, energy diversification and climate-resilient water systems will be critical to address the compounding nature of these challenges (De Gea Grela et al., 2024).

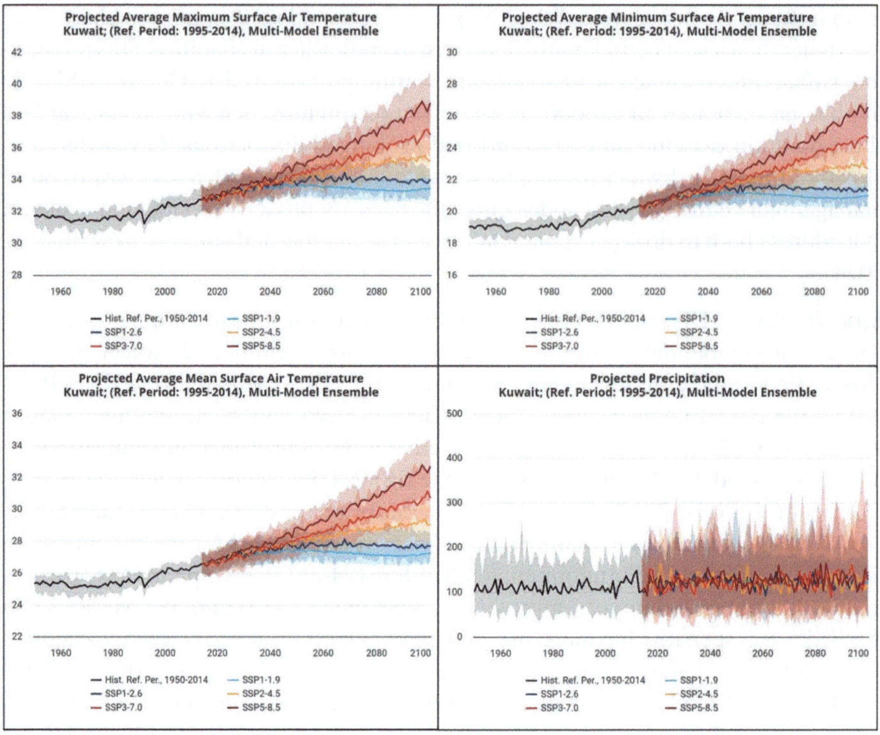

Fig. 4.5 Projected changes in maximum, minimum, mean surface air temperatures and precipitation for Kuwait under different SSP scenarios, based on a multi-model ensemble relative to the 1995–2014 reference period

Oman: Climate projections for this region, based on CMIP6 models and benchmarked against the 1995–2014 baseline, suggest a future marked by increasing heat and gradually intensifying rainfall patterns (Fig. 4.6). During the reference period, the average maximum temperature was 34.1 °C, with a mean of 27.9 °C, a minimum of 21.7 °C and average annual precipitation of 37.1 mm/year. By mid-century (2041–2060), under the SSP2-4.5 scenario pathway, all temperature indicators are projected to increase moderately. The maximum temperature is expected to rise to 35.5 °C, the mean to 29.3 °C and the minimum to 23.2 °C representing increases of 1.3 °C, 1.4 °C and 1.5 °C, respectively. In addition, rainfall is projected to rise to 41.1 mm/year, marking an increase of 4.1 mm, suggesting a slight but measurable intensification in moisture availability. Under the SSP5-8.5 high-emissions pathway during the same period, the warming intensifies. Maximum temperatures are expected to reach 36.0 °C, with mean and minimum temperatures increasing to 29.9 °C and 23.8 °C, respectively. These reflect rises of 1.8, 2.0 and 2.1 °C above the historical norms. Precipitation, meanwhile, is projected to increase more robustly to 46.4 mm/year, adding 9.3 mm compared to the baseline.

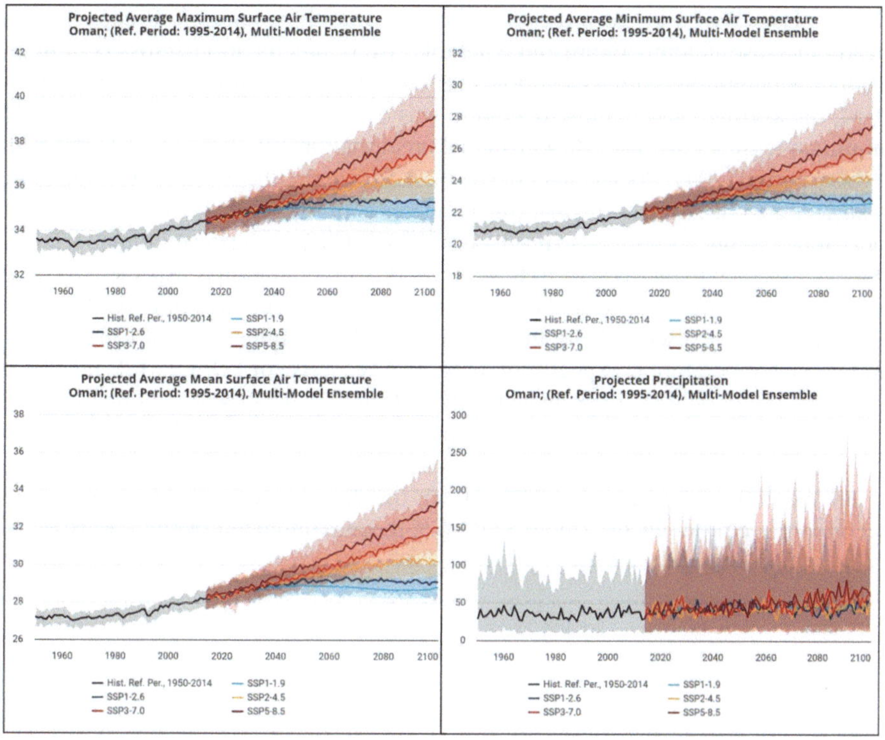

Fig. 4.6 Projected changes in maximum, minimum, mean surface air temperatures and precipitation for Oman under different SSP scenarios, based on a multi-model ensemble relative to the 1995–2014 reference period

By the end of the century (2081–2100), the gap between the two emission scenarios widens. In SSP2-4.5, the region's maximum temperature is expected to reach 36.3 °C, the mean 30.2 °C and the minimum 24.2 °C, with annual rainfall rising to 45.6 mm/year an increase of 8.6 mm from baseline. In contrast, under SSP5-8.5, the region faces more pronounced warming; maximum temperatures increase to 38.5 °C, mean to 32.6 °C and minimum to 26.7 °C, showing increases of 4.3 °C, 4.6 °C and 5.0 °C, respectively. Annual precipitation is projected to surge to 63.4 mm/year, an increase of 26.3 mm, suggesting not only a hotter future but one with potentially heavier or more frequent rainfall events. Based on the projected climate trends for Oman, there is an increasing likelihood of short-term extremes, with alternating dry periods and episodes of intense rainfall becoming more frequent. These changes carry particular relevance for Oman's socio-environmental landscape, where both water scarcity and flash flood vulnerability are pressing issues (Al-Rawas & Valeo, 2008; Purdue et al., 2021). Effective climate adaptation will require an integrated approach to water management and infrastructure resilience that accounts for the dual pressures of warming temperatures and less predictable rainfall dynamics.

Yemen: The CMIP6-based projections indicate that the country will experience a steady increase in temperature and a notable rise in precipitation across both intermediate and high-emissions scenarios (Fig. 4.7). During the baseline period of 1995–2014, the average maximum temperature was 31.6 °C, mean temperature 25.4 °C, minimum temperature 19.3 °C and annual precipitation 148.0 mm/year. In the mid-century period (2041–2060) under the SSP2-4.5 scenario, the maximum temperature is projected to rise to 32.9 °C, marking an increase of 1.4 °C above the baseline. The mean temperature is expected to reach 26.9 °C (+ 1.4 °C), while the minimum temperature is projected to increase to 20.8 °C (+ 1.5 °C). Precipitation is expected to rise to 162.5 mm/year, which represents an increase of 14.5 mm compared to the baseline. Under the SSP5-8.5 scenario for the same period, the warming is slightly higher. The maximum temperature is expected to reach 33.5 °C, a rise of 1.9 °C, while mean and minimum temperatures are projected at 27.4 °C and 21.4 °C, representing increases of 2.0 °C and 2.1 °C, respectively. Precipitation under this scenario is projected to increase more substantially to 179.6 mm/year, a rise of 31.6 mm.

By the end of the century (2081–2100), the divergence between the scenarios becomes more pronounced. Under SSP2-4.5, the maximum temperature is expected

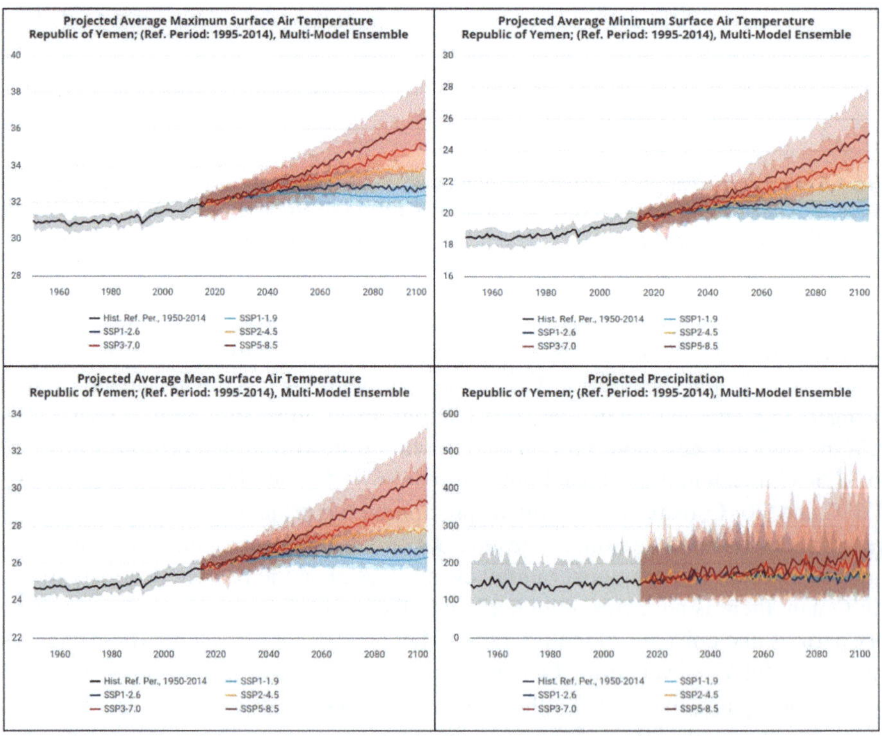

Fig. 4.7 Projected changes in maximum, minimum, mean surface air temperatures and precipitation for Yemen under different SSP scenarios, based on a multi-model ensemble relative to the 1995–2014 reference period

to increase to 33.7 °C (+ 2.2 °C), the mean temperature to 27.7 °C (+ 2.3 °C) and the minimum temperature to 21.7 °C (+ 2.4 °C). Precipitation is projected to increase to 172.0 mm/year, which is 23.9 mm higher than the baseline. In the SSP5-8.5 scenario by century's end, the country may experience more severe warming. The maximum temperature is projected to reach 36.0 °C, marking a rise of 4.4 °C. The mean and minimum temperatures are expected to reach 30.1 °C and 24.3 °C, increasing by 4.7 °C and 5.0 °C, respectively. Precipitation is projected to rise significantly to 214.3 mm/year, an increase of 66.3 mm compared to the baseline. These projections suggest that the country will not only face substantial warming, especially under high-emission scenarios, but also a notable increase in annual rainfall. While rising temperatures may intensify evapotranspiration, the increase in precipitation could offer some mitigation potential though its spatial and seasonal distribution will remain critical for effective adaptation planning. Given Yemen's topographical diversity, fragile water infrastructure and socio-economic vulnerabilities, these combined changes in temperature and precipitation emphasize the need for targeted adaptation efforts (Price, 2022). Climate-informed water governance, investment in public health preparedness and the development of drought and flood-resilient agricultural systems will be essential for long-term resilience (East & Region, 2010).

Jordan: The CMIP6-based climate projections indicate that the country will experience consistent warming across all future scenarios, with the degree of temperature rise and precipitation change varying depending on the emissions pathway (Fig. 4.8). During the baseline period of 1995–2014, the average maximum temperature was 25.7 °C, mean temperature 19.3 °C, minimum temperature 12.9 °C and annual precipitation 75.1 mm. In the mid-century period (2041–2060) under the intermediate emissions scenario SSP2-4.5, the maximum temperature is projected to rise to 27.5 °C, showing an increase of 1.8 °C from the baseline. The mean temperature is expected to reach 21.1 °C, also increasing by 1.8 °C, while the minimum temperature is likely to rise to 14.6 °C, marking a 1.7 °C increase. Precipitation is projected to decline slightly to 71.3 mm/year, which represents a decrease of 3.9 mm from the historical average. Under the high-emissions SSP5-8.5 scenario for mid-century, the warming is more pronounced. The maximum temperature is projected to reach 28.2 °C, which is 2.5 °C higher than the baseline. Mean and minimum temperatures are expected to increase to 21.7 and 15.3 °C, reflecting rises of 2.4 °C each. Precipitation under this scenario is also projected to decline slightly to 71.4 mm/year, a reduction of 3.7 mm compared to the baseline.

By the end of the century (2081–2100), the differences between emission scenarios become more evident. Under SSP2-4.5, the maximum temperature is expected to rise to 28.6 °C, representing an increase of 2.9 °C from the baseline. The mean and minimum temperatures are projected to reach 22.1 °C and 15.6 °C, showing increases of 2.8 °C and 2.7 °C, respectively. Annual precipitation is projected to decline further to 69.4 mm/year, which is 5.8 mm below the baseline average. In the SSP5-8.5 end-century scenario, the country is projected to face intense warming. The maximum temperature is expected to reach 31.2 °C, indicating a rise of 5.5 °C. The mean temperature is projected to increase to 24.7 °C (+ 5.4 °C) and the minimum

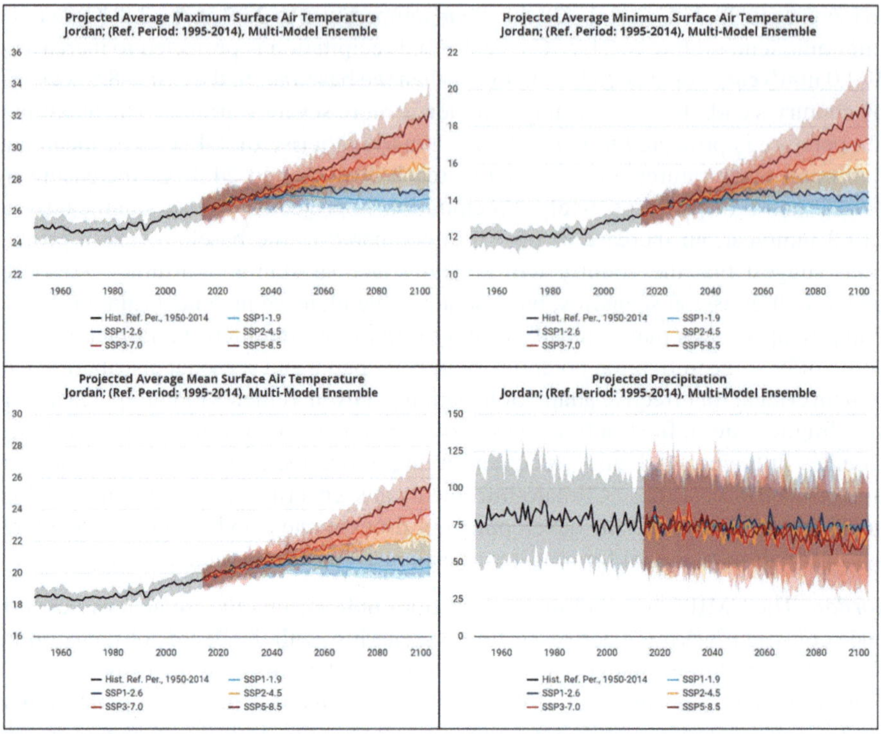

Fig. 4.8 Projected changes in maximum, minimum, mean surface air temperatures and precipitation for Jordan under different SSP scenarios, based on a multi-model ensemble relative to the 1995–2014 reference period

temperature to 18.2 °C (+ 5.3 °C). Precipitation is projected to decline more substantially under this scenario, falling to 62.8 mm/year, which is a reduction of 12.4 mm compared to the baseline. These projections highlight a clear trajectory of increasing temperature across all scenarios, with significantly higher warming under SSP5-8.5. While precipitation changes appear moderate to substantial depending on the scenario and period, the combination of higher temperatures and reduced rainfall is likely to intensify water stress and climatic challenges in the region. The convergence of increasing temperatures and more irregular precipitation patterns will likely exert additional pressure on Jordan's agriculture, water availability and overall climate resilience (Mahmoud et al., 2023).

Iraq: The climate projections for Iraq, based on CMIP6 ensemble simulations, indicate a consistent and marked warming trend throughout the twenty-first century, with significantly stronger temperature increases under the high-emissions scenario (Fig. 4.9). Using the 1995–2014 period as the baseline, during which the average maximum temperature was 29.3 °C, mean temperature 22.8 °C, minimum temperature 16.3 °C and precipitation 204.7 mm/year, the following changes are projected.

In the mid-century period (2041–2060) under the intermediate emissions scenario SSP2-4.5, the maximum temperature is expected to reach 31.2 °C, an increase of 1.9 °C above the baseline. The mean temperature is projected to be 24.7 °C, rising by 1.9 °C, while the minimum temperature is likely to reach 18.1 °C, marking an increase of 1.8 °C. Precipitation under this scenario is projected to remain relatively stable, slightly decreasing to 207.1 mm/year, which is a minor reduction of 1.6 mm. In contrast, under the high-emissions SSP5-8.5 scenario for mid-century, the warming is more substantial. The maximum temperature is projected to rise to 31.9 °C, which is 2.6 °C higher than the baseline. The mean temperature is expected to reach 25.4 °C (+ 2.6 °C) and the minimum temperature to 18.9 °C (+ 2.6 °C). Precipitation is projected to decline to 203.6 mm/year, about 1.1 mm less than the baseline.

By the end of the century (2081–2100), the differences between emission pathways become more pronounced. Under SSP2-4.5, the maximum temperature is projected to rise to 32.4 °C, an increase of 3.1 °C from the baseline. The mean temperature will increase to 25.8 °C (+ 3.0 °C) and the minimum temperature to 19.2 °C (+ 2.9 °C). Annual precipitation is projected to decrease to 202.3 mm/year,

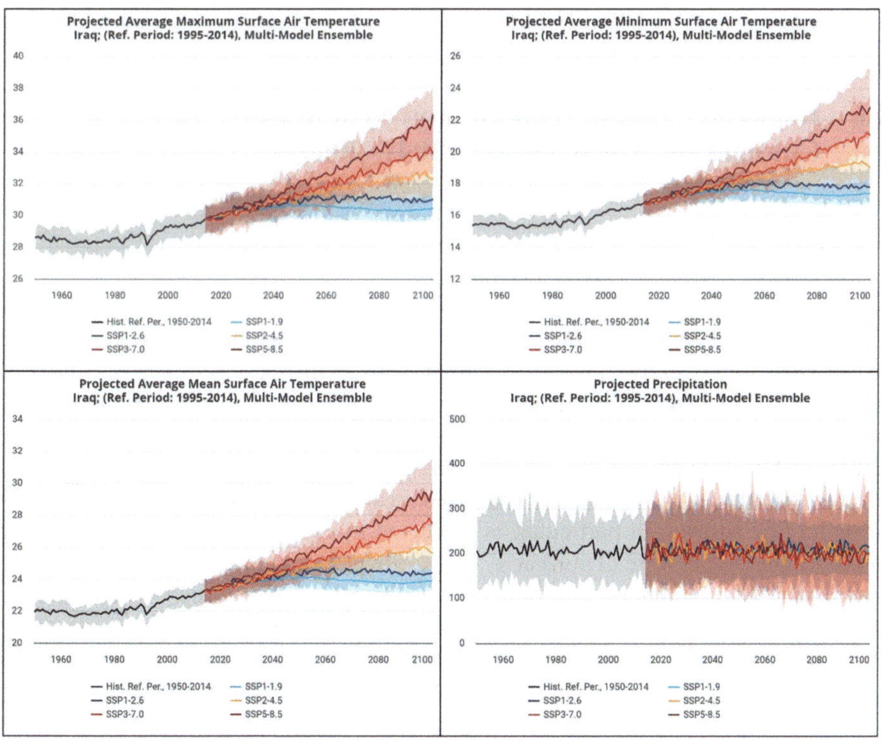

Fig. 4.9 Projected changes in maximum, minimum, mean surface air temperatures and precipitation for Iraq under different SSP scenarios, based on a multi-model ensemble relative to the 1995–2014 reference period

a reduction of 2.4 mm compared to the baseline. Under the SSP5-8.5 scenario, Iraq is expected to face extreme warming by century's end. The maximum temperature is projected to reach 35.2 °C, which is 5.9 °C above the baseline. The mean temperature will rise to 28.6 °C (+ 5.8 °C) and the minimum temperature to 22.0 °C (+ 5.7 °C). Precipitation is expected to decline more significantly to 195.5 mm/year, representing a drop of 9.2 mm from the baseline. These results clearly indicate that Iraq is expected to undergo substantial warming under all emission scenarios, with the magnitude of change closely tied to future emissions trajectories. While projected changes in precipitation appear moderate in absolute terms, even slight declines when combined with intensified heat are likely to exacerbate aridity and place greater pressure on already stressed water resources. The projections point to a dual challenge for Iraq; sustained heat stress alongside a more variable and potentially declining precipitation regime (Hashim et al., 2022). Mukheef et al. (2024) further emphasize that changing climate patterns are expected to severely impact water availability in the Middle East, particularly in Iraq, with adverse effects on both surface and groundwater resources. These findings highlight the critical need for proactive and well-targeted adaptation planning to build resilience in key sectors such as agriculture, public health, urban development and integrated water resource management.

4.6 Regional Climate Outlook and Synthesis of Projected Changes for Mid-Century (2041–2060)

Temperature Changes

Under the SSP2-4.5 scenario, a moderate yet regionally significant warming trend is projected for mid-century (Table 4.3). The maximum temperature (Tmax) is expected to rise most significantly in Iraq and Jordan, with an increase of + 1.8 °C, followed by Saudi Arabia, Qatar, Bahrain and Kuwait, each with an increase of approximately + 1.7 °C. Mean temperatures (Tmean) also reflect similar increases of + 1.8 °C in Saudi Arabia, Kuwait, Jordan and Iraq, suggesting consistent warming throughout the day. In addition, it was observed that UAE is also showing high increase in Tmean of + 1.7 °C. Kuwait, in particular, is projected to experience the highest increase in minimum temperature (Tmin) at + 1.9 °C, indicating intensified night-time heat, which may have compounding effects on human health and energy demands for cooling. Other countries projected to experience substantial increases in minimum temperature include Saudi Arabia, UAE and Iraq, each with a rise of + 1.8 °C, highlighting a broader regional trend of intensifying night-time warming. In contrast, the SSP5-8.5 scenario, which assumes a higher-emission trajectory, predicts a steeper rise in temperature by mid-century. Iraq leads with the largest projected increase in Tmax at + 2.7 °C, followed closely by Kuwait and Jordan at + 2.5 °C. Iraq showed the highest increase in mean temperature (+ 2.5 °C), a rise that is also matched by Kuwait and Saudi Arabia, indicating widespread daytime warming across these countries. In Tmin rise at + 2.6 °C was observed in Saudi Arabia, Kuwait and Iraq, suggesting persistent and exacerbated heat stress, especially in urban environments with heat island effects.

Precipitation Changes

Projected changes in precipitation patterns reveal considerable variability across the region. In the SSP2-4.5 scenario, Yemen is expected to experience the most significant increase in annual precipitation, with a projected rise of + 14.5 mm, possibly improving water availability and recharge potential. Kuwait also shows a substantial positive anomaly of + 11.0 mm, which could help alleviate its current water scarcity challenges. However, some countries are likely to experience a decline; Jordan and Iraq are projected to see reductions of − 3.8 mm and − 2.4 mm, respectively, indicating growing aridity risks. Under the SSP5-8.5 scenario, these precipitation trends are amplified. Yemen is projected to receive an even larger increase in rainfall, rising by + 31.6 mm, the highest among all countries assessed. This could bring both opportunities for improved water resources and challenges such as increased flood risk, especially if precipitation becomes more intense and episodic. Kuwait, Oman and Saudi Arabia are also projected to experience meaningful gains of + 10.5 mm, + 9.3 mm and + 8.4 mm, respectively. On the other hand, Jordan and Iraq continue to exhibit declining trends in precipitation, with projected decreases of − 3.7 mm/year and − 1.1 mm, respectively, compounding their vulnerability to drought.

Synthesis and Implications

The mid-century climate outlook under both SSP2-4.5 and SSP5-8.5 scenarios reveals a clear intensification of warming across the region, with Iraq and Kuwait consistently emerging as the most affected by rising temperatures. These thermal shifts, particularly in Tmin, point to prolonged heat exposure, reduced diurnal temperature variability and greater stress on human health, agriculture and energy infrastructure. The increase in mean temperatures (Tmean) by a projected of + 2.5 °C in Iraq, Kuwait and Saudi Arabia, indicates intensified daily heat exposure that may significantly impact agricultural productivity, labor efficiency and infrastructure resilience (Sheng et al., 2024; Waha et al., 2017). The precipitation outlook adds complexity to the regional narrative. While Yemen and Kuwait may benefit from increased rainfall under both scenarios, the drying trends projected for Jordan and Iraq suggest emerging hotspots of water stress. These diverging hydrological responses call for localized adaptation strategies such as enhanced water harvesting in wetter zones and robust drought preparedness in drying regions. Given these projections, a regionally tailored adaptation strategy becomes essential. Countries experiencing

Table 4.3 Projected Mid- and End-Century changes under SSP2-4.5 and SSP5-8.5 scenarios across the Arabian Peninsula

Country	Metric	Mid-century (2041–2100)		End-century (2081–2100)	
		Δ 4.5	Δ 8.5	Δ 4.5	Δ 8.5
Saudi Arabia	Tmax (°C)	1.7	2.4	2.8	5.5
	Tmean (°C)	1.8	2.5	2.8	5.7
	Tmin (°C)	1.8	2.6	2.9	5.9
	Precip (mm)	6.4	8.4	8.1	16
UAE	Tmax (°C)	1.6	2.2	2.5	5.1
	Tmean (°C)	1.7	2.3	2.7	5.3
	Tmin (°C)	1.8	2.4	2.8	5.5
	Precip (mm)	3.8	4.9	2.3	10.4
Qatar	Tmax (°C)	1.7	2.3	2.6	5.1
	Tmean (°C)	1.6	2.3	2.6	5.2
	Tmin (°C)	1.7	2.3	2.6	5.3
	Precip (mm)	6.8	6.3	6.5	9.3
Bahrain	Tmax (°C)	1.7	2.3	2.7	5.1
	Tmean (°C)	1.6	2.3	2.6	5.1
	Tmin (°C)	1.6	2.3	2.6	5.1
	Precip (mm)	5.5	7.3	6.7	9.4
Kuwait	Tmax (°C)	1.7	2.5	2.9	5.6
	Tmean (°C)	1.8	2.5	2.8	5.7
	Tmin (°C)	1.9	2.6	2.9	5.8
	Precip (mm)	11	10.5	7.7	16.3
Oman	Tmax (°C)	1.4	1.9	2.2	4.4
	Tmean (°C)	1.4	2	2.3	4.7
	Tmin (°C)	1.5	2.1	2.5	5
	Precip (mm)	4	9.3	8.5	26.3
Yemen	Tmax (°C)	1.3	1.9	2.1	4.4
	Tmean (°C)	1.5	2	2.3	4.7
	Tmin (°C)	1.5	2.1	2.4	5
	Precip (mm)	14.5	31.6	24	66.3
Jordon	Tmax (°C)	1.8	2.5	2.9	5.5
	Tmean (°C)	1.8	2.4	2.8	5.4
	Tmin (°C)	1.7	2.4	2.7	5.3
	Precip (mm)	− 3.8	− 3.7	− 5.7	− 12.3
Iran	Tmax (°C)	1.8	2.7	3	5.9
	Tmean (°C)	1.8	2.5	2.9	5.8
	Tmin (°C)	1.8	2.6	2.9	5.7
	Precip (mm)	− 2.4	− 1.1	− 2.4	− 9.2

significant warming, particularly Iraq, Kuwait, Saudi Arabia and UAE, will need to prioritize heat-health action plans, including early warning systems, public cooling centers and urban greening to mitigate the urban heat island effect (Matzarakis, 2024). Simultaneously, countries like Yemen, Kuwait, Oman and Saudi Arabia that are projected to receive increased rainfall must invest in water harvesting, flood control and groundwater recharge infrastructure. In contrast, Jordan and Iraq require robust drought preparedness, including demand-side water management, climate-resilient crop selection and regional cooperation on transboundary water resources. The contrast in outcomes between SSP2-4.5 and SSP5-8.5 scenarios highlights the importance of mitigation efforts in curbing future climate extremes. Reducing emissions can substantially temper future warming by about 1 °C in many locations and moderate extremes in precipitation, particularly in Yemen. Scenario-based planning will therefore be critical for ensuring infrastructure, agriculture and public health systems are resilient under a wide range of possible future climate conditions.

4.7 Regional Climate Outlook and Synthesis of Projected Changes for End-Century (2081–2100)

Temperature Changes

By the end of the century, both scenarios SSP2-4.5 and SSP5-8.5 project a marked escalation in regional warming, with more intense and widespread heat extremes than mid-century projections (Table 4.3). Under the SSP2-4.5 scenario, Iraq is projected to experience the most significant increase in maximum temperature (Tmax), with a rise of + 3.0 °C. This is followed closely by Jordan and Kuwait, each with increases of + 2.9 °C and Saudi Arabia, which is projected to see a slightly lower rise of + 2.8 °C. Mean temperatures (Tmean) show the highest increases in Iraq (+ 2.9 °C) and Kuwait, Jordan and Saudi Arabia (+ 2.8 °C), highlighting the extent of daily warming across these countries. Minimum temperatures (Tmin) are also expected to rise most in Kuwait, Saudi Arabia and Iraq, with each country projected to see an increase of + 2.9 °C, followed by UAE (+ 2.8 °C) reflecting more persistent night-time heat that will likely intensify heat stress and reduce thermal comfort. The SSP5-8.5 scenario presents an even more severe picture. Iraq again leads with the highest projected increase in Tmax at + 5.9 °C, closely followed by Kuwait (+ 5.6 °C) and Saudi Arabia and Jordan at + 5.5 °C. For Tmean, Iraq continues to record the highest rise (+ 5.8 °C), with Kuwait and Saudi Arabia not far behind at + 5.7 °C. The steepest increases in Tmin are projected in Saudi Arabia (+ 5.9 °C), Kuwait (+ 5.8 °C) and Iraq (+ 5.7 °C), pointing to the likelihood of extreme night-time heat persisting with little cooling relief.

Precipitation Changes

Projected changes in precipitation show a strong divergence between countries, with some expected to see significant gains and others facing serious declines. Under SSP2-4.5, Yemen shows the most pronounced increase in annual precipitation, with an expected rise of + 24.0 mm, suggesting improved water availability and recharge potential. This is followed by Oman (+ 8.5 mm) and Saudi Arabia (+ 8.1 mm), where enhanced rainfall may partially alleviate water stress. Conversely, Jordan and Iraq are projected to experience further drying, with declines of − 5.7 mm and − 2.4 mm, respectively. This persistent decrease raises concerns about sustained aridification and increased drought vulnerability in these regions. In the SSP5-8.5 scenario, rainfall anomalies become more extreme. Yemen exhibits a dramatic increase in precipitation of + 66.3 mm, the highest among all countries assessed, which may significantly alter local hydrological regimes. Oman (+ 26.3 mm) and Kuwait (+ 16.3 mm) also show substantial increases, potentially offering opportunities for water resource development. However, Jordan (− 12.3 mm) and Iraq (− 9.2 mm) continue on a downward trend, further compounding long-term water scarcity concerns.

Synthesis and Implications

Taken together, the end-century projections illustrate an intensifying climate regime across the region, characterized by extreme heat and diverging precipitation trends. Iran consistently emerges as the region's thermal hotspot across all metrics and scenarios, particularly under SSP5-8.5, where it records the highest increases in Tmax, Tmean and Tmin. Kuwait and Saudi Arabia also show substantial warming, especially in Tmin, indicating prolonged heat exposure and reduced night-time cooling. These conditions are likely to exacerbate heat-related health issues, increase cooling energy demand and challenge the resilience of infrastructure and agriculture. Yemen, on the other hand, shows the most substantial increases in precipitation, especially under SSP5-8.5. While this may improve water availability and support rain-fed agriculture, it also introduces the risk of flooding and water management challenges if rainfall events become more intense and erratic. Oman and Kuwait may also benefit from enhanced precipitation, providing some relief in otherwise arid conditions. In contrast, Jordan and Iran appear particularly vulnerable due to the dual threats of significant warming and declining precipitation. This combination of heat and drying increases the urgency for adaptive planning focused on water conservation, drought resilience and sustainable resource management. Ultimately, the differences between the SSP2-4.5 and SSP5-8.5 outcomes underscore the importance of global mitigation efforts. Limiting emissions can significantly reduce the severity of temperature increases and moderate precipitation extremes, providing a more manageable future climate for the region (Shukla et al., 2016). For decision-makers, this highlights the need for scenario-based adaptation strategies that strengthen regional resilience while supporting global climate commitments.

4.8 Implications of Projected Climate Change for the Arabian Peninsula

The projected climate trends analyzed highlighted a future marked by escalating thermal and hydrological stress across the Arabian Peninsula under both SSP2-4.5 and SSP5-8.5 scenarios. While the magnitude and intensity of projected changes are greater under the high-emissions pathway (SSP5-8.5), both scenarios point to region-wide increases in temperature, altered precipitation regimes and increased climate variability. Critical sectors such as health, water, agriculture, energy, infrastructure and urban systems are projected to be impacted, with more severe consequences anticipated under SSP5-8.5 and ongoing challenges expected even under intermediate-emission scenarios.

Public Health and Human Well-Being

Under both scenarios, rising maximum and minimum temperatures especially in Kuwait, Saudi Arabia, UAE and Iraq are expected to exceed human thermal comfort thresholds, with more frequent and intense heatwaves under SSP5-8.5 (Koheji, 2025). Increased nighttime temperatures, particularly under high-emission conditions, will reduce the body's capacity for physiological recovery, increasing the incidence of heat-related illnesses, sleep disturbances and cardiovascular stress. These risks are further exacerbated in urban areas experiencing the heat island effect. Addressing these impacts requires targeted investments in heat-health warning systems, climate-resilient healthcare infrastructure and public awareness campaigns tailored to vulnerable populations.

Water Resources and Sanitation

Yemen, Oman and Kuwait are projected to receive increased precipitation under both scenarios, although the gains are larger and more erratic under SSP5-8.5. These increases may provide temporary relief in water-scarce settings but are accompanied by a higher likelihood of extreme rainfall events and flooding. In contrast, Jordan and Iraq may experience declining rainfall, especially under SSP5-8.5, intensifying their prevailing water stress (Alsalal et al., 2024). These dynamics will challenge water governance and infrastructure, necessitating expanded desalination capacity, improved groundwater regulation and adaptive storage solutions, regardless of the emissions pathway.

Food Security and Agriculture

Projected increases in temperature and altered precipitation patterns will influence agricultural productivity, particularly in Iraq, Jordan and Yemen (Adamo et al., 2022). Under SSP5-8.5, the combined effects of heat stress, higher evapotranspiration and intensified drought risk are likely to be more severe. Even under SSP2-4.5, warming trends and moderate precipitation shifts can alter crop cycles and reduce rainfed yields. Adaptive responses must include the promotion of drought-resilient crop varieties, climate-smart irrigation practices and diversification of livelihoods in vulnerable agro-ecological zones.

Energy and Infrastructure

Rising daytime and nighttime temperatures will result in increased demand for cooling under both scenarios, straining power grids and increasing dependency on energy-intensive technologies. SSP5-8.5 projections indicate a more pronounced rise in cooling needs, especially in urban and coastal regions. Infrastructure will be exposed to thermal stress, degradation and localized flooding from intense rainfall. Expanding renewable energy capacity, upgrading grid resilience and integrating passive cooling design into new developments will be critical strategies under both emission futures.

Urban Systems and Built Environments

Urban areas especially in countries such as Kuwait, UAE and Saudi Arabia are projected to face sustained heat-related stress alongside an increased likelihood of extreme precipitation events. In scenarios characterized by more frequent extreme heat and intense precipitation as reflected in high-emission pathways urban systems may be increasingly vulnerable to risks such as flooding, infrastructure strain and service disruptions. Even under moderate warming trajectories, there remains a credible probability of escalating pressure on municipal infrastructure and public services, particularly in densely populated and low-lying urban areas. Enhancing urban resilience will require integrated approaches, including the adoption of green and blue infrastructure, the revision of building codes for climate compatibility and the upgrading of drainage systems (Siehr et al., 2022). Informed and forward-looking urban planning, grounded in localized climate information, will be essential to reducing exposure and building adaptive capacity under both high- and intermediate-emission scenarios.

Regional Coordination and Adaptation Planning

Despite variation in projected impacts across the Peninsula, both scenarios emphasize the importance of regional cooperation. A shared climate risk information platform, harmonized early warning systems and coordinated adaptation funding mechanisms will strengthen collective resilience. Transboundary collaboration on water management, health surveillance and infrastructure development will be essential in responding to both gradual and abrupt climate-related stressors. While the urgency is more pronounced under SSP5-8.5, the need for long-term, integrated adaptation strategies persists even under moderate warming conditions projected by SSP2-4.5.

4.9 Conclusion

Climate projections are vital tools for anticipating future climate conditions and informing policy, planning and adaptation efforts at regional and global scales. In the context of the Arabian Peninsula a region already experiencing climatic extremes these projections are indispensable for guiding infrastructure development, water resource planning, agricultural resilience, public health preparedness and disaster risk reduction. The comprehensive projections presented in this chapter highlight the scale and complexity of climate change challenges facing the Arabian Peninsula. While the quantitative analysis offers clear indications of increasing temperatures and shifting precipitation regimes, the broader implication lies in how these changes intersect with existing socio-environmental vulnerabilities in the region. Countries across the Peninsula are already struggling with water scarcity, urban heat stress and growing energy demands pressures that are likely to be amplified by the projected climate trajectories. What emerges from the mid- and end-century scenarios is not only a rising temperature curve but also a widening disparity in water resource availability across the region. The uneven distribution of projected precipitation change underscores the importance of regional coordination, especially in managing transboundary aquifers, shared river basins and regional food supply chains. Some countries will need to focus on maximizing the benefits of increasing rainfall through improved water storage and managed aquifer recharge, while others must prepare for prolonged aridity through demand-side interventions and diversification of water sources, including desalination and reuse.

The temperature trends, particularly the sharp increases in minimum temperatures, suggest a fundamental shift in the region's thermal regime, with implications for daily life, building design, occupational safety and ecological functioning. These findings demand a re-evaluation of infrastructure standards, agricultural calendars and energy system planning to account for sustained heat exposure and altered climatic baselines. Furthermore, the divergence in outcomes between the two emissions scenarios indicates a narrowing window for meaningful mitigation. While adaptation is essential, these projections reinforce that mitigation efforts at the global scale remain critical in determining the severity of regional impacts. The region's rapidly expanding urban areas, energy transition agendas and development plans offer opportunities to embed climate resilience if informed by robust, scenario-based planning.

In light of these findings, the chapter calls for an integrated approach that combines forward-looking adaptation, evidence-based planning and strategic mitigation. Future research should emphasize localized impact assessments, stakeholder engagement and the development of climate services tailored to the needs of policymakers and practitioners across the Arabian Peninsula. The results underscore the importance of incorporating high-resolution climate data into long-term planning, particularly in sectors sensitive to thermal and hydrological stress, including water resource management, agriculture, public health and urban infrastructure. Future research should prioritize improved downscaling approaches, enhanced representation of land–atmosphere feedbacks in regional climate models and the integration of observational datasets for model evaluation. As the region undergoes rapid socioeconomic development, understanding the spatial variability and sectoral implications of projected climate change will be essential to reduce exposure and enhance adaptive capacity. Equally critical is the translation of scientific insight into actionable policy. Embedding climate-risk metrics within national development plans, infrastructure codes and financial appraisal frameworks will help ensure that new investments remain robust under a range of future conditions. Regional collaboration particularly on transboundary water resources, heat-health surveillance and integrated early-warning systems can amplify the effectiveness of country-level efforts and avoid maladaptive outcomes. Finally, aligning adaptation priorities with global mitigation commitments will not only temper the most extreme projections but also unlock co-benefits for energy security, food systems and public health, thereby supporting a sustainable and climate-resilient future for the Arabian Peninsula.

References

Adamo, N., Al-Ansari, N., Sissakian, V., Jehad Fahmi, K., & Ali Abed, S. (2022). Climate change: Droughts and increasing desertification in the Middle East, with special reference to Iraq. *Engineering, 14*(7), 235–273.

Ajay, P., Pathak, B., Solmon, F., Bhuyan, P. K., & Giorgi, F. (2019). Obtaining best parameterization scheme of RegCM 4.4 for aerosols and chemistry simulations over the CORDEX South Asia. *Climate Dynamics, 53*, 329–352.

Ajjur, S. B., & Al-Ghamdi, S. G. (2022, January). Assessment of the climate change impact on the Arabian Peninsula evapotranspiration losses. In *World environmental and water resources congress 2022* (pp. 492–498).

Ajjur, S. B., & Al-Ghamdi, S. G. (2023). Assessing and projecting climatic changes in the Middle East and North Africa (MENA) region: Insights from Regional Climate Model (RCM) simulations and future projections. *Sustainable Cities in a Changing Climate: Enhancing Urban Resilience*, 29–38.

Aleisa, E. (2024). Navigating Kuwait's water scarcity challenges: A holistic analysis of water resources, environmental impact, cost, and policy implications. *Desalination, 586*, Article 117827.

Almazroui, M. (2018). Assessment of CMIP5 global climate models and projected changes in surface air temperature over the Arabian Peninsula in the twenty-first century. *Arabian Journal of Geosciences, 11*, 1–15.

Almazroui, M. (2019). Climate extremes over the Arabian Peninsula using RegCM4 for present conditions forced by several CMIP5 models. *Atmosphere, 10*(11), 675.

Almazroui, M., Abid, M. A., Athar, H., Islam, M. N., & Ehsan, M. A. (2013). Interannual variability of rainfall over the Arabian Peninsula using the IPCC AR4 Global Climate Models. *International Journal of Climatology, 33*(10), 2328–2340.

Almazroui, M., Islam, M. N., Al-Khalaf, A. K., & Saeed, F. (2016). Best convective parameterization scheme within RegCM4 to downscale CMIP5 multi-model data for the CORDEX-MENA/Arab domain. *Theoretical and Applied Climatology, 124*, 807–823.

Almazroui, M., Islam, M. N., Saeed, S., Saeed, F., & Ismail, M. (2020). Future changes in climate over the Arabian Peninsula based on CMIP6 multimodel simulations. *Earth Systems and Environment, 4*, 611–630.

Al-Rawas, G. A., & Valeo, C. (2008). Issues with flash flood modeling in the capital region of Sultanate Oman. *Geomatics Engineering at Schulich School of Engineering University of Calgary*, 1–11.

Alrowaili, S., Aldulaimi, S. H., Abdeldayem, M., & Hassan, H. I. (2024, December). Strategic water innovations for sustainable development and future resilience in Bahrain. In *2024 International conference on decision aid sciences and applications (DASA)* (pp. 1–7). IEEE.

Alsalal, S., Tan, M. L., Samat, N., Al-Bakri, J. T., & Zhang, F. (2024). Temperature and precipitation changes under CMIP6 projections in the Mujib Basin, Jordan. *Theoretical and Applied Climatology, 155*(8), 7703–7720.

Alsarraf, H., & Broeke, M. V. D. (2015). Using the WRF regional climate model to simulate future summertime wind speed changes over the Arabian Peninsula. *Journal of Climatology & Weather Forecasting, 3*, 144.

Asmamaw, M., Ambellu, A., & Tiku, S. (2015). Resilience of ecosystems to climate change. *American Journal of Environmental Protection, 4*(6), 325–333.

Attada, R., Attada, R., Kunchala, R. K., Dasari, H. P., Sivareddy, S., Yesubabu, V., Knio, O. M., & Hoteit, I. (2021). Representation of Arabian Peninsula summer climate in a regional atmospheric model using spectral nudging. *Theoretical and Applied Climatology, 145*(1), 13–30. https://doi.org/10.1007/S00704-021-03617-W

Barfus, K., & Bernhofer, C. (2014). Assessment of GCM performances for the Arabian Peninsula, Brazil, and Ukraine and indications of regional climate change. *Environmental Earth Sciences, 72*, 4689–4703.

De Gea Grela, P., Sánchez-González, D., & Gallardo Peralta, L. P. (2024). Urban and rural environments and their implications for older adults' adaptation to heat waves: A systematic review. *Land, 13*(9), 1378.

Deser, C. (2020). Certain uncertainty: The role of internal climate variability in projections of regional climate change and risk management. *Earth's Future, 8*(12), e2020EF001854.

Dessai, S., & Hulme, M. (2004). Does climate adaptation policy need probabilities? *Climate Policy, 4*(2), 107–128.

East, M., & Region, N. A. (2010). Assessing the impacts of climate change and variability on the water and agricultural sectors and the policy implications.

Giorgi, F. (2019). Thirty years of regional climate modeling: Where are we and where are we going next? *Journal of Geophysical Research: Atmospheres, 124*(11), 5696–5723.

Guo, D., & Wang, H. (2016). Comparison of a very-fine-resolution GCM with RCM dynamical downscaling in simulating climate in China. *Advances in Atmospheric Sciences, 33*, 559–570.

Gurriaran, L., Tanaka, K., Bayram, I. S., Proestos, Y., Lelieveld, J., & Ciais, P. (2022). Climate change induced increase on power demand and CO_2 emissions in the Middle East (Qatar). *Authorea Preprints.*

Gutowski, W. J., Ullrich, P. A., Hall, A., Leung, L. R., O'Brien, T. A., Patricola, C. M., Arritt, R. W., Bukovsky, M. S., Calvin, K. V., Feng, Z., Jones, A. D., Kooperman, G. J., Monier, E., Pritchard, M. S., Pryor, S. C., Qian, Y., Rhoades, A. M., Roberts, A. F., …, Sakaguchi, K. (2020). High-resolution regional climate models. *Bulletin of the American Meteorological Society, 101*(8), 693–697.

Hansen, J., Fung, I., Lacis, A., Rind, D., Lebedeff, S., Ruedy, R., Russell, G., & Stone, P. (1988). Global climate changes as forecast by Goddard Institute for Space Studies three-dimensional model. *Journal of geophysical research: Atmospheres, 93*(D8), 9341–9364.

Hashim, B. M., Al Maliki, A., Alraheem, E. A., Al-Janabi, A. M. S., Halder, B., & Yaseen, Z. M. (2022). Temperature and precipitation trend analysis of the Iraq Region under SRES scenarios during the twenty-first century. *Theoretical and Applied Climatology, 148*(3), 881–898.

Mobarak, H. E., Fatahi, E., Ranjbar, S. A.. A., & Hossein, H. (2023). Investigating the RegCM model ability in simulation of summer Khorasan dust. *Journal of Applied Researches in Geographical Sciences, 23*(71), 39–59.

Koheji, M. (2025). On cooling and comfort: The engineering of thermal spaces in Bahrain. *Engineering Studies*, 1–21.

Komurcu, M., Schlosser, C. A., Alshehri, I., Alshahrani, T., Alhayaza, W., AlSaati, A., & Strzepek, K. (2020). Mid-century changes in the mean and extreme climate in the kingdom of Saudi Arabia and implications for water harvesting and climate adaptation. *Atmosphere, 11*(10), 1068.

Kumar, A. (2018). Global warming, climate change and greenhouse gas mitigation. In *Biofuels: Greenhouse gas mitigation and global warming: Next generation biofuels and role of biotechnology*, 1–16.

Lau, N. C. (1992). Climate variability simulated in GCMs. *Climate System Modeling, 617*, 642.

Lemos, M. C., & Rood, R. B. (2010). Climate projections and their impact on policy and practice. *Wiley Interdisciplinary Reviews: Climate Change, 1*(5), 670–682.

Mahmoud, S. H., Gan, T. Y., & Zhu, D. Z. (2023). Impacts of climate change and climate variability on water resources and drought in an arid region and possible resiliency and adaptation measures against climate warming. *Climate Dynamics, 61*(9), 4079–4105.

Malik, A., Stenchikov, G., Mostamandi, S., Parajuli, S., Lelieveld, J., Zittis, G., Ahsan, M. S., Atique, L., & Usman, M. (2024). Accelerated historical and future warming in the Middle East and North Africa. *Journal of Geophysical Research: Atmospheres, 129*(22), e2024JD041625.

Matzarakis, A. (2024). *Heat health action plans-the role of human biometeorology and urban climatology* (No. EMS2024-271). Copernicus Meetings.

Mukheef, R. A., Hassan, W. H., & Alquzweeni, S. J. A. R. (2024). Projections of temperature and precipitation trends using CMhyd under CMIP6 scenarios: A case study of Iraq's Middle and West. *Atmospheric Research, 306*, Article 107470.

Mushtaq, F., Haq, A. U., Bharti, S., Alhems, L. M., & Farooq, M. (2025). *Regional climate dynamics in the Arabian Peninsula: A study of temperature trends and precipitation uncertainty* (No. EGU25-3014). Copernicus Meetings.

Patlakas, P., Stathopoulos, C., Flocas, H., Kalogeri, C., & Kallos, G. (2019). Regional climatic features of the Arabian Peninsula. *Atmosphere, 10*(4), 220.

Patlakas, P., Stathopoulos, C., Flocas, H., Bartsotas, N. S., & Kallos, G. (2021). Precipitation climatology for the arid region of the Arabian Peninsula—Variability, trends and extremes. *Climate, 9*(7), 103.

Price, R. (2022). Climate change risks and opportunities in Yemen.

Purdue, L., Kennet, D., Garnier, A., Parton, A., Djerbi, H., Botan, S., Herveux, L., Davtian, G., Carré, A., Moger, D., & Al-Jahwari, N. (2021). Ancient agriculture in Southeast Arabia: A three thousand year record of runoff farming from central Oman (Rustaq). *Catena, 204*, 105406.

Rao, K. K., Al Mandous, A., Al Ebri, M., Al Hameli, N., Rakib, M., & Al Kaabi, S. (2024). Future changes in the precipitation regime over the Arabian Peninsula with special emphasis on UAE: Insights from NEX-GDDP CMIP6 model simulations. *Scientific Reports, 14*(1), 151.

Roeckner, E. (1992). Past, present and future levels of greenhouse gases in the atmosphere and model projections of related climatic changes. *Journal of Experimental Botany, 43*(8), 1097–1109.

Rummukainen, M. (2010). State-of-the-art with regional climate models. *Wiley Interdisciplinary Reviews: Climate Change, 1*(1), 82–96.

Russo, E., Sørland, S. L., Kirchner, I., Schaap, M., Raible, C. C., & Cubasch, U. (2020). Exploring the parameters space of the Regional Climate Model COSMO-CLM 5.0 for the CORDEX Central Asia Domain. *Geoscientific Model Development Discussions, 2020*, 1–33.

Saharwardi, M. S., Dasari, H. P., Gandham, H., Ashok, K., & Hoteit, I. (2024). Spatiotemporal variability of hydro-meteorological droughts over the Arabian Peninsula and associated mechanisms. *Scientific Reports, 14*(1), 20296.

Sharif, M. (2015). Analysis of projected temperature changes over Saudi Arabia in the twenty-first century. *Arabian Journal of Geosciences, 8*, 8795–8809.

Sheng, D., Zhao, X., Edmonds, J., Waldhoff, S., Patel, P., O'Neill, B., Tebaldi, C., & Wise, M. (2024). Omitting labor responses to heat stress underestimates future climate impact on agriculture.

Shukla, P. R., Dhar, S., Pathak, M., Mahadevia, D., & Garg, A. (2016). Regional low-emission pathways from global models. *Social Science Research Network*. https://doi.org/10.2139/ssrn.2718125

Siehr, S. A., Sun, M., & Aranda Nucamendi, J. L. (2022). Blue-green infrastructure for climate resilience and urban multifunctionality in Chinese cities. *Wiley Interdisciplinary Reviews: Energy and Environment, 11*(5), Article e447.

Skelly, W. C., & Henderson-Sellers, A. N. N. (1996). Grid box or grid point: What type of data do GCMs deliver to climate impacts researchers? *International Journal of Climatology, 16*(10), 1079–1086.

Syed, F. S., Latif, M., Al-Maashi, A., & Ghulam, A. (2019a). Regional climate model RCA4 simulations of temperature and precipitation over the Arabian Peninsula: Sensitivity to CORDEX domain and lateral boundary conditions. *Climate Dynamics, 53*(11), 7045–7064.

Syed, F. S., Latif, M., Al-Maashi, A., & Ghulam, A. (2019b). Regional climate model RCA4 simulations of temperature and precipitation over the Arabian Peninsula: Sensitivity to CORDEX domain and lateral boundary conditions. *Climate Dynamics, 53*(11), 7045–7064. https://doi.org/10.1007/S00382-019-04974-Z

Tareq, A., Rajesh, G., & Bilal, H. (2021). Long term temperature and precipitation trends assessment in the state of Qatar and its implication to energy water and food nexus.

United Nations Economic and Social Commission for Western Asia. (2022). *Regional climate modelling outputs for Saudi Arabia*. https://www.unescwa.org/sites/default/files/pubs/pdf/regional-climate-modelling-outputs-saudi-arabia-english.pdf

Vinodhkumar, B., Ullah, S., Kumar, T. L., & Al-Ghamdi, S. G. (2024). Amplification of temperature extremes in Arabian Peninsula under warmer worlds. *Scientific Reports, 14*(1), 16604.

Waha, K., Krummenauer, L., Adams, S., Aich, V., Baarsch, F., Coumou, D., Fader, M., Hoff, H., Jobbins, G., Marcus, R., Mengel, M., & Schleussner, C. F. (2017). Climate change impacts in the Middle East and Northern Africa (MENA) region and their implications for vulnerable population groups. *Regional Environmental Change, 17*, 1623–1638.

World Bank Group. (2023). *Climate change knowledge portal (CCKP)*. https://climateknowledgeportal.worldbank.org/

Zhu, H., Jiang, Z., Li, J., Li, W., Sun, C., & Li, L. (2020). Does CMIP6 inspire more confidence in simulating climate extremes over China? *Advances in Atmospheric Sciences, 37*, 1119–1132.

Chapter 5
Hazards and Extreme Climate Events in the Arabian Peninsula: Patterns, Drivers and Resilience Pathways

Fayma Mushtaq, Akhouri Pramod Krishna, Manikandan Karuppasamy Ponnambalam, and Thadickal V. Joydas

5.1 Introduction

Extreme climate and weather events are unusual and rare occurrences that stand out due to their intensity or unexpected locations. These include heatwaves, extreme rainfall, floods, thunderstorms, typhoons, hurricanes, tornadoes, tropical cyclones, hailstorms, storm surges, droughts and wildfires (Valavanidis, 2022) (Fig. 5.1). Such events often lead to substantial damage to infrastructure, human health and ecosystems. Some, like wildfires and floods, can cause extensive financial losses, disrupt communities and result in casualties. However, not all extreme events have purely negative consequences; for instance, extreme rainfall can replenish water supplies in arid regions, highlighting the complex interactions between climate and environmental systems. Climate change has significantly influenced the frequency, duration and severity of extreme events (Delmonaco et al., 1999). Over the past 150 years, human activities have released large amounts of greenhouse gases into the atmosphere, trapping heat and causing global warming. As a result, certain extreme events are occurring more frequently, lasting longer and becoming more intense, reinforcing the urgent need for climate action and resilience planning. With every increment of warming, climate extremes are worsening, leading to more frequent and intense heatwaves, heavier rainfall, prolonged droughts and stronger tropical storms (Kumar et al., 2024). These compounding and cascading hazards amplify risks for human

F. Mushtaq (✉) · M. K. Ponnambalam · T. V. Joydas
Applied Research Centre for Environment and Marine Studies, King Fahd University of Petroleum and Minerals, Dhahran, Saudi Arabia
e-mail: fayma.mushtaq@kfupm.edu.sa

A. P. Krishna
Department of Remote Sensing, Birla Institute of Technology, Mesra, Ranchi, India

F. Mushtaq et al. (eds.), *Climate Change in the Arabian Peninsula*,
https://doi.org/10.1007/978-3-032-02481-7_5

health, infrastructure, food security and ecosystems, causing unprecedented losses and damages across regions (Fig. 5.2).

The Intergovernmental Panel on Climate Change (IPCC) defines extreme weather events (EWE) as occurrences that fall at or beyond the 10th or 90th percentile of a probability density function estimated from observations. The increasing frequency of these events due to climate change poses severe threats to ecosystems, societies and economies (Malhi et al., 2020). Human-induced climate change has led to widespread adverse impacts, causing losses and damage beyond natural climate variability. As both natural and human systems are pushed beyond their adaptive capacity, some impacts such as biodiversity loss, ecosystem degradation and infrastructure failure have become irreversible. The observed rise in extreme weather and climate events has already resulted in widespread disruptions (Seneviratne et al.,

Fig. 5.1 Climate change impacts: a visual representation of extreme weather events and environmental disasters

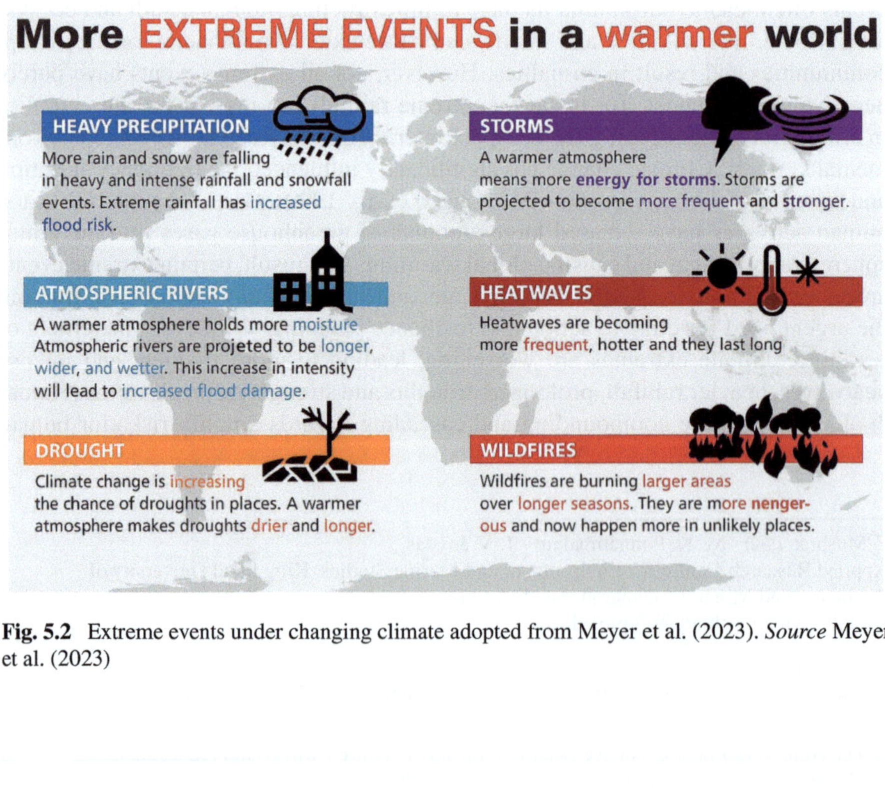

Fig. 5.2 Extreme events under changing climate adopted from Meyer et al. (2023). *Source* Meyer et al. (2023)

2021). These include hot extremes on land and in the ocean, heavy precipitation, droughts and fire weather. Since the Fifth Assessment Report (AR5), growing evidence has attributed these impacts to human-induced climate change, particularly through increased frequency and severity of extreme events. Significant consequences include higher heat-related mortality (medium confidence), increased warm-water coral bleaching and mortality (high confidence) and greater drought-related tree mortality (high confidence). Additionally, the expansion of wildfire-affected areas has been linked to human-induced climate change in several regions (medium to high confidence). The adverse effects of tropical cyclones have also intensified due to rising sea levels and increased heavy precipitation (medium confidence). Beyond rapid-onset disasters, slow-onset climate processes such as ocean acidification, sea level rise and regional decreases in precipitation have been linked to anthropogenic activities (high confidence). These gradual changes exacerbate the risk of extreme climate events, compounding vulnerabilities and making long-term adaptation efforts more challenging.

Extreme climate events encompass various hazards, each with distinct physical characteristics and impacts (McPhillips et al., 2018). Temperature extremes, such as heatwaves and cold spells, have become more pronounced, with strong evidence linking them to anthropogenic global warming. Heavy precipitation has intensified in many regions, increasing the risk of flooding, while droughts have worsened due to rising temperatures and shifting precipitation patterns, affecting water availability, agriculture and ecosystems (Tabari, 2020). Changes in oceanic and atmospheric conditions have influenced the intensity, duration and spatial distribution of tropical and extratropical cyclones, as well as severe convective storms. Additionally, compound events such as heatwaves occurring alongside droughts or heavy rainfall coinciding with storm surges have become more frequent, complicating disaster response and preparedness efforts (Jain, 2022).

The IPCC AR6 report confirms that human influence has unequivocally contributed to the observed rise in extreme weather events (Fig. 5.3). Since the mid-twentieth century, greenhouse gas emissions, land-use changes and other anthropogenic activities have altered the Earth's climate system, amplifying natural variability and increasing the probability of extreme events. Many recent heatwaves, for example, would have been highly unlikely without human-induced climate change. Similarly, the intensification of heavy precipitation and related flooding can be attributed to the increased moisture-holding capacity of a warmer atmosphere (Yin et al., 2021). Future projections indicate that extreme events will continue to escalate in intensity and frequency as global temperatures rise. Heatwaves are expected to become more severe and prolonged, while cold extremes will decline (Domeisen et al., 2023). Heavy precipitation will intensify in many regions, increasing the risk of floods, while drought-prone areas will likely experience prolonged dry spells. Coastal and low-lying areas will face greater risks from rising sea levels and storm surges (Hsiao et al., 2021). Additionally, low-probability but high-impact events such as unprecedented weather extremes may become more likely as climate change progresses, leading to potentially severe and unexpected disruptions. Understanding these risks is crucial for effective adaptation and mitigation. Strengthening early warning systems, improving disaster preparedness, implementing sustainable land and water management practices and developing climate-resilient infrastructure are essential for reducing vulnerabilities. The findings of the IPCC AR6 report provide a robust scientific foundation for informed decision-making and proactive measures to address the growing threats posed by extreme weather and climate hazards.

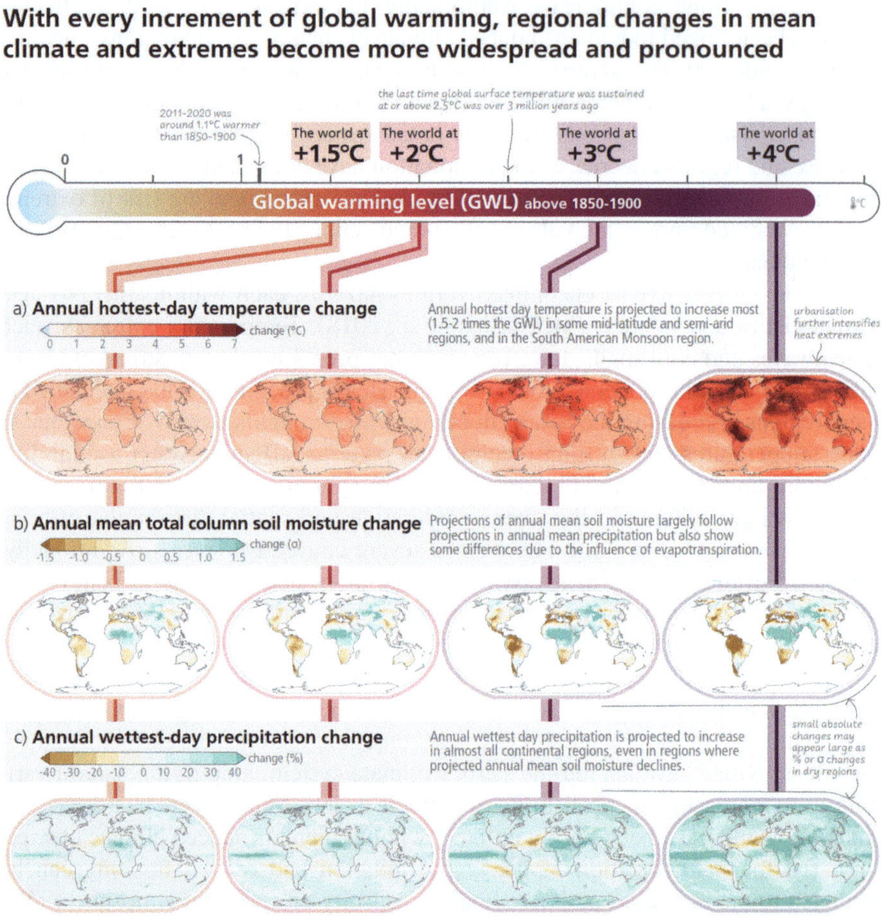

Fig. 5.3 This IPCC AR6 Synthesis Report figure illustrates how climate extremes intensify with rising global temperatures. As warming surpasses 1.5, 2 °C and beyond, heatwaves become more severe, soil moisture declines and extreme rainfall events intensify, increasing risks of droughts, floods and ecosystem disruptions. The figure underscores the urgent need for deep emissions cuts and adaptation measures to mitigate worsening climate impacts. *Source* IPCC (2023), Figure SPM.2, Sixth Assessment Report Synthesis Report, https://www.ipcc.ch/report/ar6/syr/)

5.2 Overview of Extreme Events in Arabian Peninsula

The Arabian Peninsula, one of the regions most exposed to climate change, has witnessed a growing number of severe extreme events over recent decades, including heavy floods, prolonged droughts and drying lakes (World Bank, 2014). These climate-induced challenges have significantly exacerbated environmental, agricultural and public health issues, making the region particularly vulnerable to climate variability and extreme weather events (Pal & Eltahir, 2016; World Bank, 2019). Rising temperatures, shifting precipitation patterns and increasing atmospheric and oceanic changes have intensified these hazards, affecting both coastal and inland areas. The region's warming trend has been particularly alarming, with Saudi Arabia, which covers more than 80% of the Arabian Peninsula, identified as a global hotspot for temperature increase, experiencing a warming rate nearly three times the global average.

Pal and Eltahir (2016) demonstrated in their study that wet-bulb temperature (TW), which combines temperature and humidity, is a crucial indicator of human heat stress, with a TW of 35 °C marking the threshold beyond which even healthy individuals cannot effectively thermoregulate. Their projections indicate that by 2071–2100, under high-emission scenarios, several Gulf cities, including Abu Dhabi, Dubai, Doha, Dhahran and Bandar Abbas, are expected to exceed this critical threshold multiple times, posing severe risks to human survival and habitability in the region. Meanwhile, interior desert regions are projected to experience extreme dry-bulb temperatures, with maximum temperature (Tmax) exceeding 60 °C in some years, making these areas increasingly uninhabitable. The Muslim pilgrimage (Hajj) in Mecca will also face significant heat hazards, with maximum wet-bulb temperature (TWmax) projected to reach 32–33 °C and Tmax exceeding 55 °C, creating dangerous conditions for millions of outdoor pilgrims. The Arabian Gulf and Red Sea further amplify these extreme conditions due to their low albedo, absorbing high amounts of solar radiation, which leads to increased evaporation and high humidity. Moist air transported inland exacerbates heat stress in coastal cities, compounding the risks of extreme heat.

Furthermore, water stress remains a critical issue in the Arabian Peninsula, with minimal precipitation and growing reliance on desalinated water from the Arabian Gulf (Odhiambo, 2017). While desalination has provided a crucial water supply, its extensive use has led to additional environmental stress on marine ecosystems, affecting water quality and biodiversity (Dawoud & Al Mulla, 2012). Moreover, precipitation patterns across the region exhibit significant spatial variability, influenced by multiple climatic factors, including the Mediterranean climate, the Indian monsoon and orographic variations (Patlakas et al., 2021). Climate change has intensified these patterns, leading to more erratic rainfall, increasing the risks of both flooding and prolonged droughts in different parts of the region. The combination of extreme heat and fluctuating precipitation further strains water resources, making sustainable management more challenging. Desertification is another major concern, as rising temperatures and declining precipitation are expected to accelerate the expansion of arid regions (Le Houérou, 1996). The Arabian Peninsula is already one of the most arid regions in the world and continued warming trends will likely make large portions of land even more inhospitable. The loss of arable land will significantly impact agriculture, food security and biodiversity, placing additional pressure on already vulnerable ecosystems and communities. Future climate projections suggest that the northeastern Arabian Peninsula will experience a significantly drier climate from 2016 until the mid-century, which could shift to a relatively wetter climate towards the end of the century (Almazroui et al., 2017; El-Samra et al., 2017).

The economic and social impacts of climate change will be severe, with rising temperatures and extreme weather events disrupting economies and increasing human vulnerability (Kompas et al., 2018). Episodes of heavy rainfall and subsequent flash floods have already resulted in significant economic losses and loss of life, as seen in Jeddah's historic floods in 2009 and 2011 (Almazroui, 2020; Atif et al., 2020; Krishna, 2014). The Arabian Peninsula has also recorded some of the highest temperatures globally, with Jeddah reaching a record-breaking 52 °C on June 22, 2010 (Almazroui et al., 2012). With the growing urgency to understand and mitigate climate change impacts, various studies have used Regional Climate Models (RCMs) like RegCM4 to downscale Global Climate Models (GCMs) from Coupled Model Intercomparison Project Phase 5 (CMIP5) and Phase 6 (CMIP6), improving their applicability for assessing climate extremes in the Arabian Peninsula (Almazroui et al., 2021). While some studies have analyzed climate extremes using observational data, others have incorporated downscaled climate models to enhance regional assessments (Ntoumos et al., 2020). The increasing frequency and intensity of extreme events underscore the urgent need for climate adaptation and mitigation strategies to safeguard the Arabian Peninsula against escalating climate risks. As climate change accelerates, the region will continue to experience compounding challenges, necessitating long-term planning and resilient infrastructure to minimize the socio-economic and environmental consequences of extreme weather events.

Climate Extremes

The definition of extreme events varies across regions, as climatic extremes are context-dependent. Many extreme climate events are increasing in intensity and frequency due to human-induced climate change, with their impacts further exacerbated by urbanization and the expansion of infrastructure into vulnerable areas (McPhillips et al., 2018). In general, an extreme event is characterized by a climate variable exceeding or falling below a threshold near the upper or lower limits of its historical range. Some climate extremes, such as droughts and floods, may result from the accumulation of multiple moderate weather events, which, while not individually extreme, collectively lead to severe conditions. Additionally, certain weather or climate events, even if not extreme in a purely statistical sense, can still cause significant disruptions by surpassing critical thresholds in social, ecological, or physical systems or by coinciding with other extreme events (Seneviratne et al., 2021).

Climate extremes, particularly in temperature and precipitation, play a critical role in understanding climate variability and long-term changes. To systematically monitor these extremes, the Expert Team on Climate Change Detection and Indices

(ETCCDI) developed a set of standardized climate indices. Established as a UN-backed working group, the ETCCDI aimed to define best practices for characterizing climate variability and change. The ETCCDI indices consist of 27 core indicators, derived from daily temperature and precipitation observations. While all indices are calculated on an annual basis, a subset is also analyzed seasonally and monthly to capture variations across different timescales. Of these, 16 indices focus on temperature extremes, assessing both cold and warm events using daily minimum (TN) and daily maximum (TX) temperatures. These indices account for absolute maximum and minimum values, exceedances of fixed thresholds and percentile-based thresholds to identify significant deviations from normal climate patterns. The remaining 11 indices concentrate on precipitation extremes, measuring total rainfall accumulation, threshold-based rainfall events and extreme precipitation events that contribute to floods and droughts. Together, these indices provide a comprehensive framework for analyzing climate extremes and their evolving patterns in response to climate change. These indices have been used worldwide by various researchers to study the extreme trends in rainfall and precipitation data (Brown et al., 2010; Gujree et al., 2017; Kalita et al., 2023; Li et al., 2018; Mushtaq & Nee Lala, 2017; Yun et al., 2012).

In the Arabian Peninsula, temperature and precipitation extreme indices (Table 5.1) have been analyzed by Almazroui et al. (2023) using regional climate model simulations to assess future climate extremes for the mid-century (2036–2065) and end-century (2071–2100), relative to the reference period (1976–2005). Simulations from three Global Climate Models (GCMs) under the World Climate Research Programme (WCRP) CMIP5, were dynamically downscaled over the Arabian Peninsula using the International Centre for Theoretical Physics (ICTP) Regional Climate Model (RegCM4). The analysis of climate extreme indices reveals significant changes under high-emission scenarios, particularly in temperature extremes. The frequency and intensity of warm days and nights are projected to increase substantially, with the Warm Spell Duration Index (WSDI) indicating prolonged heatwaves and more days exceeding the 90th percentile threshold. In contrast, cold extremes are expected to decline, with a marked decrease in the occurrence of cold days and nights and the Cold Spell Duration Index (CSDI) projections showing that prolonged cold spells will become increasingly rare. Precipitation patterns also exhibit greater variability, with an increase in very wet days (R95p) and extremely wet days (R99p), suggesting a higher frequency of intense rainfall events. The number of heavy precipitation days (R10mm, R20mm) is projected to rise, increasing the risk of flash floods, particularly in urban areas. At the same time, Consecutive Dry Days (CDD) are expected to increase, indicating a likelihood of longer drought periods. These projected changes in climate extremes indicate a growing vulnerability to both heat stress and hydrological hazards.

Recently, the changing precipitation regime has been analyzed by Rao et al. (2024) with a particular focus on the United Arab Emirates (UAE) (Fig. 5.4). Their study employs a subset of high-resolution NASA Earth Exchange Global Daily Downscaled Projections (NEX-GDDP) data, which are derived from CMIP6 Global

Table 5.1 Extreme temperature and precipitation indices used by Almazroui et al. (2023) using regional climate model simulations to assess future climate extremes for the mid-century (2036–2065) and end-century (2071–2100), relative to the reference period (1976–2005)

Index name	Definition	Units	ID
Warm days	Percentage of days when daily maximum temperature (TX) exceeds the 90th percentile	Percentage (%)	TX90p
Warm nights	Percentage of nights when daily minimum temperature (TN) exceeds the 90th percentile	Percentage (%)	TN90p
Warm spell duration index (WSDI)	Number of days per year part of warm spells, defined as periods of at least six consecutive days with TX > 90th percentile	Days	WSDI
Cold days	Percentage of days when daily maximum temperature (TX) falls below the 10th percentile	Percentage (%)	TX10p
Cold nights	Percentage of nights when daily minimum temperature (TN) falls below the 10th percentile	Percentage (%)	TN10p
Cold spell duration index (CSDI)	Number of days per year part of cold spells, defined as periods of at least six consecutive days with TN < 10th percentile	Days	CSDI
Very wet days	Annual precipitation amount from days with precipitation exceeding the 95th percentile	mm	R95p
Extremely wet days	Annual precipitation amount from days with precipitation exceeding the 99th percentile	mm	R99p
Heavy Rainfall Days	Number of days with daily precipitation greater than 10 mm	Days	R10
Very heavy precipitation days	Number of days with daily precipitation greater than 20 mm	Days	R20
Consecutive dry days (CDD)	Maximum number of consecutive dry days with daily precipitation < 1 mm	Days	CDD

Source Almazroui et al. (2023)

Climate Models (GCMs). In contrast to Almazroui et al. (2023), who used CMIP5-based regional downscaling, Rao et al. (2024) utilized CMIP6 data, which incorporates the latest Shared Socioeconomic Pathway (SSP) scenarios i.e., SSP1-2.6, SSP2-4.5 and SSP5-8.5 to assess future precipitation trends. The analysis was conducted over three future time periods: near future (2021–2050), mid-future (2051–2080) and far future (2081–2100), with the baseline period set as 1985–2014 (Fig. 5.4). The study examines spatiotemporal changes in extreme precipitation indices, including Simple Daily Intensity Index (SDII), 1-day maximum precipitation (RX1DAY) and the number of days with precipitation exceeding 10 mm (PD10MM) (Table 5.2). The findings indicate a progressive wetting trend across the UAE, particularly in the mid and far future, with an increase of up to 30% in annual precipitation. Additionally, CDD are projected to decrease, indicating shorter dry spells compared to historical conditions. Both studies highlight the significant impact of climate change on extreme weather patterns over the Arabian Peninsula, with a specific focus on increasing heavy rainfall events and changes in dry spells. The findings suggest a progressive wetting trend in parts of the region, particularly the UAE, where annual precipitation is projected to increase by up to 30% in the mid to late twenty-first century. This increase is driven by higher precipitation intensity, more frequent extreme rainfall events (e.g., RX1DAY, R10mm, R20mm) and a general shift toward wetter conditions under high-emission scenarios (SSP5-8.5, RCP8.5). However, a key difference between the studies lies in the projection of CDD. While Almazroui et al. (2023) indicate an increase in CDD, suggesting a rise in drought frequency in some parts of the Arabian Peninsula, Rao et al. (2024) find a decline in CDD, implying shorter dry spells in the UAE towards the end of the century. This discrepancy highlights the regional variability in precipitation trends, which may depend on factors such as spatial resolution of models, dataset selection (CMIP5 vs. CMIP6) and geographical focus. Overall, both studies emphasize the growing risks associated with extreme precipitation variability, including more frequent flash floods, increasing water management challenges and increased risks to infrastructure and human settlements. The findings underscore the importance of using high-resolution climate data to refine regional projections, ensuring effective climate adaptation strategies to address the escalating impacts of extreme weather events across the Arabian Peninsula.

Vinodhkumar et al. (2024) investigated the amplification of temperature extremes in the Arabian Peninsula under different global warming levels (GWLs) of 1.5, 2.0, 3.0 and 4.0 °C, using three Shared Socioeconomic Pathways (SSP1-2.6, SSP2-4.5 and SSP5-8.5) (Fig. 5.5). Their study utilized high-resolution datasets from 27 models of the NASA Earth Exchange Global Daily Downscaled Projections under the Coupled Model Intercomparison Project Phase 6 (NEX-GDDP-CMIP6). The findings indicate that summer maximum temperatures are projected to rise by 0.11–0.67 °C per decade, while winter minimum temperatures are expected to increase by 0.09–0.70 °C per decade across the selected SSPs. Additionally, warm temperature extremes are projected to intensify significantly, whereas cold extremes are expected to decline across all GWLs and SSP scenarios, highlighting the increasing climate vulnerability of the region under future warming conditions.

Sea Level Rise

Climate change encompasses various aspects of changing weather patterns, but one of its critical impacts is sea-level rise (SLR). SLR is a global phenomenon affecting coastlines worldwide, with its effects varying based on local geophysical factors (York et al., 2003). In the Gulf region, the low-lying, hyper-arid coastal areas are particularly vulnerable, especially along the eastern Arabian Peninsula. This includes major cities and critical infrastructure in Iraq, Kuwait, Saudi Arabia, Bahrain, Qatar and UAE (Bakhamis et al., 2024). The Gulf region hosts vital economic assets such as airports, seaports, power plants, refineries and desalination plants, which are crucial for both the local economy and global trade. Given that the Gulf is a semi-enclosed, shallow waterbody, it is highly sensitive to rising sea levels and increasing sea surface temperatures, further amplifying its vulnerability to climate

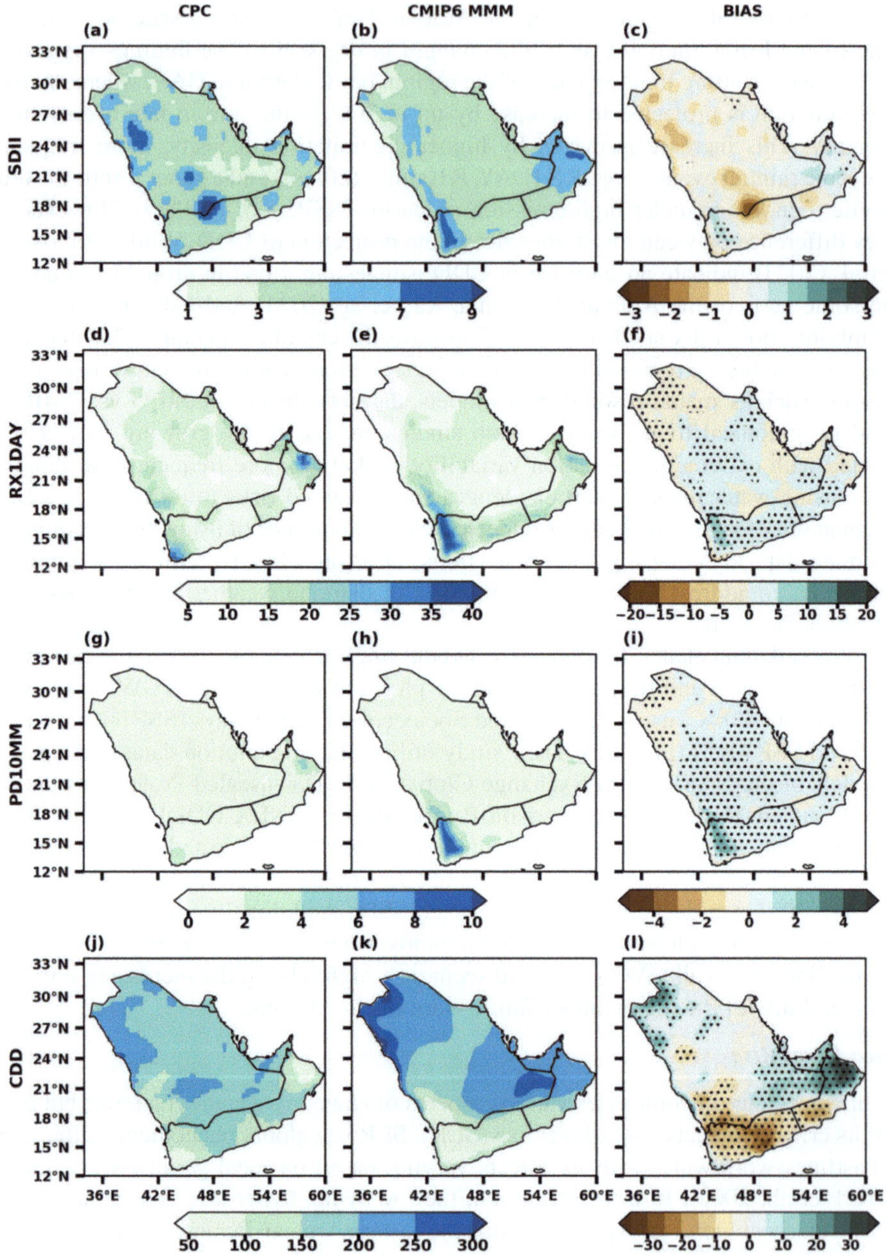

Fig. 5.4 Spatial distribution of extreme precipitation indices over the Arabian Peninsula, as analysed by Rao et al. (2024), based on Climate Prediction Center (CPC) observations, CMIP6 Multi-Model Mean (MMM) simulations and Bias during the period 1985–2014. The indices displayed include SDII, RX1DAY, PD10MM and CDD. *Source* Rao et al. (2024)

Table 5.2 Overview of the study by Rao et al. (2024) on projected changes in precipitation extremes over the Arabian Peninsula, with a focus on the UAE

Aspect	Details
Study focus	Changes in precipitation extremes over the Arabian Peninsula, with emphasis on the UAE
Climate model used	NASA Earth Exchange Global Daily Downscaled Projections (NEX-GDDP) from CMIP6
Downscaling method	High-resolution downscaling of CMIP6 data
Emission scenarios	SSP1-2.6, SSP2-4.5, SSP5-8.5
Time periods analyzed	Near future (2021–2050), Mid-future (2051–2080), Far future (2081–2100), with baseline 1985–2014
Precipitation indices used	Simple daily intensity index (SDII), 1-day maximum precipitation (RX1DAY), days with precipitation exceeding 10mm (PD10MM), consecutive dry days (CDD)
Geographical focus	Arabian Peninsula, specifically UAE

Source Rao et al. (2024)

change. The main driver of accelerated SLR is human-induced climate change, which leads to rising sea temperatures and the melting of land ice. Thermal expansion of seawater due to increasing temperatures is one of the primary contributors to SLR, accounting for approximately 41.7% of the global increase (Kopp et al., 2016). Studies have shown that thermal expansion contributes between 1.1 mm/year and 1.9 mm/year to SLR, with a more recent estimate indicating a contribution of at least 35% from this factor alone (Frederikse et al., 2020; IPCC, 2013). Melting ice sheets and glaciers are another significant contributor to SLR. The Greenland Ice Sheet (GIS) and West Antarctic Ice Sheet (WAIS) have seen accelerated melting in recent decades, contributing around 0.765 mm/year to SLR on average (Cazenave & Llovel, 2010). Recent data show that the GIS contributed 0.73 mm/year between 2002 and 2011, while the WAIS contributed 0.55 mm/yr during the same period (IPCC, 2013). These contributions have increased over time, with estimates showing a rise from 0.4 mm/yr to 1 mm/year between 1961 and 2004 (Kaser et al., 2006). Terrestrial groundwater storage changes also play a role in SLR, although they are less significant compared to thermal expansion and ice melt. Groundwater pumping for agricultural and other purposes can lead to land subsidence, indirectly contributing to SLR. However, terrestrial groundwater storage only accounts for less than 10% of total drivers affecting SLR (Frederikse et al., 2020; Wöppelmann et al., 2013).

In the Arabian Peninsula, regional drivers of SLR include unique local factors such as spatiotemporal differences in air pressure, atmospheric temperature, sea surface temperature, freshwater fluxes and ocean currents (Bakhamis et al., 2024; Mushtaq et al., 2024). The Arabian Gulf, being a semi-enclosed shallow waterbody, is particularly sensitive to these drivers. Increasing temperatures in the region fuel

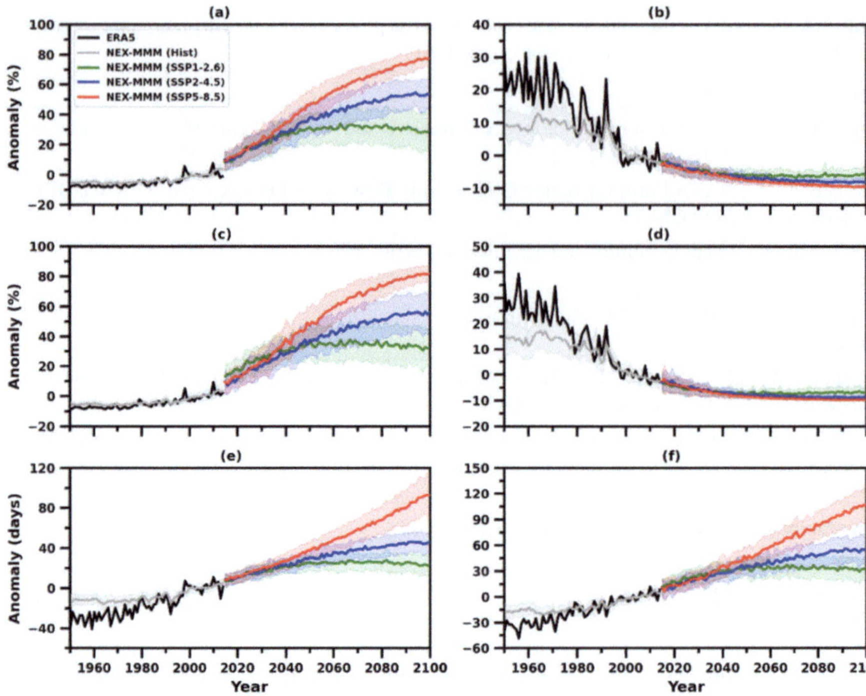

Fig. 5.5 Historical (1951–2014) and projected (2015–2100) changes in annual anomalies of extreme temperature indices over the Arabian Peninsula using NEX-GDDP-CMIP6 MMM and ERA5, relative to the reference period (1995–2014) studied by Vinodhkumar et al. (2024). The selected SSPs show trends for: **a** warm days (TX90p), **b** cold days (TX10p), **c** warm nights (TN90p), **d** cold nights (TN10p), **e** hot days (HDs) and **f** tropical nights (TR). *Source* Vinodhkumar et al. (2024)

thermal expansion, which is a critical factor for SLR (IPCC, 2013). The salinity levels in the Arabian Gulf are also influenced by desalination activities, which can affect the density and volume of seawater (Dawoud, 2012; Mushtaq, 2025). Additionally, the low precipitation rates in the region, with most rainfall occurring along the northeastern coast, contribute to the overall vulnerability (Hereher, 2020). Regional projections indicate that the Arabian Gulf has an average SLR rate of 4.3 mm/yr with an error margin of 0.4 mm, higher than the Indian Ocean's rate of 3.3 mm/year (NOAA, 2022). This accelerated rise suggests increased vulnerability to coastal flooding and infrastructure risks in the region, likely driven by its distinct geophysical characteristics and regional climatic influences. The impacts of SLR in this region are expected to be severe, including coastal erosion, saltwater intrusion and damage to critical infrastructure such as airports, seaports, power plants, refineries and desalination plants (Bakhamis et al., 2024). Figures 5.6 and 5.7 from (Bakhamis et al., 2024) provides specific details about the global drivers of sea-level rise (SLR) and the land area losses due to SLR in the Arabian Gulf, respectively.

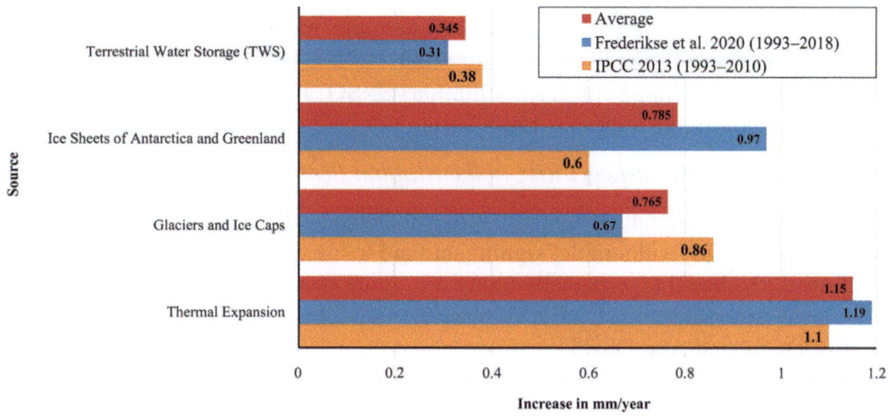

Fig. 5.6 Global SLR drivers adopted from Bakhamis et al. (2024). *Source* Bakhamis et al. (2024)

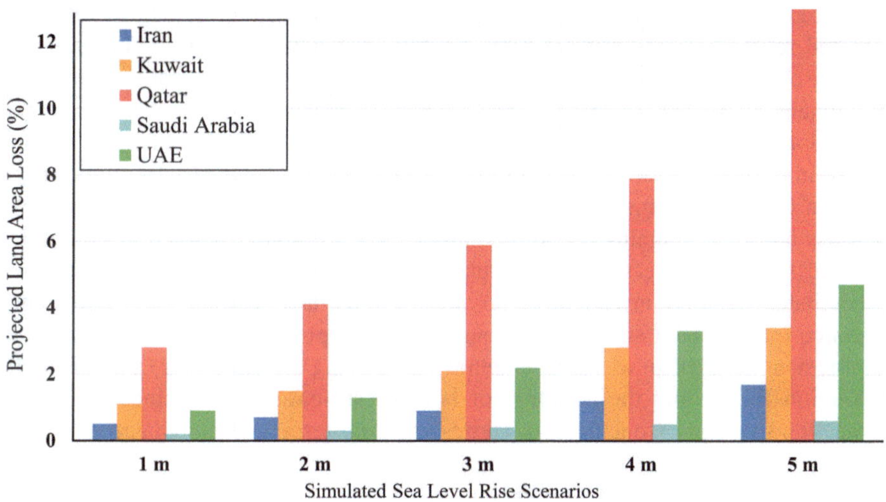

Fig. 5.7 Projected land area loss by Country due to SLR in the Arabian Gulf studied by Bakhamis et al. (2024). *Source* Bakhamis et al. (2024)

Global Drivers of Sea-Level Rise

1. *Thermal Expansion*: This is one of the prime contributors to SLR, responsible for at least 35% of the total increase. Studies show that thermal expansion contributes between 1.1 and 1.9 mm/year to SLR.
2. *Ice Sheets and Glaciers*: Ice sheets and glaciers contribute approximately 0.765 mm/year to SLR. The melting of these ice bodies has accelerated over recent decades, with contributions increasing from 0.4 to 1 mm/year between 1961 and 2004.

3. *Water Storage*: Changes in surface water due to the construction of reservoirs and dams, groundwater consumption, vertical land movement and weather patterns like El Niño-Southern Oscillation (ENSO) also affect SLR. Water storage changes contribute around 0.15–0.24 mm per year over the previous decade.

Regional Drivers of Sea-Level Rise in the Arabian Peninsula

1. Qatar: Most vulnerable country, could lose between 2.70 and 13.25% of its land.
2. Bahrain: Highly vulnerable, significant land loss expected.
3. Kuwait: Vulnerable, especially along low-lying coastal areas.
4. United Arab Emirates (UAE): Vulnerable, particularly in urbanized coastal areas like Dubai and Abu Dhabi.
5. Saudi Arabia: Least vulnerable among the mentioned countries, with less than 0.5% land loss in the worst-case scenario (5-m increase).
6. Iraq: Vulnerable, especially along the Shatt Al-Arab River system and coastal cities like Basra.

To reduce uncertainties and formulate effective mitigation strategies for SLR, targeted research is essential at both regional and national scales. This includes the consolidation of historical records, enhancement of observational networks and refinement of coupled climate–ocean models to generate high-precision regional projections (World Bank, 2022). Although desalination infrastructure has significantly improved freshwater availability for domestic and industrial sectors in the Arabian Gulf (Bakhamis et al., 2024), its cumulative impact on nearshore salinity regimes and SLR remains poorly quantified. The interplay between desalination-induced salinity shifts, saltwater intrusion and climate-driven sea level dynamics necessitates detailed investigation to inform integrated coastal zone and water resource management strategies. Strengthening adaptive capacity will require the deployment of real-time environmental monitoring systems, the use of high-resolution RCMs and the adoption of both nature-based and engineered coastal protection solutions. Concurrently, addressing long-term risks associated with SLR will depend on the decarbonization of energy systems and enhanced regional cooperation. A coordinated, data-driven and interdisciplinary approach is vital for advancing climate-resilient coastal governance and safeguarding socio-ecological systems in the Arabian Gulf.

Extreme Heatwaves

Climate change is intensifying extreme heat events across the Arabian Peninsula, posing a significant threat to human health and well-being (Hamed et al., 2024). Rising air temperatures have substantially amplified the intensity and frequency of heatwaves and other temperature-related extremes (Houmsi et al., 2023). Future projections indicate an increasing frequency and duration of extreme heat events, further exacerbating regional climate vulnerabilities (Shukla et al., 2019). The impacts of increasing temperatures intensify susceptibility to heat-related illnesses such as heat exhaustion, cardiovascular complications, renal disorders and dehydration (Almazroui et al., 2016a, 2016b; Johnson et al., 2019). Moreover, extreme heat significantly affects labor productivity, with adverse socioeconomic consequences (Dasgupta et al., 2021). As the frequency and severity of extreme heat events escalate, challenges to human well-being, economic activities and environmental systems become more profound. The Arabian Peninsula is particularly vulnerable to climate change, with an observed rise in temperature extremes and thermal bioclimatic indices (Hamed et al., 2022). The region has been termed the "global ground zero" for climate change due to its acute exposure to extreme heat and associated impacts (Al-Delaimy, 2020). Hamed et al. (2024) conducted a study mapping population exposure to varying Universal Thermal Climate Index (UTCI) levels across North Africa and the Arabian Peninsula (NAAP) under future climate change scenarios using machine learning (ML) techniques. To determine the key meteorological drivers influencing UTCI, Sobol's global sensitivity analysis was applied, identifying air temperature, wind speed, surface pressure, solar radiation and humidity as the most influential factors. The study found that air temperature had a strong positive correlation with UTCI, underscoring its critical role in heat stress, while wind speed and surface pressure exhibited negative correlations, indicating their role in reducing thermal stress. Relative humidity showed a mixed effect, with a negative correlation in most of the Arabian Peninsula but a positive correlation in the humid southern regions.

The findings indicate that even a 0.5 °C rise in global temperatures could lead to 20 additional days of moderate thermal stress in Egypt and Sudan, while strong thermal stress days are projected to increase significantly in the Arabian Peninsula. Under the SSP1-1.9 scenario (1.5 °C global temperature rise), population exposure to UTCI was found to be highly concentrated, affecting over 10 million people-°C, particularly in urban centers and densely populated regions such as the Arabian Peninsula, South Sudan, Cairo, the Nile Delta, Algiers and Tunisia. These areas face higher vulnerability due to increasing population densities, which exacerbate heat stress risks. The urban heat island effect, combined with rising temperatures, intensifies public health and infrastructure challenges, further compounding the effects of extreme heat. This risk becomes even more pronounced with a further 0.5 °C temperature increase from SSP1-1.9 to SSP1-2.6, which is projected to increase population exposure by up to 1.5 million person-°C. The most affected regions include the Arabian Peninsula, Iraq, Syria, Southern Sudan and the Mediterranean coastal areas, where thermal stress is expected to rise sharply. Future projections of UTCI categories suggest a shift toward higher thermal stress levels across the NAAP region. Under SSP1-1.9, strong cold stress and extreme thermal stress (Table 5.3) remain negligible, while moderate and strong thermal stress levels exhibit significant variability. Regions such as South Sudan, Oman and Yemen are projected to experience 100–200 days of moderate thermal stress, while Libya and Egypt's Mediterranean coast could see 100–150 days in this category. The sensitivity of UTCI to slight temperature increases indicates a considerable intensification of thermal stress risks, leading to a reduction in slight cold stress days along the Mediterranean coast and Syria, while moderate thermal stress days will increase in north-western NAAP, Libya, Egypt and northern Sudan. In the far future, spatial patterns suggest a broader and more intense UTCI impact, with regions such as Oman, Sudan and parts of the Arabian Peninsula experiencing the highest rise in strong thermal stress days. Meanwhile,

Table 5.3 Universal thermal climate index (UTCI) stress classification

UTCI (°C)	Stress category
< -13	Strong cold stress
-13 to 0	Moderate cold stress
0–9	Slight Cold Stress
9–26	No thermal stress
26–32	Moderate thermal stress
32–38	Strong thermal stress
38–46	Very strong thermal stress
> 46	Extreme thermal stress

Source Hamed et al. (2024)

Mauritania, Morocco, Iraq, Algeria and Tunisia are likely to gain up to 12 additional days of strong thermal stress, while the Red Sea coast is projected to see the highest decrease in stress days. These results underscore the urgent need for targeted climate adaptation measures such as heat warning systems, climate-resilient urban planning and integrated water resource management to address the escalating thermal risks in the region. The findings also highlight the nonlinear nature of heat stress intensification, even with marginal temperature increases and call for focused interventions in economically significant and high-risk urban areas where extreme heat poses serious threats to livelihoods, energy infrastructure and public health systems.

Desertification and Sandstorms

Desertification, a major driver of environmental challenges, transforms productive land into barren, unproductive landscapes due to poor land management. It reduces the land's ability to support life, negatively affecting wild species, domestic animals, agricultural crops and human populations (Amin, 2004). The reduction in plant cover accelerates soil erosion by wind and water, further exacerbating land degradation. This issue is particularly severe in semi-arid regions and drylands, such as the Arab States of the Gulf, where land degradation has become a critical ecological concern (Abahussain et al., 2002). Recognized as early as 1992 by the United Nations Conference on Environment and Development (UNCED) in Chapter 12 of Agenda 21, desertification remains a pressing challenge. Globally, drylands account for 41% of the total land area, covering approximately 150 million km^2. Of this, around 60 million km^2 are classified as degraded drylands and nearly two-thirds suffer from low fertility and severe ecological damage, effectively becoming desert-like landscapes. Human activities are responsible for approximately 35% of this degradation equivalent to 14 million km^2 of land (Al-Awadhi & Alkandary, 2024). The consequences of desertification extend beyond the environment, making it a socio-economic crisis affecting nearly one billion people across 110 countries.

In Arabian Peninsula the severity of desertification was studied by Rajendran et al. (2023) by examining land degradation and agricultural changes in the Abu Samra region of Qatar over two decades (2000–2021) using Advanced Spaceborne

Thermal Emission and Reflection Radiometer (ASTER) satellite data. Aeolian (wind-driven) sand deposits were mapped to understand their movement and impact on the landscape. By analyzing the spectral properties of sand, particularly in the 8–9.50 μm wavelength range, the study identified significant shifts in the distribution of these deposits. The findings revealed a notable decrease in sand deposit coverage from 9.70% in 2000 to just 2.94% in 2021, indicating that wind erosion and transportation played a major role in reshaping the region. Satellite images further revealed drastic land-use changes, including the disappearance of sabkhas (salt flats) by 2015 and a decline in agricultural land, which was present in 2000 but no longer visible after 2005. The extent of land degradation in the area was further substantiated through field surveys and soil sample analyses. The study highlights how desertification is transforming the landscape, with wind erosion redistributing sand and accelerating the decline of fertile land. The loss of farmland and the disappearance of sabkhas point to a broader transformation of the region's ecosystem, driven by both natural processes and human activities.

In addition, Salih et al. (2021) underscored the severity of desertification in Saudi Arabia through their assessment of the Al-Ahsa Oasis, where land degradation is particularly evident. This date palm-dominated region in the Eastern Province has been extensively mapped and analyzed to assess the extent and severity of land degradation. Using Landsat-8 OLI imagery from December 2019, the study applied the Desertification Degree Index (DDI) and found that an alarming 83% of the oasis had been affected by desertification. The study further classified the severity of degradation, revealing that 14.2% of the oasis experienced extreme desertification, 19.3% was severely degraded, 23.9% faced moderate desertification and 25.6% was classified as low-level degradation. These findings highlight the widespread impact of desertification, even in historically fertile regions, emphasizing the urgency of addressing land degradation through sustainable management practices and mitigation strategies. The situation in Al-Ahsa Oasis reflects a broader trend across Saudi Arabia, where increasing aridity, unsustainable land use and climate change are exacerbating the vulnerability of key agricultural and ecological zones. Without immediate intervention, desertification will continue to threaten vital resources, food security and economic stability in the region. Together, these studies illustrate the urgent need for proactive measures to combat desertification across the Arabian Peninsula. As climate change, land mismanagement and increasing aridity continue to drive land degradation, integrated strategies must be developed to restore degraded lands, enhance resilience and ensure long-term sustainability for both ecosystems and human societies.

Climate change plays a crucial role in accelerating desertification, which, in turn, increases the frequency and intensity of sand and dust storms (SDS) (WMO, 2022). These storms are among the most common meteorological hazards in arid and semi-arid regions, where vast amounts of mineral dust particles are lifted into the atmosphere (Awadh, 2023). Sparse vegetation, prolonged droughts and land degradation further contribute to their severity and occurrence. As a result, sand and dust storms significantly impact air quality by increasing particulate matter (PM) concentrations, often making airborne dust the primary source of PM-related pollution in many regions (Meo et al., 2021). This poses serious health risks, particularly for those exposed to these particles over extended periods. According to United Nations Convention to Combat Desertification (UNCCD) experts, approximately two billion tons of sand and dust enter the atmosphere annually equivalent to the weight of 350 Great Pyramids of Giza. More than 25% of this problem is attributed to human activities, including deforestation, land degradation, unsustainable land management and water mismanagement, all of which contribute to worsening environmental and socio-economic conditions (Atieku & Segbefia, 2024). The impact extends beyond air pollution and human health, affecting agriculture, industry, transportation and water resources. In agricultural regions, dust deposition reduces soil fertility and lowers crop yields, while in urban areas, it disrupts infrastructure and transportation networks. Globally, an estimated 330 million people are exposed daily to airborne dust particles, which can be transported by wind over thousands of kilometers from their source regions (Lian et al., 2025).

The Arabian Peninsula is the second-largest dust source in the world, surpassed only by the Sahara Desert (De Longueville et al., 2010; Tanaka & Chiba, 2006) (Fig. 5.8). Each year, dust storms impact various regions of the peninsula, causing severe environmental and health consequences. Saudi Arabia, which covers nearly four-fifths of the Arabian Peninsula, is particularly vulnerable to these storms due to its vast desert landscapes and arid climate (Fig. 5.9) that influence its atmosphere, ecosystems and human health. Dust storms in Saudi Arabia occur throughout the year but are most intense between February and June (Notaro et al., 2013). These storms bring immediate disruptions, reducing visibility and creating hazardous conditions for transportation. The deterioration in air quality leads to respiratory issues in both humans and livestock, exacerbating health problems in vulnerable populations. In addition to health impacts, dust storms compromise atmospheric radiative transfer, diminishing solar energy efficiency and accelerating soil degradation by stripping topsoil nutrients. The accumulation of dust also damages telecommunications and mechanical systems, increases dust deposition in urban and agricultural areas and contributes to the degradation of property and vegetation, particularly in ecologically fragile regions.

Beyond local sources, Saudi Arabia is also influenced by dust transported from surrounding regions (Table 5.4). The Middle East, as one of the world's largest dust-producing areas (Rezazadeh et al., 2013), experiences extensive dust storm activity, particularly in regions such as the Syria-Iraq plains, northern and eastern Saudi Arabia, the Arabian Gulf and southwestern and southeastern Iran (Hamidi et al., 2017). The primary dust sources in this region form a continuous belt extending from the upper reaches of the Tigris and Euphrates rivers to the coast of Oman (Prospero et al., 2002). Within Saudi Arabia, major dust-producing areas include the Rub' Al Khali, An-Nafud and An-Dahna deserts. However, dust storms affecting the country are not solely generated within its borders. Long-range dust transport from the Sahara Desert and Iraq significantly contributes to dust loading in western, northern and eastern Saudi Arabia (Fig. 5.10), depending on atmospheric circulation patterns (Prakash et al., 2015).

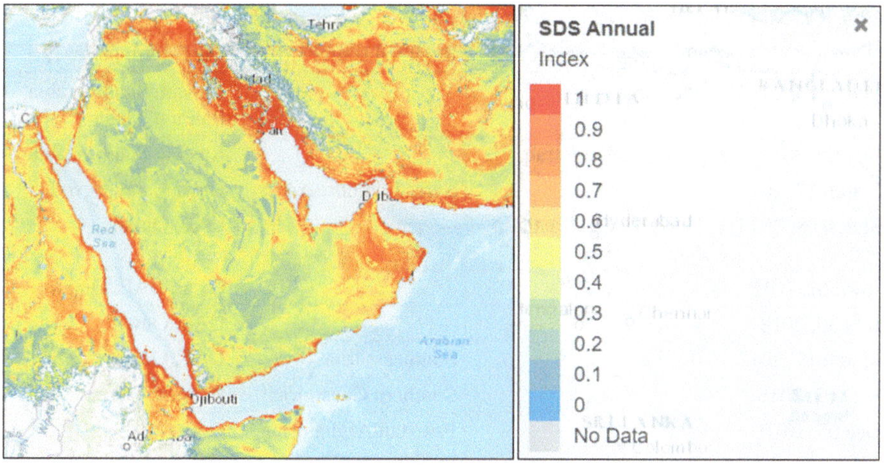

Fig. 5.8 Annual sand and dust storms source base-map from 2000 to 2018. *Source* UNCCD/UNEP/WMO, 2019

Fig. 5.9 Satellite view of a major dust plume over the Arabian Peninsula extending into the Arabian Gulf. *Source* European Space Agency (ESA), July 11th, 2008

The seasonality and frequency of dust storms vary due to climatic factors such as drought and precipitation patterns (UNEP, 2013). While dust activity persists year-round in much of the Middle East, its intensity fluctuates with seasonal changes. During winter, most parts of the region experience reduced dust storm activity, although localized peaks may still occur, particularly in areas where strong winds mobilize fine desert sediments. These fluctuations highlight the complexity of dust storm dynamics and the need for continuous monitoring and assessment. Given the widespread impacts of dust storms on human health, infrastructure and regional

Table 5.4 Major dust storm sources and pathways over Arabian Peninsula

Date	Dust storm source	Regional source	Dust storm pathway
March 18, 2012	West Iraq	–	Kuwait, Saudi Arabia, Bahrain, Qatar
July 05, 2012	Mesopotamian Iraq	–	Kuwait, Saudi Arabia
March 17, 2012	–	Syria-Iraq border	Kuwait, Saudi Arabia
March 08, 2015	Northern and Southern Saudi Arabia	–	UAE, Oman
April 02, 2015	–	–	Qatar, UAE, Oman
July 30, 2018	–	–	UAE, Oman
March 19, 2012	–	Southeast Iran	UAE, Oman
Sept 03, 2015	–	Southern Coast, Iran	Bahrain, Qatar, UAE
Dec 05, 2010	–	Pakistan/Afghanistan border	Oman
Oct 14, 2014	–	Pakistan/India border	Oman
March 19, 2017	–	East Africa	Saudi Arabia
July 20, 2011	–	East Africa	Saudi Arabia

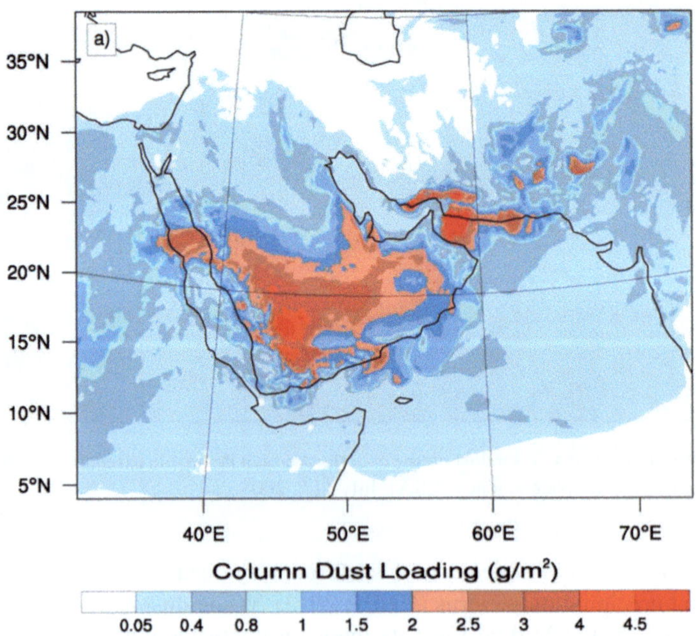

Fig. 5.10 Spatial distribution of simulated dust load (19th March 2012 over the Arabian Peninsula and Red Sea adopted from Prakash et al. 2015). *Source* Prakash et al. (2015)

climate, there is a growing need for improved forecasting and mitigation strategies. Research efforts combining satellite observations, ground-based measurements and numerical models have enhanced the understanding of dust storm patterns and their contributing factors. Addressing this challenge requires a coordinated approach at regional and international levels to develop effective early warning systems, mitigate environmental damage and safeguard communities from the adverse effects of dust storms.

Flash Floods

Floods are among the most devastating natural disasters, posing a significant threat to lives, property and economies worldwide (McCarthy, 2001). Flash floods, in particular, have severe socio-economic and environmental consequences, often resulting in loss of life, infrastructure damage and ecosystem destruction (Abdelkader et al., 2021). These floods typically occur due to intense rainfall over a short period or the sudden release of large volumes of water from rivers or dams (Murray & Ebi, 2012). With climate change intensifying extreme weather events, the frequency and severity of flash floods are projected to increase, placing many regions at increased risk (Khan et al., 2010). The Arabian Peninsula, is experiencing an increasing flood risk, primarily driven by land use changes resulting from rapid urbanization, economic expansion and population growth in recent decades (Hussein et al., 2020). Urbanization significantly alters hydrological processes, leading to greater surface runoff, elevated peak discharges, shorter response times, deeper flood depths, increased flood volumes and wider flood extents during extreme rainfall events (Booth & Bledsoe, 2009). One of the most affected areas in the region is wadis, an Arabic term meaning "valley", where flash floods have become more frequent in recent years (Abdelkader et al., 2021). These typically dry riverbeds experience prolonged drought conditions throughout the year but become highly susceptible to intense flooding during seasonal rainfall events. Studies indicate that arid and semi-arid regions are more prone to severe flooding than humid areas due to intense rainfall events and rapid runoff accumulation (Costa, 1987).

Coastal areas are also experiencing frequent and severe flash floods, with their impacts becoming more pronounced over time. Several key factors contribute to the intensity of coastal flooding, including climate change, rapid urban expansion, the proximity of critical infrastructure and assets to shorelines and ongoing urbanization (Arnous et al., 2022). The increasing prevalence of flash flood-related hazards presents significant challenges to global development efforts, leading to numerous fatalities and substantial economic losses (Dhiman et al., 2019). Moreover, the disruptive nature of floods extends beyond immediate damage, imposing a considerable financial burden on mitigation and adaptation strategies (Cotti et al., 2022). Across the Arabian Peninsula, addressing the growing risks of flash flood hazards under projected climate change scenarios requires both a comprehensive understanding of extreme rainfall events and the implementation of effective mitigation and adaptation measures (Terry et al., 2023). Several locations across the region, including

Jeddah, Makkah, Madinah, Riyadh and the UAE, have already experienced significant flooding events, highlighting need for proactive flood management strategies (Table 5.5).

Flash Floods in Jeddah: The 2009 and 2011 Events

Flash floods represent a critical hazard to the city of Jeddah, with their increasing frequency and severity being strongly associated with shifting climatic patterns (Al Saud, 2010). These extreme hydrometeorological events not only threaten human lives and infrastructure but also impose substantial disruptions on the socio-economic systems of the region (Ameur, 2016). Situated on the coastal plain of western Saudi Arabia, Jeddah has, in recent years, experienced anomalously intense rainfall events (Abdullah et al., 2019). These extremes, when coupled with topographical constraints, altered land use patterns and inadequate drainage infrastructure, have significantly increased surface runoff, leading to alarming flood incidents (Shah et al.,, 2023) (Table 5.6). The Jeddah region, has witnessed extreme short-duration rainfall events, often lasting up to three hours, culminating in high-intensity flash floods. Among these, the 2009 flood is considered one of the most devastating in the Kingdom's recent history. On November 25, 2009, Jeddah received between 70 and 90 mm of rainfall within a span of just four hours almost double the city's average annual precipitation. This unprecedented downpour triggered widespread flooding across Jeddah and adjacent coastal zones along the Red Sea, causing catastrophic damage (Al-Ahmad, 2009; Murphy, 2009; Davies, 2013; Zaigham et al., 2021). A major contributing factor to the magnitude of destruction was the absence of an operational early warning system, which left residents unprepared for the rapidly rising floodwaters (Tammar et al., 2020).

The flash floods of 2009 and 2011 clearly exemplify of the vulnerability of Jeddah's urban environment to extreme weather events (Youssef et al., 2016). Both

Table 5.5 Flash floods occurrences of Riyadh (1985–2016)

Year	Date(s)
1985	December 24
2000	November 15
2005	January 21 and February 15
2009	November 24
2010	May 3
2011	January 18
2012	April 14
2013	May 1
2013	November 18
2014	April 16
2015	November 25
2016	April 14

Source Rahman et al. (2016)

Table 5.6 Key contributing factors of Jeddah flash flood

Factor	Description	Citation
Climate change links	Increased rainfall intensity and rising sea levels	Sefry et al. (2024); Shah et al. (2023)
Environmental factors	Jeddah's topography and geomorphology, including wadis and impermeable rocks	Youssef et al. (2016); Zaigham et al. (2021)
Urban planning and infrastructure	Disruption of natural drainage systems and inadequate infrastructure	Daoudi and Niang (2019); Motasim (2018)
Socio-economic impacts	Significant human casualties, economic losses and social vulnerability	Alharbi and Mills (2022); Youssef et al. (2016)

episodes resulted in severe infrastructural damage, affecting residential buildings, vehicles, public transportation systems and telecommunications networks. The storm in 2009 resulted in the deaths of over 113 individuals and caused extensive damage to more than 10,000 residential and commercial structures. In addition to the human toll, the disaster left approximately 1,200 families homeless and led to the destruction or severe damage of over 17,000 vehicles. The flood severely inundated the city's infrastructure, paralyzing transportation networks and disrupting essential services. Two years later, in 2011, Jeddah experienced another major flood event, during which 111 mm of rainfall was recorded within just three hours. This episode, comparable in magnitude to the 2009 disaster, further exposed the city's vulnerability to extreme hydrological events. The 2011 flood not only caused widespread property and infrastructure damage but also resulted in the failure of a local dam, significantly amplifying the scale of destruction (Youssef et al., 2016).

A number of initiatives have been undertaken by the Saudi government to enhance flood control and management in the Jeddah region (Abdulrazzak et al., 2019; Al-Wathinani et al., 2023; Hijji et al., 2013). Various precautionary measures have been implemented to mitigate the impacts and risks associated with natural hazards. However, the effectiveness of these interventions has varied, influenced by factors such as scale, site selection and the applied engineering methodologies. In the aftermath of the severe flooding events in 2009 and 2011, a comprehensive assessment was carried out in Jeddah and its surrounding areas. The findings revealed that many of the existing protective measures had limited effectiveness in reducing flood-related damages. For instance, several older structures were found to lack adequate disaster-resilient features. This issue is prevalent along the western coast of Saudi Arabia, where even structurally sound buildings exhibit fissures and cracks. These shortcomings are not necessarily indicative of inadequate engineering standards but rather point to the absence or insufficiency of integrated protection and maintenance controls (Al Saud & Al Saud, 2015).

Based on the assessment of flood-affected areas and the analysis of various dimensions impacted by the events, several flood control measures have been recommended for Jeddah City (Al Saud & Al Saud, 2015). These include the construction of retaining walls along roads and waterways to stabilize slopes and reduce surface runoff, as well as the installation of flow barriers within valley systems to dissipate kinetic energy. Additional proposed interventions involve the excavation of impervious zones to promote infiltration by enhancing soil permeability, thereby reducing surface water accumulation. Modifications to existing drainage infrastructure, such as widening and geometrically reconfiguring outdated channels, have also been suggested to increase discharge capacity. Furthermore, the development of cross-canyon conveyance systems and dedicated flood diversion channels along major wadis is considered essential. To support early warning systems and flood forecasting, the formulation of site-specific hyetographs has also been proposed (Elfeki et al., 2014).

UAE Floods: 2024 Extreme Event

In April 2024, the United Arab Emirates experienced an unprecedented rainfall event that caused severe flooding and widespread disruption across several emirates. The rainfall on April 16 was particularly intense, with 24-h precipitation totals exceeding 250 mm in some locations making it the heaviest recorded since meteorological observations began in 1949 (Francis et al., 2025; Hussein et al., 2025). Although the region typically exhibits low runoff characteristics, the intensity of the rainfall led to extensive flooding in urbanized areas, highlighting the complexities of urban hydrology and the challenges associated with flood forecasting in arid environments (Alhosani et al., 2024). This extreme event was largely concentrated in the densely populated northeastern part of the UAE, extending from Abu Dhabi to Ras Al Khaimah. In contrast, the southwestern and southern regions recorded minimal to no rainfall. Spatial analysis of data from eight airport monitoring stations confirmed that the highest rainfall intensities were observed in cities such as Dubai, Sharjah, Ajman, Fujairah, Ras Al Khaimah and Umm Al Quwain areas that bore the brunt of the flooding (Bangar, 2024). Scientifically, the April 2024 storm is categorized as the maximum daily precipitation event across the UAE during the spring season, a period during which mesoscale convective systems (MCSs) occur more frequently (Francis et al., 2025). These organized convective systems played a central role in producing the exceptional rainfall and its spatial variability throughout the country. The event resulted in a confirmed total of 24 fatalities 20 in neighboring Oman and 4 in the UAE underscoring the transboundary nature and scale of the disaster (Abuhani et al., 2025; Zhang et al., 2025). While the UAE is internationally recognized for its arid climate and limited annual rainfall, recent years have shown a rising trend in extreme precipitation events. This increase is attributed to a combination of land use and land cover changes, climatic anomalies and the broader influence of anthropogenic climate change (Alhosani et al., 2024).

5.3 Resilience-Building Approaches Against Extreme Events

The Arabian Peninsula is increasingly vulnerable to extreme climate events, including intense rainfall, extreme heatwaves, SLR, desertification, sandstorms and flash floods. These hazards, exacerbated by climate change and rapid urbanization, pose significant risks to human settlements, infrastructure and ecosystems. Building resilience to these events necessitates a multi-faceted approach encompassing infrastructure development, policy interventions, advanced technology, community engagement and nature-based solutions.

Urban Flooding and Flash Flood Resilience

Enhancing resilience to flash floods and extreme rainfall requires integrated stormwater management systems. Investments in robust flood defenses such as expanded drainage networks, retention basins and underground reservoirs are critical for managing excess rainwater. Nature-based solutions like permeable pavements, rain gardens, urban wetlands and rooftop harvesting systems reduce runoff and promote groundwater recharge. Real-time meteorological monitoring, powered by remote sensing and artificial intelligence, strengthens early warning systems. These technologies, combined with community-based flood preparedness and water conservation strategies, can significantly reduce urban flood vulnerability.

Adapting to Extreme Heatwaves

With temperatures in the Arabian Peninsula frequently reaching higher levels, adaptive strategies must address both environmental and public health dimensions. Urban cooling measures including green spaces, cool roofs, reflective surfaces and urban forests help combat the urban heat island effect. Sustainable building designs using passive cooling techniques, high-albedo materials and district cooling systems improve indoor comfort. Public awareness campaigns, heat action plans and the provision of cooling centers and hydration stations are essential for protecting vulnerable populations during extreme heat events.

Sea Level Rise and Coastal Vulnerability

Rising sea levels are increasing the risk of coastal erosion, saltwater intrusion and tidal flooding across the region. Resilience strategies include constructing seawalls and reinforcing coastal barriers, alongside restoring natural coastal defenses like mangroves, coral reefs and sand dunes. Integrated Coastal Zone Management (ICZM) approaches that consider ecological, social and economic factors are essential. Urban planning policies should restrict development in high-risk areas and prioritize the elevation of critical infrastructure. Long-term solutions may also include the strategic relocation of vulnerable communities.

Combating Desertification and *Land Degradation*

To reverse land degradation trends, large-scale initiatives such as the Saudi Green Initiative and the UAE's afforestation efforts are critical. These projects contribute to soil stabilization, biodiversity enhancement and carbon sequestration. Sustainable land management practices like agroforestry, rotational grazing and soil conservation techniques further reinforce land resilience. Community-driven conservation programs ensure sustained impact, while satellite-based land monitoring systems provide data for targeted restoration efforts.

Sandstorm Mitigation and Dust Resilience

Frequent sandstorms in arid zones disrupt health, infrastructure and visibility. Vegetative barriers, windbreaks and dune stabilization using drought-tolerant species can reduce the intensity and spread of dust. In urban settings, resilience can be improved by using dust-resistant building materials and installing high-efficiency air filtration systems. Satellite-supported dust monitoring and predictive models aid in early warning, enabling communities and industries to take timely protective actions.

In the face of escalating climate extremes, resilience-building in the Arabian Peninsula must be grounded in both traditional knowledge and modern science. The integration of engineered and nature-based solutions, along with community participation and policy support, offers a pathway to sustainable adaptation across different hazards. A regionally coordinated, proactive approach is vital for safeguarding ecosystems, infrastructure and human well-being from the growing threats of climate variability.

5.4 Conclusion

The increasing occurrence and intensity of extreme climate events present a pressing global challenge, calling for sustained attention and proactive resilience-building efforts. As discussed in this chapter, phenomena such as prolonged heatwaves, irregular precipitation, rising sea levels, land degradation and frequent dust and sandstorms are becoming more evident under changing climatic conditions. These shifts carry implications not only for environmental systems but also for societal well-being, economic stability and long-term sustainability. The impacts of extreme climate events extend across multiple sectors, with consequences that are increasingly interlinked. Rising temperatures exert sustained pressure on public health systems, contributing to greater incidence of heat-related illnesses and reduced labor efficiency, particularly in outdoor environments. Thermal extremes, combined with rapid urban expansion, intensify stress on infrastructure, where inadequate drainage systems often fail to manage flash floods, leading to widespread urban disruption. In coastal areas, the combined effects of sea level rise and extreme precipitation result in erosion, saltwater intrusion and increased vulnerability of shoreline infrastructure. These developments further complicate water resource management, especially in regions already facing scarcity. Simultaneously, prolonged droughts and advancing desertification reduce agricultural productivity, posing challenges to food security and ecosystem stability. As vegetative cover declines, the frequency and intensity of sand and dust storms increase, degrading air quality, disrupting transportation and affecting respiratory health. Together, these cascading impacts underscore the systemic nature of climate risks and the need for integrated, cross-sectoral strategies that strengthen resilience and promote long-term sustainability.

In response to these evolving challenges, resilience must be approached as a continuous, inclusive and adaptive process. Strengthening early warning systems, promoting nature-based and context-specific solutions, encouraging sustainable land and water management and supporting climate-informed infrastructure are key pillars of effective adaptation. Equally important is the role of education, community engagement and knowledge sharing in fostering a culture of preparedness and mutual responsibility. While the impacts of climate extremes may differ by location, the shared need for coordinated, evidence-based and forward-looking strategies is universal. Efforts to enhance resilience are most effective when guided by scientific insights, supported by enabling policies and aligned with local realities and capacities. Moreover, integrating traditional ecological knowledge with modern technologies can provide holistic pathways to resilience that are both practical and culturally grounded. Ultimately, addressing the risks associated with climate extremes is not only viewed as a means to minimize potential harm, but also as an opportunity through which social cohesion, environmental stewardship and adaptive governance may be strengthened. Collaborative action across sectors and scales is encouraged and investment in resilient systems is regarded as essential to ensuring a secure, sustainable and equitable future for all.

References

Abahussain, A. A., Abdu, A. S., Al-Zubari, W. K., El-Deen, N. A., & Abdul-Raheem, M. (2002). Desertification in the Arab Region: Analysis of current status and trends. *Journal of Arid Environments, 51*(4), 521–545.

Abdelkader, M. M., Al-Amoud, A. I., El Alfy, M., El-Feky, A., & Saber, M. (2021). Assessment of flash flood hazard based on morphometric aspects and rainfall-runoff modeling in Wadi Nisah, central Saudi Arabia. *Remote Sensing Applications: Society and Environment, 23*, Article 100562.

Abdullah, M. A., Youssef, A. M., Nashar, F., & AlFadail, E. A. (2019). Statistical analysis of rainfall patterns in Jeddah City, KSA: future impacts. *Rainfall-Extremes Distribution and Properties*, 1–17.

Abdulrazzak, M., Elfeki, A., Kamis, A., Kassab, M., Alamri, N., Chaabani, A., & Noor, K. (2019). Flash flood risk assessment in urban arid environment: Case study of Taibah and Islamic universities' campuses, Medina, Kingdom of Saudi Arabia. *Geomatics, Natural Hazards and Risk, 10*(1), 780–796.

Abuhani, D. A., Aldamani, R., Zualkernan, I., AlAli, R., & Grover, H. (2025). Floods' Impact on Cities' Critical Infrastructure: A Case Study on Dubai. *IEEE Access*.

Al Saud, M. (2010). Assessment of flood hazard of Jeddah area 2009, Saudi Arabia. *Journal of Water Resource and Protection, 2*(9), 839–847.

Al Saud, M. M., & Al Saud, M. M. (2015). Required controls for floods and torrents in Jeddah Region. *Flood Control Management for the City and Surroundings of Jeddah, Saudi Arabia*, 141–161.

Al-Ahmed, A. (2009). Jeddah flood deaths shame Saudi royals. *The Guardian, 3*.

Al-Awadhi, J. M., & Alkandary, D. S. (2024). Desertification and land degradation in the Arab States of the Gulf. In *Environmental hazards in the Arabian Gulf Region* (pp. 355–376). Springer.

Al-Delaimy, W. K. (2020). Vulnerable populations and regions: Middle East as a case study. In Health of people, health of planet and our responsibility: Climate change, air pollution and health (pp. 121–133).

Alharbi, S., & Mills, G. (2022). Assessment of exposure to flash flooding in an arid environment: a case study of the Jeddah city neighborhood Abruq ar rughamah, Saudi Arabia. In: *Wadi flash floods: Challenges and advanced approaches for disaster risk reduction* (pp. 383–397).

Al-Hemoud, A., Al-Dashti, H., Al-Saleh, A., Petrov, P., Malek, M., Elhamoud, E., Al-Khafaji, S., Li, J., Koutrakis, P., Doronzo, D., & Middleton, N. (2022). Dust storm 'hot spots' and transport pathways affecting the Arabian Peninsula. *Journal of Atmospheric and Solar-Terrestrial Physics, 238*, 105932.

Alhosani, N., Hussein, K., Al-Areeq, A. M., Al Aghbari, A. A., Elkamali, M., Alsumaiti, T.,.Almurshidi, A. M. G., Sharif, H. O., & Abdalati, W. (2024). From sandstorms to deluges: assessing flood risks in Al Ain City, United Arab Emirates. *Geomatics, Natural Hazards and Risk, 15*(1), 2429735.

Almazroui, M. (2020). Changes in temperature trends and extremes over Saudi Arabia for the period 1978–2019. *Advances in Meteorology, 2020*(1), 8828421.

Almazroui, M., Islam, M. N., Alkhalaf, A. K., Saeed, F., Dambul, R., & Rahman, M. A. (2016a). Simulation of temperature and precipitation climatology for the CORDEX-MENA/Arab domain using RegCM4. *Arabian Journal of Geosciences, 9*, 1–14.

Almazroui, M., Nazrul Islam, M., Athar, H., Jones, P. D., & Rahman, M. A. (2012). Recent climate change in the Arabian Peninsula: Annual rainfall and temperature analysis of Saudi Arabia for 1978–2009. *International Journal of Climatology, 32*(6), 953–966.

Almazroui, M., Nazrul Islam, M., Saeed, S., Alkhalaf, A. K., & Dambul, R. (2017). Assessment of uncertainties in projected temperature and precipitation over the Arabian Peninsula using three categories of Cmip5 multimodel ensembles. *Earth Systems and Environment, 1*, 1–20.

Almazroui, M., Saeed, F., Islam, M. N., & Alkhalaf, A. K. (2016b). Assessing the robustness and uncertainties of projected changes in temperature and precipitation in AR4 Global Climate Models over the Arabian Peninsula. *Atmospheric Research, 182*, 163–175.

Almazroui, M., Saeed, F., Saeed, S., Ismail, M., Ehsan, M. A., Islam, M. N., Abid, M. A., O'Brien, E., Kamil, S., Rashid, I. U., & Nadeem, I. (2021). Projected changes in climate extremes using CMIP6 simulations over SREX regions. *Earth Systems and Environment, 5*(3), 481–497.

Almazroui, M., Yaqub, A., Nadeem, I., Islam, M. N., Saeed, S., Kamil, S.,... & Hassanen, H. (2023). Analysis of climate extremes over the Arabian Peninsula using downscaled RCM data. *Journal of King Abdulaziz University: Meteorology, Environment & Arid Land Agriculture Sciences, 32*(2).

Al-Wathinani, A. M., Barten, D. G., Borowska-Stefańska, M., Gołda, P., AlDulijan, N. A., Alhallaf, M. A., Samarkandi, L. O., Almuhaidly, A. S., Goniewicz, M., Samarkandi, W. O., & Goniewicz, K. (2023). Driving sustainable disaster risk reduction: a rapid review of the policies and strategies in Saudi Arabia. *Sustainability, 15*(14), 10976.

Ameur, F. (2016). Floods in Jeddah, Saudi Arabia: unusual phenomenon and huge losses. What prognoses. In *E3S Web of Conferences* (Vol. 7, p. 04019). EDP Sciences.

Amin, A. A. (2004). The extent of desertification on Saudi Arabia. *Environmental Geology, 46*, 22–31.

Arnous, M. O., El-Rayes, A. E., El-Nady, H., & Helmy, A. M. (2022). Flash flooding hazard assessment, modeling, and management in the coastal zone of Ras Ghareb City, Gulf of Suez, Egypt. *Journal of Coastal Conservation, 26*(6), 77.

Atieku, J. N., & Segbefia, S. K. (2024). Land Degradation: a global challenge and its effects on humankind and the environment, with a special focus on Ghana. *Indiana Journal of Humanities and Social Sciences, 5*(7), 58–64.

Atif, R. M., Almazroui, M., Saeed, S., Abid, M. A., Islam, M. N., & Ismail, M. (2020). Extreme precipitation events over Saudi Arabia during the wet season and their associated teleconnections. *Atmospheric Research, 231*, Article 104655.

Awadh, S. M. (2023). Impact of North African sand and dust storms on the Middle East using Iraq as an example: Causes, sources, and mitigation. *Atmosphere, 14*(1), 180.

Bakhamis, A. N., Bilal, H., Heggy, E., Al-Kuwari, M. S., & Al-Ansari, T. (2024). On the drivers, forecasts, and uncertainties of relative sea level rise in the Eastern Arabian Peninsula: A review. *Regional Studies in Marine Science*, 103503.

Bangar, N. (2024, April 17). *Look: Top UAE places hit by floods after record rains.* Khaleej Times. https://www.khaleejtimes.com/uae/dubai-rains-6-most-flood-hit-areas-after-record-downpour

Booth, D. B., & Bledsoe, B. P. (2009). Streams and urbanization. In *The water environment of cities* (pp. 93–123).

Brown, P. J., Bradley, R. S., & Keimig, F. T. (2010). Changes in extreme climate indices for the northeastern United States, 1870–2005. *Journal of Climate, 23*(24), 6555–6572.

Cazenave, A., & Llovel, W. (2010). Contemporary sea level rise. *Annual Review of Marine Science, 2*(1), 145–173.

Costa, J. E. (1987). Hydraulics and basin morphometry of the largest flash floods in the conterminous United States. *Journal of Hydrology, 93*(3–4), 313–338.

Cotti, D., Harb, M., Hadri, A., Aboufirass, M., Chaham, K. R., Libertino, A., Campo, L., Trasforini, E., Krätzschmar, E., Bellert, F., & Hagenlocher, M. (2022). An integrated multi-risk assessment for floods and drought in the Marrakech-Safi region (Morocco). *Frontiers in Water, 4*, 886648.

Daoudi, M., & Niang, A. J. (2019). Flood risk and vulnerability of Jeddah city, Saudi Arabia. *Recent Advances in Flood Risk Management*, 634–654.

Dasgupta, S., van Maanen, N., Gosling, S. N., Piontek, F., Otto, C., & Schleussner, C. F. (2021). Effects of climate change on combined labour productivity and supply: An empirical, multi-model study. *The Lancet Planetary Health, 5*(7), e455–e465.

Davies, R. (2013). Jeddah Flood Defenses. Floodlis July, 28, 2013.

Dawoud, M. A. (2012). Environmental impacts of seawater desalination: Arabian Gulf case study. *International Journal of Environment and Sustainability, 1*(3).

Dawoud, M. A., & Al Mulla, M. M. (2012). Environmental impacts of seawater desalination: Arabian Gulf case study. *International Journal of Environment and Sustainability, 1*(3), 22–37.

De Longueville, F., Hountondji, Y. C., Henry, S., & Ozer, P. (2010). What do we know about effects of desert dust on air quality and human health in West Africa compared to other regions? *Science of the Total Environment, 409*(1), 1–8.

Delmonaco, G., Margottini, C., & Serafini, S. (1999). Climate change impact on frequency and distribution of natural extreme events: an overview. In *Floods and landslides: integrated Risk Assessment* (pp. 45–66)

Dhiman, R., VishnuRadhan, R., Eldho, T. I., & Inamdar, A. (2019). Flood risk and adaptation in Indian coastal cities: Recent scenarios. *Applied Water Science, 9*(1), 5.

Domeisen, D. I., Eltahir, E. A., Fischer, E. M., Knutti, R., Perkins-Kirkpatrick, S. E., Schär, C., Seneviratne, S. I., Weisheimer, A., & Wernli, H. (2023). Prediction and projection of heatwaves. *Nature Reviews Earth & Environment, 4*(1), 36–50

Elfeki, A. M., Ewea, H. A., & Al-Amri, N. S. (2014). Development of storm hyetographs for flood forecasting in the Kingdom of Saudi Arabia. *Arabian Journal of Geosciences, 7*, 4387–4398.

El-Samra, R., Bou-Zeid, E., Bangalath, H. K., Stenchikov, G., & El-Fadel, M. (2017). Future intensification of hydro-meteorological extremes: Downscaling using the weather research and forecasting model. *Climate Dynamics, 49*, 3765–3785.

Francis, D., Fonseca, R., Nelli, N., Cherif, C., Yarragunta, Y., Zittis, G., & Jan de Vries, A. (2025). From cause to consequence: examining the historic April 2024 rainstorm in the United Arab Emirates through the lens of climate change. *npj Climate and Atmospheric Science, 8*(1), 1–14.

Frederikse, T., Landerer, F., Caron, L., Adhikari, S., Parkes, D., Humphrey, V. W., Dangendorf, S., Hogarth, P., Zanna, L., & Wu, Y. H. (2020). The causes of sea-level rise since 1900. *Nature, 584*(7821), 393–397.

Gujree, I., Wani, I., Muslim, M., Farooq, M., & Meraj, G. (2017). Evaluating the variability and trends in extreme climate events in the Kashmir Valley using PRECIS RCM simulations. *Modeling Earth Systems and Environment, 3*(4), 1647–1662.

Hamed, M. M., Al-Hasani, A. A. J., Nashwan, M. S., Sa'adi, Z., & Shahid, S. (2024). Assessing the growing threat of heat stress in the North Africa and Arabian Peninsula region connected to climate change. *Journal of Cleaner Production, 447*, Article 141639.

Hamed, M. M., Nashwan, M. S., Shiru, M. S., & Shahid, S. (2022). Comparison between CMIP5 and CMIP6 models over MENA region using historical simulations and future projections. *Sustainability, 14*(16), 10375.

Hamidi, M., Kavianpour, M. R., & Shao, Y. (2017). A quantitative evaluation of the 3–8 July 2009 Shamal dust storm. *Aeolian Research, 24*, 133–143.

Hereher, M. E. (2020). Assessment of climate change impacts on sea surface temperatures and sea level rise—The Arabian Gulf. *Climate, 8*(4), 50.

Hijji, M., Amin, S., Iqbal, R., & Harrop, W. (2013, December). A critical evaluation of the rational need for an IT management system for flash flood events in Jeddah, Saudi Arabia. In *2013 Sixth International Conference on Developments in eSystems Engineering* (pp. 209–214). IEEE.

Houmsi, M.R., Ismail, Z., Othman, K., Ishak, D., Hamed, M.M., Iqbal, Z., Syamsunur, D., Shahid, S., 2023. Spatiotemporal changes in hourly Wet bulb globe temperature in peninsular Malaysia. Stoch. Environ. Res. Risk Assess. 292 (1 September 2023), 106838–106838.

Hsiao, S. C., Chiang, W. S., Jang, J. H., Wu, H. L., Lu, W. S., Chen, W. B., & Wu, Y. T. (2021). Flood risk influenced by the compound effect of storm surge and rainfall under climate change for low-lying coastal areas. *Science of the Total Environment, 764*, Article 144439.

Hussein, K., Alhosani, N., Al-Areeq, A. M., Al Aghbari, A. A., Elkamali, M., Alsumaiti, T., Sharif, H. O., Almurshidi, A. M. G., & Abdalati, W. (2025). Unprecedented rainfall in the United Arab Emirates: Hydrologic and flood impact analysis of the April 2024 event. *Natural Hazards*, 1–23.

Hussein, K., Alkaabi, K., Ghebreyesus, D., Liaqat, M. U., & Sharif, H. O. (2020). Land use/land cover change along the Eastern Coast of the UAE and its impact on flooding risk. *Geomatics, Natural Hazards and Risk, 11*(1), 112–130.

IPCC, I., 2013. The physical science basis. In *Contribution of working group I to the fifth assessment report of the intergovernmental panel on climate change, 1535.*

IPCC. (2023). *Figure SPM.2: With every increment of global warming, regional changes in mean climate and extremes become more widespread and pronounced.* In Climate Change 2023: Synthesis Report. Intergovernmental Panel on Climate Change. https://www.ipcc.ch/report/ar6/syr/

Jain, M. (2022). Increasing atmospheric extreme events and role of disaster risk management: Dimensions and approaches. In *Extremes in atmospheric processes and phenomenon: Assessment, impacts and mitigation* (pp. 303–328). Springer Nature Singapore.

Johnson, R. J., Sánchez-Lozada, L. G., Newman, L. S., Lanaspa, M. A., Diaz, H. F., Lemery, J.,... Rodriguez-Iturbe, B., Tolan, D. R., Butler-Dawson, J., Sato, J, Garcia. G., Hernando, A. A., & Roncal-Jimenez, C. A. (2019). Climate change and the kidney. *Annals of Nutrition and Metabolism, 74*(Suppl. 3), 38–44.

Kalita, R., Kalita, D., & Saxena, A. (2023). Trends in extreme climate indices in Cherrapunji for the period 1979 to 2020. *Journal of Earth System Science, 132*(2), 74.

Kaser, G., Cogley, J. G., Dyurgerov, M. B., Meier, M. F., & Ohmura, A. (2006). Mass balance of glaciers and ice caps: consensus estimates for 1961–2004. *Geophysical Research Letters, 33*(19).

Khan, S. I., Hong, Y., Wang, J., Yilmaz, K. K., Gourley, J. J., Adler, R. F., Robert Brakenridge, G., Policelli, F., Habib, S., & Irwin, D. (2010). Satellite remote sensing and hydrologic modeling for flood inundation mapping in Lake Victoria basin: Implications for hydrologic prediction in ungauged basins. *IEEE Transactions on Geoscience and Remote Sensing, 49*(1), 85–95.

Kompas, T., Pham, V. H., & Che, T. N. (2018). The effects of climate change on GDP by country and the global economic gains from complying with the Paris climate accord. *Earth's Future, 6*(8), 1153–1173.

Kopp, R. E., Kemp, A. C., Bittermann, K., Horton, B. P., Donnelly, J. P., Gehrels, W. R., Hay, C. C., Mitrovica, J. X., Morrow, E. D, & Rahmstorf, S. (2016). Temperature-driven global sea-level variability in the Common Era. *Proceedings of the National Academy of Sciences, 113*(11), E1434–E1441.

Krishna, L. V. (2014). Long term temperature trends in four different climatic zones of Saudi Arabia. *International Journal of Applied Science and Technology, 4*(5).

Kumar, S., Chatterjee, U., David Raj, A., & Sooryamol, K. R. (2024). Global warming and climate crisis/extreme events. In *Climate crisis: Adaptive approaches and sustainability* (pp. 3–18). Springer Nature Switzerland.

Le Houérou, H. N. (1996). Climate change, drought and desertification. *Journal of Arid Environments, 34*(2), 133–185.

Li, C., Wang, J., Hu, R., Yin, S., Bao, Y., & Ayal, D. Y. (2018). Relationship between vegetation change and extreme climate indices on the Inner Mongolia Plateau, China, from 1982 to 2013. *Ecological Indicators, 89*, 101–109.

Lian, L., Huang, J., Chen, S., Du, S., Zhang, L., & Yang, J. (2025). A Comprehensive review of dust events: Characteristics, climate feedbacks, and public health risks. *Current Pollution Reports, 11*(1), 1–19.

Malhi, Y., Franklin, J., Seddon, N., Solan, M., Turner, M. G., Field, C. B., & Knowlton, N. (2020). Climate change and ecosystems: Threats, opportunities and solutions. *Philosophical Transactions of the Royal Society B, 375*(1794), 20190104.

McCarthy, J. J. (2001). *Climate dhange 2001: Impacts, adaptation, and vulnerability: contribution of Working Group II to the Third Assessment Report of the Intergovernmental Panel on Climate Change*. Cambridge University Press.

McPhillips, L. E., Chang, H., Chester, M. V., Depietri, Y., Friedman, E., Grimm, N. B., Kominoski, J. S., McPhearson, T., Méndez-Lázaro, P., Rosi, E. J., Shafiei Shiva, J. (2018). Defining extreme events: A cross-disciplinary review. *Earth's Future, 6*(3), 441–455.

Meo, S. A., Almutairi, F. J., Abukhalaf, A. A., Alessa, O. M., Al-Khlaiwi, T., & Meo, A. S. (2021). Sandstorm and its effect on particulate matter PM 2.5, carbon monoxide, nitrogen dioxide, ozone pollutants and SARS-CoV-2 cases and deaths. *Science of the Total Environment, 795*, 148764.

Meyer, A., Bresson, H., Gorodetskaya, I. V., Harris, R. M., & Perkins-Kirkpatrick, S. E. (2023). Extreme climate and weather events in a warmer world. *Climate Change, 44.*

Motasim, H. (2018). [Re] covering Jeddah's Wadis-Building the City's Resilience through Green Infrastructure. *Journal of Sustainable Development, 11*(4), 228–228.

Murphy, C. (2009). Saudi Arabians use Facebook to vent fury over Jeddah flood deaths. *Christian Science Monitor, 6.*

Murray, V., & Ebi, K.L. (2012). IPCC Special report on managing the risks of extreme events and disasters to advance climate change adaptation (SREX). BMJ Publishing Group Ltd., London, UK. https://doi.org/10.1136/jech-2012-201045.

Mushtaq, F. (2025). Quantitative analysis of oceanographic parameters in Arabian Gulf: Google Earth Engine Approach. *Authorea Preprints.*

Mushtaq, F., & Nee Lala, M. G. (2017). Assessment of climatic variability in the catchments of Himalayan Lake, Jammu & Kashmir. *Geocarto International, 32*(10), 1090–1104.

Mushtaq, F., Haq, A. U., Bharti, S., Farooq, M., Ponnambalam, M. K., & Alhems, L. M. (2024, December). Exploring SST and Chlorophyll-A Anomalies in the Arabian Gulf: A Google Earth Engine Analysis. In 2024 IEEE India Geoscience and Remote Sensing Symposium (InGARSS) (pp. 1–4). IEEE.

NOAA. (2022). Ocean Topography and Cryosphere Branch/Laboratory for Satellite Altimetry/Sea Level Rise. https://www.star.nesdis.noaa.gov/socd/lsa/SeaLevelRise/LSA_SLR_timeseries_regional.php. Accessed 07 Feb 25

Notaro, M., Alkolibi, F., Fadda, E., & Bakhrjy, F. (2013). Trajectory analysis of Saudi Arabian dust storms. *Journal of Geophysical Research: Atmospheres, 118*(12), 6028–6043.

Ntoumos, A., Hadjinicolaou, P., Zittis, G., & Lelieveld, J. (2020). Updated assessment of temperature extremes over the Middle East-North Africa (MENA) region from observational and CMIP5 data. *Atmosphere, 11*(8), 813.

Odhiambo, G. O. (2017). Water scarcity in the Arabian Peninsula and socio-economic implications. *Applied Water Science, 7*(5), 2479–2492.

Pal, J. S., & Eltahir, E. A. (2016). Future temperature in southwest Asia projected to exceed a threshold for human adaptability. *Nature Climate Change, 6*(2), 197–200.

Patlakas, P., Stathopoulos, C., Flocas, H., Bartsotas, N. S., & Kallos, G. (2021). Precipitation climatology for the arid region of the Arabian Peninsula—Variability, trends and extremes. *Climate, 9*(7), 103.

Prakash, P. J., Stenchikov, G., Kalenderski, S., Osipov, S., & Bangalath, H. (2015). The impact of dust storms on the Arabian Peninsula and the Red Sea. *Atmospheric Chemistry and Physics, 15*(1), 199–222.

Prospero, J. M., Ginoux, P., Torres, O., Nicholson, S. E., & Gill, T. E. (2002). Environmental characterization of global sources of atmospheric soil dust identified with the Nimbus 7 Total Ozone Mapping Spectrometer (TOMS) absorbing aerosol product. *Reviews of Geophysics, 40*(1), 1–2.

Rahman, M. T., Aldosary, A. S., Nahiduzzaman, K. M., & Reza, I. (2016). Vulnerability of flash flooding in Riyadh, Saudi Arabia. *Natural Hazards, 84*, 1807–1830.

Rajendran, S., Al Kuwari, H. A. S., Sadooni, F. N., Nasir, S., Govil, H., & Ghrefat, H. (2023). Remote sensing of desertification and study of temporal variability of Aeolian deposits in parts of the Arabian Desert for sustainable development in an arid environment. *Environmental Research, 232*, Article 116279.

Rao, K. K., Al Mandous, A., Al Ebri, M., Al Hameli, N., Rakib, M., & Al Kaabi, S. (2024). Future changes in the precipitation regime over the Arabian Peninsula with special emphasis on UAE: Insights from NEX-GDDP CMIP6 model simulations. *Scientific Reports, 14*(1), 151.

Rezazadeh, M., Irannejad, P., & Shao, Y. (2013). Climatology of the Middle East dust events. *Aeolian Research, 10*, 103–109.

Salih, A., Hassaballa, A. A., & Ganawa, E. (2021). Mapping desertification degree and assessing its severity in Al-Ahsa Oasis, Saudi Arabia, using remote sensing-based indicators. *Arabian Journal of Geosciences, 14*, 1–15.

Sefry, S. A., El-Haddad, B. A., AbuAlfadael, E. Y., & Youssef, A. M. (2024). Extreme rainfall events in Saudi Arabia as a result of climate change: A case study: The devastating Jeddah Flood on November 25, 2022. In *Environmental hazards in the Arabian Gulf Region* (pp. 519-542). Springer.

Seneviratne, S. I., Zhang, X., Adnan, M., Badi, W., Dereczynski, C., Luca, A. D., Ghosh, S., Iskandar, I., Kossin, J., Lewis, S., Otto, F., Pinto, I., Satoh, M., Vicente-Serrano, S. M., & Allan, R. (2021). Weather and climate extreme events in a changing climate.

Shah, S. M. H., Yassin, M. A., Abba, S. I., Lawal, D. U., Al-Qadami, E. H. H., Teo, F. Y., Mustaffa, Z., & Aljundi, I. H. (2023). Flood risk and vulnerability from a changing climate perspective: An overview focusing on flash floods and associated hazards in Jeddah. *Water, 15*(20), 3641. https://doi.org/10.3390/w15203641

Shukla, P.R., Skea, J., Calvo Buendia, E., Masson-Delmotte, V., P¨ortner, H.O., Roberts, D., Zhai, P., Slade, R., Connors, S., & Van Diemen, R. (2019). IPCC, 2019: Climate Change and Land: an IPCC Special Report on Climate Change, Desertification, Land Degradation, Sustainable Land Management, Food Security, and Greenhouse Gas Fluxes in Terrestrial Ecosystems.

Tabari, H. (2020). Climate change impact on flood and extreme precipitation increases with water availability. *Scientific Reports, 10*(1), 13768.

Tammar, A., Abosuliman, S. S., & Rahaman, K. R. (2020). Social capital and disaster resilience nexus: A study of flash flood recovery in Jeddah City. *Sustainability, 12*(11), 4668.

Tanaka, T. Y., & Chiba, M. (2006). A numerical study of the contributions of dust source regions to the global dust budget. *Global and Planetary Change, 52*(1–4), 88–104.

Terry, J. P., Al Ruheili, A., Almarzooqi, M. A., Almheiri, R. Y., & Alshehhi, A. K. (2023). The rain deluge and flash floods of summer 2022 in the United Arab Emirates: Causes, analysis and perspectives on flood-risk reduction. *Journal of Arid Environments, 215*, Article 105013.
UNEP. (2013). Forecasting and early warning of dust storms. Article reproduced from United Nations environment Programme (UNEP) Global Environmental Alert Service (GEAS). *Environmental Development, 6*, 117–129.
Valavanidis, A. (2022). Extreme weather events exacerbated by the global impact of climate change. Available online: Chem-toxecotox.org/ScientificReviews. Accessed on 28 May 2023.
Vinodhkumar, B., Ullah, S., Kumar, T. L., & Al-Ghamdi, S. G. (2024). Amplification of temperature extremes in Arabian Peninsula under warmer worlds. *Scientific Reports, 14*(1), 16604.
Wöppelmann, G., Le Cozannet, G., De Michele, M., Raucoules, D., Cazenave, A., Garcin, M., Hanson, S., & Santamaría-Gómez, A. (2013). Is land subsidence increasing the exposure to sea level rise in Alexandria, Egypt?. *Geophysical Research Letters, 40* (12), 2953–2957.
World Bank Publications. (2014). *Turn down the heat: Confronting the new climate normal*. World Bank Publications.
World Bank, 2022, Sea Level Rise Dataset. https://datacatalog.worldbank.org/search/dataset/0041449/World-Sea-Level-Rise-Dataset. Accessed 06 Feb 2025
World Bank. (2019). Sustainable land management and restoration in the Middle East and North Africa Region: Issues, challenges, and recommendations.
World Meteorological Organization. (2022). New sand and dust forecast website launched. https://wmo.int/media/news/new-sand-and-dust-forecast-website-launched
Yin, J., Guo, S., Gentine, P., Sullivan, S. C., Gu, L., He, S., Chen, J., & Liu, P. (2021). Does the hook structure constrain future flood intensification under anthropogenic climate warming? *Water Resources Research, 57*(2), e2020WR028491.
York, R., Rosa, E. A., & Dietz, T. (2003). A rift in modernity? Assessing the anthropogenic sources of global climate change with the STIRPAT model. *International Journal of Sociology and Social Policy, 23*(10), 31–51.
Youssef, A. M., Sefry, S. A., Pradhan, B., & Alfadail, E. A. (2016). Analysis on causes of flash flood in Jeddah city (Kingdom of Saudi Arabia) of 2009 and 2011 using multi-sensor remote sensing data and GIS. *Geomatics, Natural Hazards and Risk, 7*(3), 1018–1042.
Yun, K. S., Heo, K. Y., Chu, J. E., Ha, K. J., Lee, E. J., Choi, Y., & Kitoh, A. (2012). Changes in climate classification and extreme climate indices from a high-resolution future projection in Korea. *Asia-Pacific Journal of Atmospheric Sciences, 48*, 213–226.
Zaigham, N. A., Mahar, G. A., Aburizaiza, O. S., & Nayyar, Z. A. (2021). Historical flood imprints reveal factors causing catastrophic impacts on modern Jeddah, Saudi Arabia: A case of flash floods versus rapid urbanization. *Arabian Journal of Geosciences, 14*(5), 342.
Zhang, W., Zhou, T., Ye, W., Zhang, T., Zhang, L., Wolski, P., Risbey, J., Wang, Z., Min, S.-K., Ramsay, H., Brody, M., Grimm, A., Clark, R., Ren, K., Jiang, J., Chen, X., Fu, S., Li, L., Tang, S., & Hu, S. (2025). A year marked by extreme precipitation and floods: Weather and climate extremes in 2024.

Chapter 6
Socio-Economic Dimensions of Climate Change in the Arabian Peninsula: Challenges and Strategic Responses

Zaiema Rouf and Aasimah Tanveer

6.1 Introduction

Climate change poses a significant threat to global development by disrupting ecological balance and exerting growing pressure on human societies and economies. The intensification of climate-related changes such as rising temperatures, altered precipitation patterns and an increase in the frequency and severity of extreme weather events is no longer a projected risk but an observable reality with significant socio-economic consequences (Patel et al., 2021). According to the Intergovernmental Panel on Climate Change (IPCC), global average temperatures increased by 1.09 °C between 2011 and 2020 compared to pre-industrial levels, with a 50% chance of surpassing the 1.5 °C threshold even under low-emission pathways. These evolving climate dynamics are increasingly disrupting livelihoods, infrastructure and development planning across diverse contexts.

The agricultural sector is particularly vulnerable to these climatic disturbances. Changes in temperature and precipitation are already altering cropping patterns, reducing yields and increasing the frequency of pest and disease outbreaks (Lobell & Gourdji, 2012; Skendžić et al., 2021). These impacts are compounded by the intensification of extreme weather events, which threaten harvest cycles and long-term agricultural sustainability. Livestock systems are similarly affected, as animals experience reduced productivity due to heat stress, water scarcity and deteriorating pasture conditions. The fisheries sector, especially in coastal areas, is also under stress from ocean warming, habitat degradation and changes in species distribution, placing pressure on food systems and those whose livelihoods depend on aquatic resources

Z. Rouf
Department of Environmental Science, University of Kashmir, Srinagar, Jammu & Kashmir, India

A. Tanveer (✉)
Department of Water Management, Government Degree College, Baramulla, Jammu & Kashmir, India
e-mail: aasimaht@gmail.com

F. Mushtaq et al. (eds.), *Climate Change in the Arabian Peninsula*,
https://doi.org/10.1007/978-3-032-02481-7_6

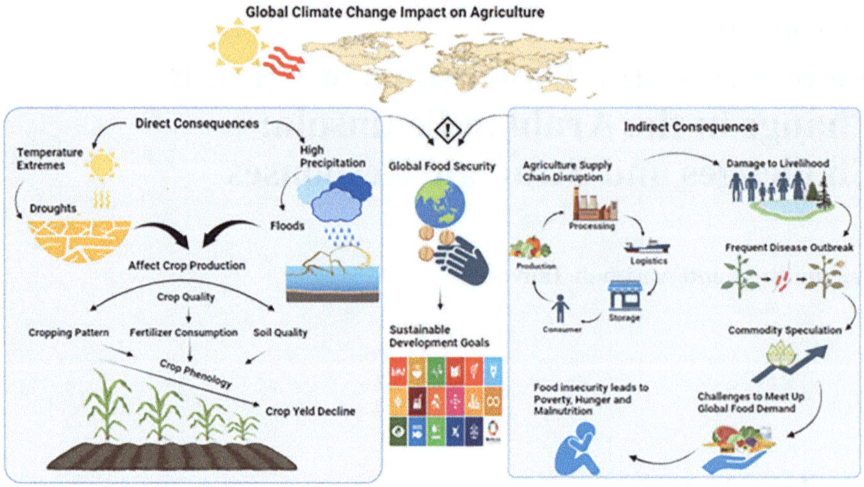

Fig. 6.1 Direct and indirect consequences of global climate change on agriculture and food security, highlighting impacts on crop production, supply chains and livelihoods adopted from Saleem et al. (2024). *Source* Saleem et al. (2024)

(Weatherdon et al., 2016) (Fig. 6.1). These agricultural and food security challenges are closely linked to rising water stress. With declining rainfall, increased evaporation and overexploited groundwater reserves, many regions are facing worsening water scarcity (Gosling & Arnell, 2016). This not only jeopardizes agricultural productivity but also undermines access to safe drinking water, sanitation services and industrial operations. The compounding effect of climate change on water availability directly affects both rural and urban populations, reinforcing socio-economic vulnerabilities and limiting development prospects.

Concurrently, climate-induced disruptions are increasingly impacting national economies through effects on labor productivity, industrial activity and energy systems (Bijnens et al., 2024). Rising temperatures contribute to heat-related work slowdowns and infrastructure failures, while floods and storms can interrupt production, transportation and supply chains (Radford et al., 2022). In addition, demand for energy particularly for cooling is increasing, placing further strain on power generation and distribution systems already vulnerable to climatic shocks (Van Vliet et al., 2016). The tourism industry, heavily dependent on predictable climate and ecosystem integrity, is also at risk. Changing weather conditions, biodiversity loss and water shortages can diminish the attractiveness of destinations, reduce tourist inflows and erode economic returns from one of the world's most employment-intensive sectors.

The consequences of climate change are not experienced uniformly across populations. Socially and economically marginalized groups, including smallholder farmers, informal workers, women-headed households, indigenous communities and people with disabilities, are more exposed to climate-related risks and possess fewer resources to adapt or recover (Wagner, 2020). Their vulnerability is shaped by a combination of location, livelihood dependence on climate-sensitive resources and systemic barriers to access, be it financial, institutional, or political. As a result, climate change often reinforces existing inequalities and creates new layers of socio-economic insecurity. Moreover, efforts to address climate change through mitigation and adaptation policies must be carefully designed to avoid unintended adverse effects (Hallegatte, 2009). Carbon pricing mechanisms, if implemented without adequate safeguards, can increase the cost of energy and transportation for low-income households (Samuel, 2025). Restrictions on natural resource use, when introduced without participatory planning, may displace traditional practices and disrupt livelihoods. Ensuring social justice within climate policy frameworks requires deliberate attention to equity, inclusion and local context in both planning and implementation.

To enhance resilience and promote inclusive development, it is imperative to integrate community knowledge and participation into climate action (Yadav, 2025). Local actors possess invaluable insights into environmental trends and adaptive practices and can play a central role in designing solutions that are both effective and contextually relevant. Recent IPCC assessments emphasize the importance of combining scientific expertise with indigenous and local knowledge systems to improve the effectiveness and social legitimacy of climate responses. Strengthening access to climate finance, enhancing institutional capacities and facilitating community engagement are essential to ensure that adaptation and mitigation efforts reach the most vulnerable (Van Aalst et al., 2008). Climate change thus represents a complex and multifaceted socio-economic challenge. Its impacts span agriculture, water security, industry and public services, with disproportionate burdens on already disadvantaged groups. A coherent response demands integrated strategies that combine environmental sustainability with social and economic resilience, ensuring that climate action contributes to equitable and sustainable development pathways (Pelling et al., 2012). This chapter builds upon these insights to explore the socio-economic implications of climate change in the Arabian Peninsula, a region characterized by both ecological sensitivity and dynamic economic and environmental developments.

6.2 Water Scarcity and Challenges in Water Management

Water is indispensable for a multitude of applications, encompassing agriculture, residential consumption, industrial utilization, tourism and the maintenance of ecological systems. The accessibility and quality of water exert a profound influence on the environment, economy and overall development of various regions. Freshwater resources exhibit an uneven distribution globally, primarily influenced by prevailing climatic conditions (Rodell et al., 2018). Certain regions, particularly those situated within tropical and temperate zones, benefit from abundant and consistent precipitation, while others especially those classified as arid encounter significant challenges due to severely restricted freshwater availability. This geographic and climatic variability leads to circumstances in which the demand for water substantially surpasses the quantity of sustainably accessible supply (Sophocleous, 2004). In comparison to regions with relatively abundant water resources, arid zones face severe water scarcity, making water security a pressing concern. Water scarcity is a widespread issue affecting not only arid and semi-arid regions but also areas with high rainfall due to damaged infrastructure and contaminated sources (Priyan, 2021). It involves both the quantity and quality of available water and poses a significant challenge to meeting essential needs and maintaining public health standards (Alotaibi et al., 2023). The sustainability of ecosystems relies heavily on adequate water resources. In addition to drinking, water is crucial for food security, sanitation and hygiene key factors influencing the global disease burden. Vulnerable and disadvantaged populations are disproportionately affected by climate change impacts on water supply and related health outcomes (DeNicola et al., 2015). This multifaceted concept can be analyzed from various perspectives. In regions experiencing minimal or no water scarcity, water resources are plentiful relative to consumption and extractions from rivers and aquifers remain negligible. Conversely, in locales facing physical water scarcity, such as much of the Arabian Peninsula, the natural environment itself is insufficient to provide the requisite water to satisfy all demands, including those necessary for sustaining ecosystems. Economic water scarcity, however, arises when water is present but financial, infrastructural, or institutional constraints inhibit equitable access (Vallino et al., 2020). Furthermore, nations classified as water-stressed typically possess total renewable freshwater resources ranging from 1000 to 1700 cubic meters per capita per annum, a threshold that several countries within the Arabian Gulf region fail to meet (Awawdeh & Nejatian, 2016).

6.3 Water Scarcity in the Arabian Peninsula

Water scarcity in the Arabian Peninsula has historically represented a significant challenge owing to the region's intrinsically limited freshwater resources, which have profoundly influenced the existence and livelihoods of its populace (Odhiambo, 2017a, 2017b). In recent decades, this predicament has been further aggravated by rapid socio-economic advancement, pronounced population surges, increasing urbanization and the escalation of agricultural practices, all of which have imposed mounting pressures on already precarious water systems. The Arabian Peninsula is predominantly characterized by arid to hyper-arid conditions, with the majority of its land area comprising desert terrains (Gaur & Squires, 2018). This otherwise dry expanse is interspersed with narrow coastal zones and isolated mountainous regions. Precipitation throughout the Peninsula is both minimal and highly erratic, with average annual rainfall typically fluctuating between 70 and 130 mm (Patlakas et al., 2021). Nonetheless, in specific coastal regions affected by orographic influences, precipitation levels may surpass 500 mm annually, thereby providing limited yet essential recharge opportunities for local aquifers.

Renewable groundwater resources within the region are exceedingly constrained. Shallow alluvial aquifers yield some renewable groundwater in targeted coastal locales, yet their spatial extent and recharge capabilities are rather limited. Conversely, deep aquifers harbor considerable volumes of non-renewable fossil water, particularly in nations such as Saudi Arabia (Sherif et al., 2023). Historically, these deep aquifers have facilitated extensive water extraction to satisfy escalating demands. However, their reserves are being depleted at an alarming rate as a consequence of unsustainable extraction methodologies. Despite the Arabian Gulf region comprising merely a small fraction of the global land area and population, it encompasses only approximately 1.1% of the world's total renewable water resources (Table 6.1). This disparity between natural resource endowment and demand is further exemplified by the exceedingly low per capita availability of renewable water in several countries within the region. For instance, in Kuwait, Bahrain and Qatar, the annual renewable water availability per capita is below 100 cubic meters a figure significantly beneath the threshold indicative of severe water stress (Odhiambo, 2017a, 2017b).

The primary renewable water resources in the region are shallow aquifers, which are particularly abundant within wadi channels and alluvial fans. These aquifers predominantly depend on indirect recharge from intermittent water flows, rendering their sustainability susceptible to climatic fluctuations and anthropogenic pressures. While the quality of groundwater in these shallow systems is generally favorable, extensive extraction particularly in coastal areas has precipitated the incursion of saltwater into freshwater lenses, consequently jeopardizing water quality and restricting its applicability for certain uses. Estimates indicate that the reserves of alluvial aquifers in the Arabian Peninsula approximate 131 billion cubic meters, with Saudi Arabia constituting a significant proportion of this aggregate (Mazzoni et al., 2018). Nevertheless, in numerous downstream regions, the poor quality of groundwater characterized by elevated salinity and various contaminants may hinder the utility of these reserves for domestic consumption without prior treatment. Building upon these challenges, the Arabian Peninsula is now facing mounting hydrological stress driven by the overexploitation of its fossil groundwater reserves (Hussain et al., 2025). These reserves are primarily stored within extensive sandstone and limestone formations that span nearly two-thirds of Saudi Arabia and extend into neighboring

Table 6.1 Availability of water resources in the Arabian Peninsula

Country	Groundwater recharge (M m^3)	Non-renewable reserve (M m^3)	Annual renewable water resources (km^3/year)	Total renewable water per capita (m^3/year/person)
Bahrain	110	Negligible	11.6	7.65
Kuwait	160	n/a	58.3	6.154
Oman	900	102,000	0.6	462.84
Qatar	50	Negligible	29	29.30
Saudi Arabia	3850	428,400	33.7	86.45
UAE	190	n/a	64.5	16

Source Odhiambo (2017a, 2017b)

countries. Classified as primary or secondary aquifers, these geological formations serve as the backbone of the region's water supply. Among the most significant are the Saq, Tabuk and Wajid aquifers in Saudi Arabia and the Dammam aquifer, which extends into Bahrain and Kuwait (Ali, 1983). While some of these deep aquifers contain relatively high-quality water, others yield brackish water that necessitates considerable treatment prior to use in domestic or agricultural sectors.

An ongoing challenge throughout the region is the widespread unregulated extraction of groundwater, which has become a principal contributor to the rapid depletion of these valuable aquifer reserves. This lack of oversight not only jeopardizes the sustainability of both rural and urban communities but also hastens the exhaustion of nonrenewable groundwater stocks. In response to growing shortages, several nations in the region have increasingly resorted to seawater desalination as a supplemental source of freshwater (Saeed et al., 2019). However, while desalination provides a technological lifeline, it is both energy-intensive and accompanied by environmental and economic challenges, including brine disposal and high operational costs. The severity of groundwater depletion is further intensified by interconnected drivers such as escalating population growth, inefficient water resource governance and persistent reliance on fossil water. In many instances, national water consumption has already surpassed sustainable thresholds, placing significant strain on aquifer systems and risking long-term ecological damage. Yemen presents a clear example of this emerging challenges, the country suffers from extreme water scarcity and is depleting its groundwater reserves at a rapid pace, leading to the serious possibility of total water depletion in several of its major urban centers (Mohamed et al., 2017). This trajectory underscores the urgent need for sustainable water management practices, robust regulatory frameworks and a shift away from dependence on finite groundwater resources. In parallel, rapid population growth and rising per capita water consumption have intensified pressure on already limited water resources (DeNicola et al., 2015). This demographic expansion, coupled with urbanization and improved living standards, has led to heightened water demand across domestic, agricultural and industrial sectors further accelerating aquifer depletion and complicating efforts toward sustainable water management.

This rising demand is distributed unevenly across sectors, with agriculture being the dominant consumer of water resources in the region (Al-Zubari, 2016). In countries such as Saudi Arabia, Oman and Yemen, irrigation accounts for more than 85% of total water withdrawals, whereas in Bahrain, Kuwait and Qatar, the share is lower, though still substantial, at under 60% (Odhiambo, 2017a, 2017b) (Table 6.2). Groundwater remains the primary source of irrigation, but its overexploitation exacerbated by low natural recharge rates has led to saltwater intrusion, declining water tables and concerns over long-term agricultural viability (Mogren, 2015). Water use efficiency in agriculture is particularly low, averaging around 40%, raising alarms over sustainability.

Municipal and industrial water use also varies significantly across the Peninsula. For example, municipal consumption ranges from as low as 13 m^3 per capita in Yemen to as high as 245 m^3 in Bahrain (Ansari, 2013). Industrial withdrawals are comparatively modest but variable, averaging around 70 m^3 per capita with Yemen consuming

Table 6.2 Freshwater withdrawal by sector in Arabian Peninsula Countries

Country	Domestic use (%)	Industrial use (%)	Agricultural use (%)	Domestic use (m^3/p/year)	Industrial use (m^3/p/year)	Agricultural use (m^3/p/year)
Bahrain	50	6	45	221	27	199
Kuwait	44	2	54	132	6	162
Oman	10	1	88	45	5	400
Qatar	39	2	59	115	6	174
Saudi Arabia	9	3	88	81	27	793
UAE	15	2	83	127	17	705
Yemen	8	2	90	11	3	126

Source Odhiambo (2017a, 2017b)

only 3 m^3 per capita for industrial purposes. The growing competition for groundwater between agriculture, cities and industries is intensifying water stress across the region. In response to these mounting pressures, desalination has become increasingly critical to meet urban and industrial water needs, especially in coastal nations. However, reliance on desalination is not uniform and is shaped by each country's access to technology, coastline and financial capacity. Simultaneously, government-driven agricultural expansion has further increased demand, often outpacing the ability of natural and artificial systems to provide sustainable supply. Altogether, water demand in the region increased nearly fourfold between 1980 and 1990, with projections pointing to continued growth across all sectors (Wehbe & Temimi, 2021) (Table 6.3). Combined with rapid urbanization and shifting consumption patterns, this sectoral imbalance underscores the urgent need for integrated water planning, efficiency improvements and diversification of water sources to prevent a deepening of the region's water crisis.

Future Scenarios of Water Scarcity and Socio-economic Implications under Climate Change

The Arabian Peninsula, already one of the most water-stressed regions in the world, faces escalating hydrological challenges due to the compounding impacts of climate change. The region's arid climate, high evaporation rates, low groundwater recharge and limited renewable freshwater resources provide a precarious baseline that is now further exacerbated by global warming. Climate change projections indicate a temperature rise of up to 2 °C in the next two decades (Almazroui et al., 2012) and over 4 °C by the end of the century, resulting in hotter, drier and increasingly erratic conditions, accompanied by declining precipitation trends. These climatic changes have direct consequences for the hydrological cycle in the region. Reduced precipitation leads to lower surface runoff and stream flows, thereby diminishing opportunities for aquifer recharge. Estimates suggest that by 2050, the Arabian

Table 6.3 Past and projected water demand (2000–2025) in the Arabian Peninsula

Country	2000 Domestic	2000 Agriculture	2000 Industrial	2025 Domestic	2025 Agriculture	2025 Industrial	1990 Total	2000 Total	2025 Total
Bahrain	169	124	26	230	271	73	223	319	574
Kuwait	375	110	105	670	140	160	383	590	970
Qatar	90	185	15	230	205	50	194	290	485
Oman	170	1270	85	630	1500	350	1236	1525	2480
UAE	750	1400	30	1100	2050	50	1490	2180	3200
Saudi Arabia	2350	15,000	415	6450	16,300	1450	16,300	17,765	24,200
Yemen	360	3100	60	840	3800	137	2899	3520	4777
Total	4264	21,189	736	10,150	24,266	2270	22,725	26,189	36,686

Source Odhiambo (2017a, 2017b)

Peninsula may face a reduction of 10% or more in total water availability (Odhiambo, 2017a, 2017b). Countries such as Saudi Arabia and the UAE are expected to experience pronounced declines in precipitation and corresponding increases in temperature, worsening water shortages and contributing to rising soil salinity (Rao et al., 2024). Coastal aquifers will be particularly vulnerable to saltwater intrusion. Similarly, Yemen is anticipated to suffer from severe water scarcity, deterioration in water quality, increasing drought frequency and the spread of desertification due to climate change effects (Varisco, 2019).

Looking ahead, the future water scarcity scenario is expected to become more severe due to rising population, rapid urbanization and economic development (Shahin & Shahin, 2007). Water demand is projected to rise significantly in individual countries, while current renewable and non-renewable water resources are already insufficient to meet escalating consumption (Mir et al., 2023). As a result, water shortages will likely intensify, affecting all major sectors agriculture, domestic supply and industry. Declining groundwater tables and shrinking river flows will trigger adverse social, economic and environmental repercussions. In this context, effective water management strategies, conservation initiatives and integrated water resource management schemes become vital for restoring balance between supply and demand. Furthermore, the economic feasibility of long-term water solutions is a growing concern (Birol et al., 2010). The development of low-cost desalination and wastewater treatment technologies is essential, but also capital-intensive (Trivedy, 2007). Continued dependence on mining fossil groundwater, constructing large desalination plants and expanding wastewater treatment infrastructure is expected to impose considerable financial strain on national budgets, particularly in economies with limited fiscal space. If existing water governance frameworks remain static, the reliance on unsustainable sources and inadequate desalination infrastructure will likely exacerbate the region's water security crisis.

The unsustainable use of groundwater in the Arabian Gulf region poses significant challenges to socio-economic development, with consequences for urban water demand, national budgets and the environment (Brown et al., 2018). Based on current trends, water demand in the region may exceed 50 BCM by 2030, leading to further groundwater depletion, water quality deterioration and salinization of agricultural land. Continued reliance on groundwater reserves without significant desalination or wastewater reuse efforts could severely impact water availability for social, agricultural and industrial needs. Economic diversification efforts in oil-rich countries like the UAE, Qatar, Kuwait and Oman are shifting focus to sectors like tourism, which can strain water resources through activities like golf course irrigation (Koul, 2024). The rising demand for water, compounded by climate change effects, threatens the sustainability of water resources in the region, particularly in tourist destinations reliant on water-intensive activities. Water scarcity may hinder the growth of tourism sectors in countries like the UAE, Oman and Qatar, affecting popular destinations and posing economic challenges (Shahin & Salem, 2015). Tourism-related activities such as maintaining gardens, golf courses, spas and swimming pools are projected to significantly increase water consumption, exacerbating water scarcity concerns in the region.

To address these risks, future water security in the Arabian Peninsula will depend on forward-looking strategies such as virtual water trade, efficient water resource management and demand-side interventions. Countries like the UAE and Saudi Arabia have already initiated reforms ranging from legislative changes and water pricing mechanisms to public awareness programs and expanded reuse of treated wastewater. Long-term resilience requires integrated water management planning, market-based allocation systems, conservation-oriented governance and technological innovation to ensure the sustainable use and equitable distribution of water resources under intensifying climate pressures.

Adaptation Strategies

Governments in the Arabian Peninsula are implementing strategies to mitigate the impact of climate change on water resources (Alodah, 2023). Efforts include expanding desalination plant facilities and developing cost-effective solutions like flash flood dams to recharge underground aquifers and promote water conservation (Alodah, 2023; Beaumont, 1977; Tahir et al., 2019). In the UAE, numerous recharge and storage dams have been constructed in catchment areas to collect flash flood water, with capacities amounting to millions of cubic meters per year. Similarly, Saudi Arabia has built dams throughout the country to capture surface runoff water from wadis. Despite these efforts, future drier conditions may reduce water collection from dams, increasing the reliance on desalination plants and elevating operational costs for governments. Governments are focusing on solutions like flash flood dams to replenish underground aquifers and enhance water conservation. The UAE and Saudi Arabia have constructed dams to capture flash flood and surface runoff water. The growing dependence on desalination plants, resulting from reduced water collection from dams, may lead to increased operational costs in the future.

National Initiatives for Addressing Water Scarcity in the Arabian Peninsula

Countries such as the UAE and Saudi Arabia have implemented a range of strategic initiatives to address water scarcity (Al-Zahrani & Baig, 2011; Al-Zubari et al., 2017; Odhiambo, 2017a, 2017b). These actions span legal, institutional, economic and technological dimensions. Key initiatives include:

- *Revision and Modernization of Water Legislation*: Updating existing water laws and strengthening enforcement mechanisms to enhance regulatory effectiveness and support integrated water management.
- *Institutional Reforms*: Reforming water and agriculture sector institutions to reduce fragmentation of responsibilities and improve coordination among agencies.
- *Economic Incentives for Conservation*: Introducing pricing mechanisms and incentives to encourage water and power conservation across households, commercial establishments and industries.
- *Public Awareness Campaigns*: Launching national-level programmes to raise awareness about the importance of water and energy conservation among consumers.

- *Promotion of Private Sector Participation:* Encouraging private entities to construct, operate and maintain water and energy infrastructure to reduce public expenditure and enhance efficiency.
- *Wastewater Recycling:* Reusing treated sewage water for agriculture, landscaping and industrial applications to augment non-conventional water supplies.
- *Subsidy Reforms and Realistic Pricing:* Reviewing and adjusting water and electricity tariffs to reflect actual costs and reduce wasteful consumption. For example, in January 2015, the UAE increased tariffs for expatriates and reduced subsidies for nationals an initial step viewed by experts as a move toward long-term fiscal savings and conservation.
- *Consumer-Level Rationalization Campaigns:* Introducing effective national campaigns to promote responsible water use, complementing other technical and policy measures.

These initiatives collectively reflect a proactive and integrated approach toward achieving sustainable water security in the Arabian Peninsula under growing climatic and socio-economic stress.

Case Study: Hydro-economic Assessment of Water Scarcity in the MENA Region by (Asad et al., 2024)

A recent hydro-economic assessment conducted by Asad et al. (2024) provides critical insights into the water resource challenges facing the Middle East and North Africa (MENA) region, particularly in light of ongoing climate and socio-economic transformations. The study presents a comprehensive evaluation of how climate variability, population dynamics and urban expansion are reshaping water availability and demand and underscores the urgency of adopting more efficient and sustainable water management strategies.

The hydro-economic assessment employed a robust, integrated methodological framework combining both quantitative modeling and qualitative insights. Future scenarios were developed using combinations of Shared Socio-economic Pathways (SSPs) and Representative Concentration Pathways (RCPs) to simulate climate and socio-economic impacts on water resources. The modeling approach included the use of the ECHO hydro-economic model, which integrates economic and hydrological data to evaluate resource allocation and use, alongside the hydrological model CWatM (Catchment Water Management) for simulating water availability and demand at monthly and sub-basin scales. Expert consultations and regional development visions were incorporated to contextualize the scenarios, ensuring region-specific relevance. Simulations were conducted at high spatial and temporal resolutions, enabling fine-scale analysis of water deficits and demand dynamics. This mixed-methods approach allowed for a comprehensive assessment of future water scarcity and informed targeted policy recommendations.

One of the central conclusions of the study is the continuous decline in water availability across the MENA region. This trend is largely attributed to the impacts of climate change and increased climatic variability. Existing water scarcity issues already a defining feature of the region are being further aggravated by socio-economic stressors such as rapid population growth and accelerating urbanization. The research also identifies an escalating demand for water across MENA countries. This rising demand, when coupled with the diminishing supply of renewable water resources, poses a formidable challenge for water managers and policymakers. The widening gap between supply and demand highlights the need for integrated strategies to ensure equitable and sustainable water access across sectors. In response to these pressures, the study emphasizes the need for more efficient resource allocation. Improved allocation across different sectors and sub-basins is deemed crucial for addressing growing water demands and mitigating the consequences of water scarcity. Such strategies are particularly important in regions with competing agricultural, municipal and industrial needs.

Furthermore, the study emphasizes the importance of advanced water management strategies, calling for the adoption of modern water conservation techniques and reliable water supply technologies. These approaches are essential to reduce stress on already limited water resources and to secure long-term sustainability. A key methodological feature of the study is its use of scenario simulations, combining SSPs with RCPs. These simulations offer valuable projections of future water resource conditions under varying climatic and socio-economic scenarios, thereby informing long-term planning and preparedness. The findings of the study call for cost-effective and sustainable water management practices. By implementing such measures, countries in the MENA region can reduce the severity of water scarcity and safeguard the long-term viability of their water resources. In conclusion, the hydro-economic assessment by Asad et al. (2024) highlights the urgent need for improved water governance in the MENA region. With the dual pressures of climate change and socio-economic development reshaping water systems, proactive and integrated water resource management will be essential to ensure future resilience and sustainability.

6.4 Impact on Fisheries, Agriculture, Livestock and Food Security

Climate change has emerged as a major driver of stress across global food systems, with direct implications for key sectors such as agriculture, fisheries, livestock and food security (Myers et al., 2017). Intensifying temperature extremes, shifting precipitation patterns and increasing climate variability are disrupting the stability and productivity of food supply chains worldwide. These disruptions are particularly concerning for environmentally vulnerable and import-dependent regions, including the Arabian Peninsula. In the marine sector, climate-induced changes are significantly affecting global fisheries through rising sea surface temperatures, ocean acidification and altered oceanic circulation patterns (Brander, 2010). These environmental shifts are contributing to reductions in fish stocks, shifts in species distribution and declining marine biodiversity. Coastal communities in regions such as the Arabian

Gulf, which rely heavily on fisheries for protein and livelihood, face increased socio-economic risks (Wabnitz et al., 2018). Declines in fish catch can lead to income losses and increased food insecurity, particularly in countries with limited alternative livelihood options. Adaptation strategies such as sustainable fisheries management and diversification of economic activities are essential to offset these emerging threats.

As climate change continues to impact marine systems, its consequences are also being strongly felt across terrestrial food production, particularly within the agricultural sector. In terrestrial systems, the agricultural sector is highly exposed to the impacts of rising temperatures and declining water availability (Bernacchi & VanLoocke, 2015). Rain-fed agriculture, in particular, is susceptible to reduced yields due to drought, soil degradation and shifting growing seasons. These challenges are further intensified in arid and semi-arid regions like the Arabian Peninsula, where water resources are already scarce. Regional studies project a 15–45% decrease in water availability by the end of the century, posing severe implications for agricultural productivity (Chowdhury & Al-Zahrani, 2013). The resulting decline in output could exacerbate food import dependence, increase exposure to global food price fluctuations and contribute to socio-economic vulnerability among food-insecure populations.

Beyond crops, climate change is also disrupting animal production systems, placing additional strain on food and income sources. Livestock systems are similarly vulnerable to climate change, particularly through the compounding effects of heat stress, water scarcity and declining pasture and feed quality (Nardone et al., 2010). These stressors contribute to reduced animal productivity, increased mortality and higher disease susceptibility. The economic implications include reduced income for livestock producers and rising costs associated with maintaining animal health. These pressures are particularly relevant for regions such as the Arabian Peninsula, where livestock remains a key livelihood source for certain rural and semi-arid communities. Adaptation responses such as the introduction of heat-resilient livestock breeds and improved resource management practices are increasingly critical in supporting the viability of the sector under changing climatic conditions. The combined vulnerabilities across agriculture, fisheries and livestock sectors collectively pose a growing threat to food security (Ding et al., 2017).

Food security, which is shaped by the combined performance of fisheries, agriculture and livestock systems, is increasingly threatened by climate change. Disruptions to food production, transport and pricing both locally and globally are undermining the availability, accessibility and affordability of food (Igwe et al., 2024). This is especially concerning for countries that are heavily reliant on imports, such as those in the Arabian Peninsula. For example, Kuwait imports over 90% of its food supply, making it acutely vulnerable to supply chain disruptions and global market volatility (Asem, 2024). Climate change further compounds these risks by reducing domestic production capacity and amplifying the volatility of global food prices, disproportionately affecting low-income and marginalized populations. Understanding these interconnected risks is essential for developing integrated strategies that build resilience in regional food systems. Collectively, these sectoral vulnerabilities highlight the urgent need for adaptive responses that are both robust and context-specific. While climate change presents similar challenges across global regions, the scale and nature of its impact in the Arabian Peninsula are shaped by distinct environmental, economic and social conditions. Strengthening the resilience of food systems in this region will require integrated strategies that include sustainable agricultural practices, improved water governance, investment in climate-resilient technologies and enhanced regional cooperation. The following sections examine the specific impacts of climate change on agriculture, fisheries, livestock and food security within the Arabian Peninsula, providing a focused analysis of vulnerabilities and emerging adaptation approaches.

Agriculture and Food Security in MENA Countries

Climate change poses a complex and intensifying threat to agriculture in the MENA region, where food systems are already under pressure from resource scarcity, land degradation and population growth (Waha et al., 2017). The MENA region, which includes countries of the Arabian Peninsula such as Saudi Arabia, the UAE, Oman, Qatar, Bahrain, Kuwait and Yemen, is particularly vulnerable due to its arid climate, limited freshwater availability and growing reliance on imported food. One of the most immediate and critical threats is the anticipated decline in water availability. Projections indicate that water discharge could decrease by 15–45% by the end of the century, posing a major challenge for agricultural production across the region (Chowdhury & Al-Zahrani, 2013). This threat is magnified by the fact that approximately 70% of MENA's agricultural output depends on rain-fed systems. In such an arid region, any further reduction in water supply could severely curtail irrigation potential, increase crop failure risks and destabilize rural livelihoods. In parallel, the region is experiencing steadily rising temperatures, a trend expected to intensify as climate change progresses. These higher temperatures lead to increased evapotranspiration and cause direct heat stress on crops, which reduces yields and compromises food quality. This phenomenon is particularly alarming for the MENA region, where baseline temperatures are already among the highest globally and where adaptive capacity, especially in rural zones remains uneven.

Soil degradation represents another major climate-related threat to agriculture in MENA (Ziadat et al., 2022). Processes such as salinization, erosion and desertification are being exacerbated by both climatic pressures and unsustainable farming practices. These factors are diminishing the extent and productivity of arable land, further undermining the ability of farming systems to sustain output and rural employment. In some cases, degraded soils are no longer viable for cultivation, accelerating rural decline and land abandonment. As agricultural productivity declines due to these compounding climate impacts, countries in the MENA region are projected to become increasingly dependent on food imports to meet domestic demand (Le Mouël et al., 2023). This shift presents a significant food security risk, particularly for low- and middle-income countries that lack sufficient financial buffers to absorb global food price shocks. The volatility of international food markets, combined with potential disruptions in supply chains, makes heavy import dependence a structural vulnerability for the region's food systems.

These economic stressors are closely linked with broader social impacts. Reduced agricultural viability is expected to drive rural-to-urban migration, particularly among communities whose livelihoods depend on small-scale farming (de Bruin et al., 2021). Migration in response to climate-driven agricultural collapse may place additional pressure on urban areas, exacerbate inequality and contribute to socio-political instability. Many countries in the region have already experienced environmental displacement and future trends suggest this will likely intensify. Moreover, the capacity to adapt to these challenges remains uneven across the region. Low- and middle-income countries in MENA are especially vulnerable due to limited institutional readiness, weak infrastructure and inadequate financial resources. The agricultural sector in the MENA region stands at the intersection of multiple climate risks ranging from water scarcity and soil degradation to rising heat stress and growing dependence on food imports. Without decisive action, these trends will continue to erode food security and rural resilience. Effective adaptation will require climate-smart agriculture, sustainable land and water management practices, strengthened governance and targeted support for the region's most vulnerable populations.

Case Study: Impact of Climate Change on Food Security in Saudi Arabia by (Rahman et al., 2022)

Food security in Saudi Arabia is under increasing threat due to the multifaceted impacts of climate change, population growth and limitations in natural resources. Food security, broadly defined to include the dimensions of production, commerce and nutrition, is highly sensitive to climatic variables such as temperature, precipitation and atmospheric CO_2 levels. In the context of Saudi Arabia a country characterized by arid climate, scarce freshwater availability and high dependence on food imports climate change exacerbates existing vulnerabilities, putting both production systems and the wider food supply chain under pressure. Climate change directly affects agricultural output through increased heat stress, altered rainfall patterns and soil degradation. It also indirectly influences food security by disrupting the ecological and economic conditions necessary for stable food production. These disruptions manifest in multiple ways, ranging from shifts in species behavior and ecosystem services to volatility in food prices and supply chains. Rising global temperatures, changing precipitation regimes and increased frequency of extreme weather events further complicate the challenge of ensuring a consistent and affordable food supply. These concerns are especially critical for Saudi Arabia, where limited arable land and overreliance on groundwater have constrained domestic agricultural expansion.

Saudi Arabia's extreme climatic conditions marked by soaring summer temperatures that can reach 45–54 °C and irregular rainfall are increasingly volatile under climate change. These temperature extremes contribute to high rates of evapotranspiration, reduced soil moisture retention and the acceleration of mechanical weathering. Studies spanning from 1967 to 2016 have documented a 1.9 °C increase in average temperatures, with the highest rise observed during the summer months. This trend has direct implications for agricultural productivity: for every 1 °C increase in temperature, crop yields in Saudi Arabia are estimated to decline by 7–25%. Crops such as wheat, barley, dates and vegetables all staples in Saudi agriculture are expected to suffer yield declines exceeding 30% under predicted climate scenarios. A regional study further revealed that a 1% increase in winter temperature may cause a 1.12% reduction in agricultural output, underscoring the sensitivity of Saudi agriculture to thermal stress. Rainfall variability further intensifies food production challenges. Saudi Arabia receives an average of only 52.5 mm of rainfall annually and precipitation is unevenly distributed throughout the year. Over 80% of rainfall occurs between November and May, with the driest conditions prevailing during the summer. Topography, particularly elevation, plays a vital role in precipitation patterns, with mountainous regions receiving relatively more rainfall. For example, Jeddah benefits from moderate rainfall due to its elevation and subtropical location. However, even in regions with slightly higher rainfall, the amount remains insufficient to support reliable, large-scale rain-fed agriculture. While rainfall variability affects crop cycles and productivity, the influence of temperature is ultimately more critical in determining food security outcomes in the country. Saudi Arabia's water scarcity is perhaps the most significant constraint to food production.

The Kingdom relies heavily on groundwater, both renewable and non-renewable, to meet its agricultural and domestic needs. Groundwater depletion far outpaces natural recharge, creating a situation of long-term unsustainability. The estimated volume of deep groundwater reserves in Saudi Arabia is around 1,919 billion cubic meters, but annual recharge rates are only 2.7 billion cubic meters. To cope with these limitations, Saudi Arabia supplements its water supply with desalinated water and treated wastewater. Desalination, while effective, comes with high energy and economic costs. Moreover, approximately 30% of desalinated water is lost during distribution due to aging infrastructure and inefficiencies in water management. Agricultural practices in Saudi Arabia further strain water resources. Although advanced irrigation technologies have been adopted, irrigation efficiency remains relatively low-estimated at around 50%. Poor water management and outdated irrigation systems contribute to substantial water losses. Meanwhile, the expansion of irrigated agriculture has grown rapidly, driven by state-backed initiatives aimed at achieving food security. Horticultural crops such as vegetables, melons and fruit trees have seen increased cultivation, including salt-sensitive varieties, which are particularly vulnerable to declining water quality. The annual growth rate in irrigated agriculture peaked at around 5% before 1990 and has since slowed to approximately 1.2%.

In addition to abiotic stressors like heat and drought, biotic threats such as pests and diseases are also increasing due to climate change. Altered precipitation patterns and rising temperatures have contributed to greater prevalence and expansion of pest populations and plant diseases. These outbreaks reduce income from meat and crop production, disrupt aquatic and forest ecosystems and increase the costs associated with pest and disease control. In Saudi Arabia, the cumulative impact of these stressors results in reduced crop harvests and higher yield variability. Demographic and economic trends compound these environmental pressures. Saudi Arabia's population has grown significantly, from 4 million in 1960 to over 32.5 million by 2018, with projections suggesting it could surpass 56 million by 2050. This rapid growth, coupled with increasing urbanization, puts additional strain on food and water systems. Urbanization also leads to dietary shifts and greater reliance on imported food, as local agricultural production cannot meet the demand of growing urban populations. The expansion of supermarket chains and access to international suppliers has increased food availability but has also increased dependence on external food sources.

Climate change also disrupts the broader food supply chain. Droughts, floods and extreme weather events impair agricultural production, damage infrastructure and hinder distribution. These conditions increase food insecurity risks, particularly for low-income households. Temperature variations, ozone exposure and fertilizer shortages also impact crop yields and nutritional quality. Higher CO_2 concentrations have been shown to reduce the protein and nitrogen content in essential crops like soybeans and alfalfa. Poor grain and forage quality not only diminishes yields but also affects livestock productivity in pasture and rangeland systems. Warmer and wetter conditions favor the proliferation of weeds, pests and fungi, contributing to annual yield losses in Saudi Arabia. For instance, severe insect pest outbreaks have led to substantial crop losses. Similarly, instances such as reduced maize output in the U.S. Corn Belt and cherry losses in Michigan serve as international parallels to what could occur in Saudi Arabia under unchecked climate stress. The economic dimension of food security is also closely tied to climate impacts. Economic shifts influence employment, income levels and household purchasing power, which in turn shape access to nutritious food. As Gross Domestic Product (GDP) increases, so does food demand, which places further pressure on limited domestic resources and global food supply systems. Given these mounting environmental, demographic and economic pressures, continued reliance on nonrenewable groundwater for agriculture poses a significant long-term sustainability challenge. This reinforces the view that food import dependence in Saudi Arabia should be recognized as an adaptive response to resource constraints, rather than a temporary policy decision.

Adaptation Initiatives to Climate Change Impacts on the Food System in Saudi Arabia

Given these overlapping vulnerabilities, Saudi Arabia has begun to implement a range of adaptation initiatives categorized into three primary areas: supply-side, infrastructure-based and demand-side strategies (Rahman et al., 2022). These efforts aim to safeguard national food security and build resilience across the food system.

Supply-side initiatives are aimed at enhancing agricultural productivity and supply chain stability in the face of climate stress. These include the adoption of drought-resistant crop varieties, expanded use of drip and sprinkler irrigation systems and the promotion of conservation agriculture. Further approaches involve increasing soil organic matter, integrated nutrient management and pest and disease control. Agroforestry and crop-livestock integration are encouraged as diversified farming systems that build resilience and improve income stability. Controlled-environment agriculture, such as hydroponics and aquaponics, is also gaining traction, particularly in urban settings where land and water are limited.

Infrastructure initiatives focus on addressing water needs and protecting vulnerable areas from extreme events. Dams are being built or maintained to store water and reduce flash flood risks. Enhancements to transport infrastructure such as rural roads and market access routes are essential to ensure continuity of food distribution even under adverse weather conditions.

Demand-side measures target consumption habits, food waste reduction and public awareness. Promoting plant-based diets not only improves health outcomes in a country where 71% of the population is overweight but also reduces pressure on grazing lands and helps lower greenhouse gas emissions. Strategies to reduce food loss and waste are also essential. In Saudi Arabia, food waste is estimated to cost SAR 40 billion annually. In response, the government has launched initiatives such as the National Foundation for Food Preservation (Saudi Gazette, 2022). Other actions include post-harvest innovations, mandatory waste reporting by corporations and public education to support sustainable consumption patterns.

6.5 Fisheries

Marine biodiversity, ecosystem health and fisheries are increasingly under threat from overfishing, pollution and other anthropogenic pressures. Climate change has introduced additional complexity to sustainable management and conservation efforts by altering key oceanic conditions, including water temperature and biogeochemical processes. These environmental shifts have prompted physiological, phenological and behavioral responses among marine organisms, influencing their population dynamics and geographic distributions (Donelson et al., 2019). As a result, changes in ocean atmosphere systems are projected to drive significant alterations in species richness, community composition and broader ecosystem functions. These transformations are expected to impact the availability and stability of marine goods and services that support human well-being. In the context of global marine change, the marine environment surrounding the Arabian Peninsula particularly the Arabian Gulf offers a distinctive case study due to its extreme climatic conditions and ecological sensitivity (Hereher, 2020).

Characterized by high seasonal variability in temperature and salinity (Mushtaq et al., 2025), the Gulf supports species with remarkable tolerance to environmental stress. Although relatively species-poor compared to neighboring systems, it remains an ecologically valuable component of the broader Indo-West Pacific region, with direct implications for biodiversity and food security in the Peninsula. The biodiversity and associated ecosystem services of the Gulf are increasingly threatened by a confluence of climate-related stressors including rising sea temperatures, declining oxygen levels and sea level rise compounded by intensive human activities such as oil extraction, coastal infrastructure expansion, seawater desalination and unsustainable fishing practices (Naser, 2014). While many marine organisms in the region demonstrate comparatively high thermal tolerance, the cumulative impacts of climate warming are already evident. Between 1950 and 2010, sea surface temperatures in the Gulf increased by approximately 0.57 °C, contributing to observable disruptions in vulnerable species and habitats (Shirvani et al., 2015). Among the most heavily impacted ecosystems are coral reefs, which have endured multiple episodes of thermal stress, with sea temperatures reaching between 35 and 37 °C at least five times since the late 1990s. These thermal anomalies have triggered widespread coral bleaching events and led to substantial declines in coral cover, with estimates suggesting that approximately 70% of the region's coral reefs have been lost over recent decades. This degradation has, in turn, been linked to substantial declines in fish species richness and overall ecosystem complexity.

A study conducted by Wabnitz et al. (2018) provides a comprehensive assessment of the potential impacts of climate change on marine biodiversity and fisheries catches in the Arabian Gulf. The study employed three separate niche modelling approaches under a 'business-as-usual' climate change scenario to project the future habitat suitability of the Gulf for 55 expert-identified priority species, encompassing both ecologically significant and non-fish species. In addition to these ecological projections, a vulnerability assessment was performed to evaluate the susceptibility of national economies to climate-induced changes in fisheries. The modelling outputs indicated a high rate of local extinction, with up to 35% of the initial species richness expected to be lost by 2090 relative to 2010 (Fig. 6.2). Spatially, these local extinctions are projected to be most pronounced in the southwestern part of the Gulf, particularly along the coasts of Saudi Arabia, Qatar and the UAE. While the projected patterns offer valuable insights into potential biodiversity losses, the study indicates a degree of uncertainty regarding the magnitude of changes in habitat suitability. Fisheries-specific results forecast a reduction in future catch potential for several countries along the western Gulf, with relatively consistent projections across models. Qatar and the UAE are projected to experience significant declines, with catch potential falling by more than 26%. Furthermore, when changes in catch potential were integrated with socio-economic indicators, the results indicated that the fisheries sectors in Bahrain and Iran may be the most vulnerable to the effects of climate change (Fig. 6.3).

6.6 Livestock

Climate change significantly affects the livestock food supply chain, influencing various stages from production to distribution (Cheng et al., 2022). These impacts are both direct and indirect, necessitating targeted adaptation and mitigation strategies to maintain sustainability and food security. Direct impacts on livestock include health and productivity declines due to rising temperatures, which cause heat stress, lower growth rates, reduced reproductive performance and increased disease susceptibility, resulting in reduced milk and meat yields (Kuczynski et al., 2011). Rising temperatures and shifting precipitation patterns also contribute to the spread of vector-borne diseases, increasing morbidity and mortality rates in livestock populations. Indirectly, climate change reduces the quality and availability of feed resources. Elevated

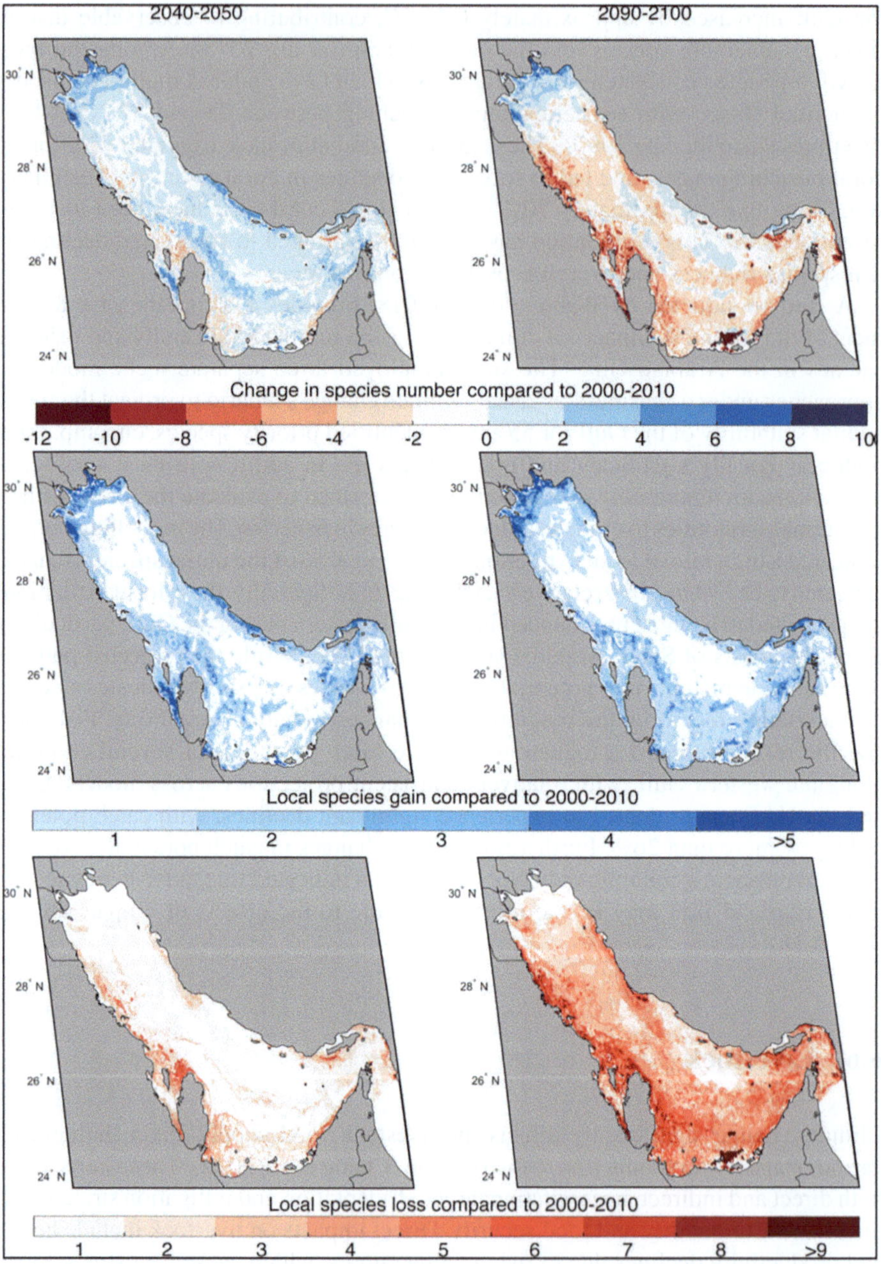

Fig. 6.2 Projected change in species number, species invasion and extinction in the Gulf by 2050 and 2090 relative to 2010 studied by Wabnitz et al. (2018). *Source* Wabnitz et al. (2018)

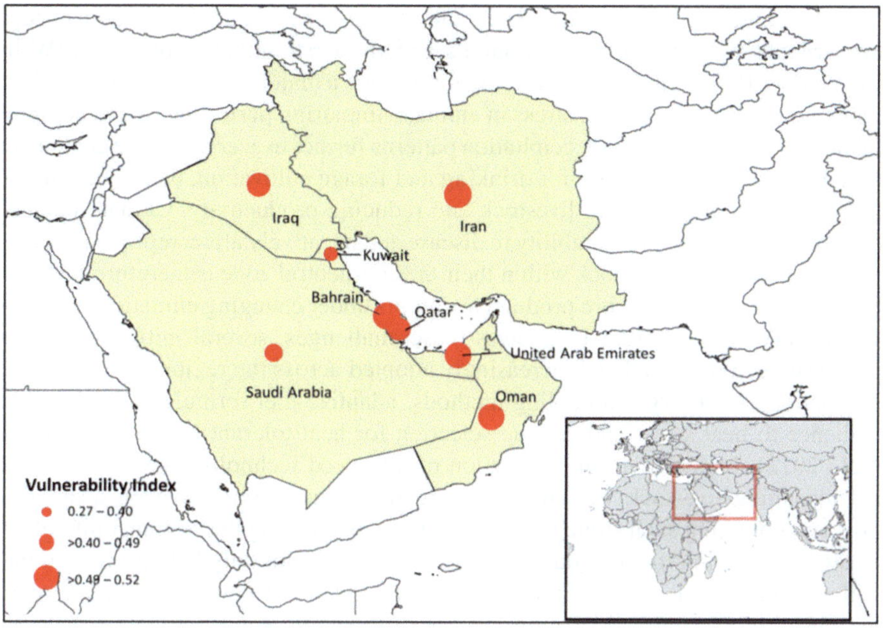

Fig. 6.3 Relative vulnerability of national economies in 2090 to climate change impacts on fisheries in the Gulf studied by Wabnitz et al. (2018). *Source* Wabnitz et al. (2018)

temperatures and altered rainfall decrease herbage yield and increase plant lignification, lowering forage digestibility. In some regions, shrub encroachment driven by climate variability has further reduced accessible and nutritious forage.

Economic repercussions include increased input costs and increased market price volatility across the livestock supply chain (Huang et al., 2011). These effects are not distributed evenly, with low-income countries facing more severe challenges due to their dependence on grassland-based systems and limited adaptive capacity. Adaptation strategies such as the use of climate-resilient livestock breeds, improved water management and feed diversification are essential to reduce vulnerability (Sejian et al., 2021). Mitigation measures include reducing greenhouse gas emissions from livestock through better manure management and dietary modifications. Sustainable livestock practices like agroforestry and regenerative agriculture enhance ecosystem resilience and promote equitable resource use. Collaborative engagement among farmers, policymakers and researchers is also vital for fostering innovation and building climate-resilient food systems. Livestock systems are not only impacted by climate change but also contribute significantly to it, particularly through methane emissions (Lassey, 2008). A balanced approach is needed that sustains livestock production while minimizing its environmental footprint. Addressing these challenges requires a comprehensive understanding of the complex interactions between climate change and livestock systems.

In the context of Arabian countries, the impacts of climate change on livestock are particularly acute due to the region's already harsh climatic conditions (Waha et al., 2017). Rising environmental temperatures are a major concern, as extreme heat intensifies the risk of thermal stress in animals, impairing performance and compromising welfare. Changes in precipitation patterns further exacerbate these challenges by limiting water availability for drinking and forage cultivation, thereby adding to the physiological burden on livestock and reducing productivity. Climate-induced stressors also increase vulnerability to disease and negatively affect reproductive efficiency. Maintaining livestock within their thermo-neutral zone is therefore essential to preserve health and ensure productive outputs under changing climatic conditions (Al-Haidary et al., 2012). To address these challenges, several mitigation strategies have been proposed and increasingly adopted across the region. These include the application of targeted cooling methods, adaptive diet formulation to improve resilience to heat stress and genetic selection for heat-tolerant breeds (Renaudeau et al., 2012). In addition, the integration of advanced technologies for monitoring and managing livestock welfare is gaining importance as a precision adaptation tool. Implementing such strategies is central to safeguarding the sustainability of the livestock sector in Arabian countries. These measures not only help protect animal welfare but also enhance the sector's capacity to adapt to evolving climate realities, ensuring its long-term viability in one of the world's most climate-vulnerable regions.

6.7 Economic Ramifications, Including Impacts on Industries Such as Energy and Tourism

Climate change poses significant economic implications for global GDP and various industry sectors (Dellink et al., 2019). Its economic impact is complex and multifaceted, influencing overall economic growth, sectoral productivity and financial stability. These consequences are not distributed evenly; some regions and sectors are expected to experience more severe disruptions than others. As such, the economic ramifications of climate change are both profound and far-reaching, necessitating urgent, coordinated global action to mitigate its effects and adapt to emerging economic realities. One of the most pressing concerns is the projected decline in global GDP due to the accelerating impacts of climate change (Newell et al., 2021). If mitigation measures are not implemented, estimates suggest that global GDP could shrink by as much as 18% by 2050. This projection reflects the cumulative burden of rising temperatures, shifting weather patterns and increasing frequency of extreme events on productive capacity and capital infrastructure. Further studies show that the loss in global GDP could range between 0.7 and 2.5% by 2060, with more pronounced effects projected for the longer term if emissions continue to rise unchecked (Dellink et al., 2014). Importantly, these economic costs are not uniform across regions. Less developed areas, including South Asia, South-East Asia, North Africa and Sub-Saharan Africa, are anticipated to bear a disproportionate share of GDP losses due to their higher exposure and lower adaptive capacity. Beyond aggregate GDP, climate change exerts varying levels of stress on individual sectors. The energy sector, for instance, is undergoing a transformative shift as nations pivot toward renewable energy sources in an effort to decarbonize their economies (Hassan et al., 2024). This transition carries substantial economic implications, especially for fossil fuel-dependent industries and economies. Moreover, energy infrastructure itself faces physical risks from climate change, including damage from rising sea levels and the increasing frequency of extreme weather events. The tourism sector is also vulnerable, as environmental degradation and ecosystem changes diminish the appeal of natural attractions, reducing tourist arrivals and leading to substantial revenue losses.

Climate change further introduces uncertainty into financial and economic stability (Battiston et al., 2021). Variability in inflation and increased financial market stress are among the anticipated consequences. The costs associated with adaptation and mitigation efforts may also place upward pressure on public debt in many countries. In addition, the insurance and broader financial sectors are increasingly exposed to the growing risks posed by climate-related natural disasters and environmental disruptions (Caselli & Figueira, 2020). These risks contribute to higher insurance premiums, reduced coverage availability and increased uncertainty in financial markets. Despite the predominantly adverse impacts, climate change also presents new economic opportunities, particularly in the realm of innovation and sustainability. The transition to a low-carbon economy can drive economic growth by fostering the development of environmentally friendly industries and technologies (Linnenluecke et al., 2019). This shift encourages innovation in renewable energy, circular economy practices and sustainable infrastructure. Such transformations offer the potential to generate new economic prospects, create green jobs and enhance competitiveness in emerging sectors. While the economic implications of climate change are largely negative, the emerging opportunities associated with sustainability transitions underscore the importance of strategic planning and innovation (Steward, 2012). The shift to a low-carbon economy can serve as a catalyst for long-term growth and resilience. However, the economic transformation required to effectively address climate change remains uncertain, as outcomes will depend heavily on the climate scenarios that unfold and the policy paths pursued by governments.

Climate change poses significant economic implications for the Arabian Peninsula, particularly influencing GDP trajectories and a broad spectrum of industry sectors (Rahman et al., 2022). The region's economic profile marked by a strong dependence on hydrocarbon exports and a high degree of vulnerability to environmental variability amplifies the complexity of its climate-related challenges (Dyduch et al., 2018). The economic effects are multifaceted, involving direct environmental impacts such as rising temperatures, erratic precipitation and sea level rise, all of which necessitate extensive investments in infrastructure resilience and energy consumption. Simultaneously, the ongoing global transition toward renewable energy threatens to diminish the oil and gas revenues that constitute a substantial portion of the Gulf Cooperation Council (GCC) economies (Darwish et al., 2018). Moreover, the region's agricultural sector, which is largely reliant on food imports, faces additional stress from potential disruptions to global food supply chains (Ali et al., 2022). Collectively, these dynamics influence the long-term economic stability and growth prospects of the Arabian Peninsula. Among the direct economic impacts, infrastructure costs stand out as a pressing concern. Rising sea levels and the increasing frequency of extreme weather events compel governments to invest heavily in coastal protection and infrastructure adaptation (Woodruff et al., 2018). This includes not only reinforcing existing structures but, in some cases, necessitates the rebuilding or relocation of entire infrastructure systems, thereby placing a considerable financial burden on national budgets. In parallel, high temperatures drive a marked increase in demand for cooling, particularly through air conditioning (Howarth et al., 2020). This surge in energy use imposes higher operational costs on both public utilities and consumers, with additional risks of electricity supply failures during peak consumption periods.

At the macroeconomic level, the oil and gas sector faces rising uncertainty (Alqahtani & Klein, 2021). As global initiatives to decarbonize accelerate, demand for fossil fuels is expected to decline, directly threatening the fiscal revenues of GCC states where hydrocarbons remain the dominant economic driver (Dargin, 2021). This decline has significant implications for GDP, employment and investment, especially in the absence of effective diversification strategies. Agriculture, although a smaller contributor to regional GDP, is also highly susceptible to climate change (Amin et al., 2016). Alterations in rainfall distribution and rising temperatures negatively affect agricultural yields, increasing the region's dependency on food imports and further straining economic resources. In addition, sectors like tourism and real estate are affected as environmental degradation and extreme heat diminish the region's attractiveness to tourists and reduce property values, particularly in coastal zones vulnerable to sea level rise. Beyond sector-specific outcomes, broader economic relationships further underscore the complexity of climate change impacts in the region. The link between economic growth and carbon emissions in the GCC remains a complex policy issue (ElMassah & Hassanein, 2023). As economies expand, energy consumption and associated emissions tend to rise, necessitating integrated strategies that reconcile economic development with environmental sustainability. Furthermore, climate change has the potential to influence global trade and market dynamics. For example, increased global food prices may have mixed implications for local economies. While higher agricultural prices may benefit some producers, they could also slow overall GDP growth and reduce household incomes, particularly in rural communities where resilience to price shocks is limited.

Adaptation and mitigation strategies are therefore essential to managing the economic risks posed by climate change. Robust policy frameworks and international cooperation are required to navigate the transition. GCC countries must prioritize investment in renewable energy infrastructure and the promotion of sustainable economic practices as central components of their climate response. Equally important is targeted investment in the agricultural sector and the expansion of social protection programs to buffer vulnerable populations from the adverse impacts of food insecurity and income instability. While the economic implications of climate change for the Arabian Peninsula are undoubtedly serious, they also present opportunities for transformation. Through sustained investment in renewable energy and the adoption of sustainability-oriented policies, the region has the potential not only to address the immediate challenges of climate change but also to reorient its development trajectory toward greater resilience and inclusiveness.

Linking Tourism, Energy Use and Climate Change: Insights from GCC Economies

A valuable case study that illustrates the interconnections between climate change, tourism, energy use and economic dynamics in the Gulf region is provided by the study titled "*The Nexus between tourism-energy-environmental degradation: Does financial development matter in GCC countries?*" *by* Farooq et al. (2024). This research addresses key issues related to climate change by analyzing the relationship between carbon dioxide (CO_2) emissions and the tourism-energy nexus within the GCC countries. The study highlights several important implications concerning climate change. Firstly, the research identifies CO_2 emissions as a significant indicator of climate change. It finds that both tourism activities and energy consumption patterns in the GCC countries have direct implications for CO_2 emissions, thereby establishing a clear link between these sectors and broader climate change concerns. This correlation underscores the environmental implications of current development trends in the region. Secondly, the study draws attention to the impact of fossil fuel energy (FFE) on environmental quality. It reveals a positive correlation between FFE consumption and CO_2 emissions, pointing to the environmental challenges posed by the region's heavy reliance on fossil fuels. Since fossil fuel combustion is a major contributor to greenhouse gas emissions, this finding reinforces the urgency of addressing the role of energy systems in accelerating climate change.

In contrast, the research identifies renewable energy as having a mitigating effect on climate-related environmental degradation. Specifically, the study finds that renewable energy sources exhibit a negative relationship with CO_2 emissions. This suggests that a transition towards renewable energy can play a significant role in reducing the adverse impacts on climate change, highlighting the importance of adopting sustainable energy practices in the GCC region to combat environmental threats. Furthermore, the study explores the role of financial development as a potential moderating factor in this nexus. It demonstrates that financial development can reduce the positive impact of tourism and fossil fuel energy consumption on CO_2 emissions. In other words, a robust and well-developed financial sector may support the implementation of strategies aimed at lowering emissions, thereby lessening the environmental footprint associated with economic activities. In light of these findings, the study puts forward recommendations for integrating financial development into policy frameworks designed to reduce the negative impacts of tourism and energy consumption on environmental quality. This integrated approach is presented as essential to addressing the complex challenges posed by climate change in the GCC countries. By promoting sustainable practices and leveraging financial development, the region can better manage CO_2 emissions and work towards long-term environmental sustainability. In summary, this study clearly illustrates the critical connections between tourism, energy consumption and CO_2 emissions and their relevance to climate change. It emphasizes the urgent need for sustainable practices and financial sector involvement in climate mitigation strategies. The findings support the argument that addressing climate change in the GCC context requires a multidimensional strategy involving renewable energy adoption, policy reform and the alignment of financial development with environmental goals.

6.8 Conclusion

Climate change poses a complex set of challenges to the socio-economic development of the Arabian Peninsula. The region's inherent ecological limitations, particularly in water availability, are magnified by accelerating climatic shifts, including rising temperatures, increased evaporation, erratic rainfall and sea-level rise. These environmental stressors converge with rapid population growth, economic development and urbanization to exert unprecedented pressure on water systems, food security and economic resilience. The water crisis is at the core of this climate vulnerability. Across the Peninsula, renewable freshwater resources are scant and unevenly distributed. Groundwater, including fossil aquifers, has been excessively exploited, leading to alarming depletion rates, salinization and ecological degradation. Despite the widespread adoption of desalination and wastewater reuse technologies, the region remains heavily reliant on energy-intensive and economically unsustainable water solutions. With projections indicating further declines in precipitation and rising water demand, the risk of long-term water insecurity looms large. This jeopardizes agricultural viability, domestic supply and industrial needs while increasing national expenditure on water infrastructure. In agriculture, the impacts of climate change are compounded by soil degradation, declining irrigation efficiency and dependency on rain-fed systems in a hyper-arid climate. The anticipated decline in water discharge by up to 45% and the intensifying heat stress are set to reduce yields and widen the food import gap. Consequently, food insecurity remains a persistent challenge, especially for lower-income populations. Countries such as Saudi Arabia and Kuwait, already importing over 90% of their food, are increasingly vulnerable to global market volatility and disruptions in supply chains. Fisheries and marine biodiversity in the Arabian Gulf are similarly under significant pressure. Rising sea surface temperatures, oxygen depletion, coral bleaching and overfishing are collectively eroding the resilience of marine ecosystems. Studies project up to a 35% decline in species richness and a significant reduction in fish catch potential by 2090, with substantial socio-economic implications for coastal communities and national economies. The livestock sector is also facing critical challenges. Increased heat stress, reduced feed and water availability and a rise in disease vectors threaten animal health, productivity and farmer livelihoods. As livestock systems are both vulnerable to and contributors to climate change, adaptation strategies such as heat-tolerant breeds, improved veterinary care and sustainable feed management are increasingly essential. Beyond these sectors, climate change exerts broader economic pressures. The region's hydrocarbon-dependent economies are exposed to the global shift towards decarbonization. At the same time, rising energy demand due to increased cooling needs and coastal infrastructure adaptation creates new fiscal burdens. Tourism, a key diversification sector for Gulf economies, is also at risk from environmental degradation and extreme weather events. Nevertheless, amidst these challenges, the Arabian Peninsula is demonstrating a proactive approach. Governments in the GCC are implementing a range of adaptation strategies across supply, infrastructure and demand domains. These include revising water laws, enhancing desalination capacity, promoting conservation agriculture, investing in climate-resilient infrastructure and launching public awareness campaigns. Saudi Arabia and the UAE, for instance, have pioneered regulatory reforms, renewable energy integration and water pricing mechanisms to reduce unsustainable consumption. The case studies reviewed ranging from hydro-economic assessments of water scarcity to climate impacts on food security and marine ecosystems underscore the need for coordinated, evidence-based and locally adapted interventions. It is evident that climate change mitigation and adaptation must be approached as integrated development challenges requiring collaboration across sectors, scales and governance systems. Ultimately, resilience in the Arabian Peninsula depends on the capacity to transform current vulnerabilities into opportunities for sustainable growth. This entails investing in research and innovation, strengthening regional cooperation, empowering communities and embedding equity and environmental justice into climate policy frameworks. The insights from this chapter provide valuable guidance for aligning climate action with socio-economic resilience, supporting a sustainable and inclusive future for the region in the face of growing environmental challenges.

References

Al-Haidary, A. A., Aljumaah, R. S., Alshaikh, M. A., Abdoun, K. A., Samara, E. M., Okab, A. B., & Alfuraiji, M. M. (2012). Thermoregulatory and physiological responses of Najdi sheep exposed to environmental heat load prevailing in Saudi Arabia.

BM Ali I Manikas B Sundarakani 2022 Food security in the United Arab Emirates: External cereal supply risks Cogent Economics & Finance 10 1 2149491

I Ali 1983 Development of Groundwater Resources of Saudi Arabia Proceedings of the Institution of Civil Engineers 74 4 885 898

M Almazroui MN Islam PD Jones H Athar MA Rahman 2012 Recent climate change in the Arabian Peninsula: Seasonal rainfall and temperature climatology of Saudi Arabia for 1979–2009 Atmospheric Research 111 29 45

A Alodah 2023 Towards sustainable water resources management considering climate change in the case of Saudi Arabia Sustainability 15 20 14674

BA Alotaibi MB Baig MM Najim AA Shah YA Alamri 2023 Water scarcity management to ensure food scarcity through sustainable water resources management in Saudi Arabia Sustainability 15 13 10648

A Alqahtani T Klein 2021 Oil price changes, uncertainty, and geopolitical risks: On the resilience of GCC countries to global tensions Energy 236 121541

KH Al-Zahrani MB Baig 2011 Water in the Kingdom of Saudi Arabia: Sustainable management options The Journal of Animal & Plant Sciences 21 3 601 604

WK Al-Zubari 2016 Status of water in the Arab region The water, energy, and food security nexus in the Arab region Springer International Publishing 1 24

Al-Zubari, W., Al-Turbak, A., Zahid, W., Al-Ruwis, K., Al-Tkhais, A., Al-Muataz, I., Abdelwahab, A., Murad, A., Al-Harbi, M., & Al-Sulaymani, Z. (2017). An overview of the GCC unified water strategy (2016–2035). *Desalination and Water Treatment*, *81*, 1–18.

MT Amin SH Mahmoud AA Alazba 2016 Observations, projections and impacts of climate change on water resources in Arabian Peninsula: Current and future scenarios Environmental Earth Sciences 75 1 17

MSA Ansari 2013 The water demand management in the Kingdom of Bahrain Journal of Engineering and Advanced Technology 2 5 544 554

Asad, S., Sahu, R. K., Fridman, D., Willaarts, B., & Kahil, T. (2024, April). Hydro-economic assessment of the impact of climate and socio-economic changes on water resources in the MENA region. In *EGU General Assembly Conference Abstracts* (p. 17546).

Asem, S. O. (2024). Prospects on Agricultural Development in Kuwait, An Overview. *Research on World Agricultural Economy*, 189–202.

Awawdeh, F., & Nejatian, A. (2016). Water Resources and Agriculture in Arab Countries. "This multi-authored compilation provides informative insights from Arab, Brazilian, Mexican, and Chilean intellectuals and illustrates the significant differences in scholarship on either side of the Atlantic." William P. Glade, Professor Emeritus, Department of Economics, University of Texas at Austin, 245.

S Battiston Y Dafermos I Monasterolo 2021 Climate risks and financial stability Journal of Financial Stability 54 100867

Beaumont, P. (1977). Water and development in Saudi Arabia. *Geographical Journal*, 42–60.

CJ Bernacchi A VanLoocke 2015 Terrestrial ecosystems in a changing environment: A dominant role for water Annual Review of Plant Biology 66 1 599 622

Bijnens, G., Anyfantaki, S., Colciago, A., De Mulder, J., Falck, E., Labhard, V., Lopez-Garcia, P., Lourenço, N., Meriküll, J., & Parker, M., Röhe, O., Schroth, J., Schulte, P., & Strobel, J. (2024). The impact of climate change and policies on productivity. *ECB Occasional Paper*, (2024/340).

E Birol P Koundouri Y Kountouris 2010 Assessing the economic viability of alternative water resources in water-scarce regions: Combining economic valuation, cost-benefit analysis and discounting Ecological Economics 69 4 839 847

K Brander 2010 Impacts of climate change on fisheries Journal of Marine Systems 79 3–4 389 402

JJ Brown P Das M Al-Saidi 2018 Sustainable agriculture in the Arabian/Persian Gulf region utilizing marginal water resources: Making the best of a bad situation Sustainability 10 5 1364

Caselli, G., & Figueira, C. (2020). The impact of climate risks on the insurance and banking industries.In *Sustainability and financial risks: The impact of climate change, environmental degradation and social inequality on financial markets* (pp. 31–62).

M Cheng B McCarl C Fei 2022 Climate change and livestock production: A literature review Atmosphere 13 1 140

S Chowdhury M Al-Zahrani 2013 Implications of climate change on water resources in Saudi Arabia Arabian Journal for Science and Engineering 38 1959 1971

J Dargin 2021 The pathway to a green Gulf: A review and analysis of the evolution of Saudi Arabia, Qatar, and the United Arab Emirates' climate change positions Carbon Climate Law Rev 15 313 341

Darwish, S., Abdo, H., & AlShuwaiee, W. M. (2018). Opportunities, challenges and risks of transition into renewable energy: the case of the Arab Gulf Cooperation Council. *International Energy Journal*, *18*(4).

S Bruin de J Dengerink J Vliet van 2021 Urbanisation as driver of food system transformation and opportunities for rural livelihoods Food Security 13 4 781 798

R Dellink E Lanzi J Chateau 2019 The sectoral and regional economic consequences of climate change to 2060 Environmental and Resource Economics 72 309 363

Dellink, R., Lanzi, E., Chateau, J., Bosello, F., Parrado, R., & De Bruin, K. (2014). Consequences of climate change damages for economic growth: a dynamic quantitative assessment.

E DeNicola OS Aburizaiza A Siddique H Khwaja DO Carpenter 2015 Climate change and water scarcity: The case of Saudi Arabia Annals of Global Health 81 3 342 353

Q Ding X Chen R Hilborn Y Chen 2017 Vulnerability to impacts of climate change on marine fisheries and food security Marine Policy 83 55 61

Donelson, J. M., Sunday, J. M., Figueira, W. F., Gaitán-Espitia, J. D., Hobday, A. J., Johnson, C. R., Leis, J.M., Ling, S.D., Marshall, D., Pandolfi, J. M., Pecl, G., Rodgers, G. G., Booth, D. J., & Munday, P. L. (2019). Understanding interactions between plasticity, adaptation and range shifts in response to marine environmental change. *Philosophical Transactions of the Royal Society B*, *374*(1768), 20180186.

Dyduch, J., Jarząbek, J., & Skorek, A. (2018). The dependence of Gulf Countries on hydrocarbons export: a perspective of regional security complex theory.

S ElMassah EA Hassanein 2023 Economic development and environmental sustainability in the GCC countries: New insights based on the economic complexity Sustainability 15 10 7987

U Farooq MI Tabash MA Saleh Al-Faryan C Işık T Dogru 2024 The Nexus between tourism-energy-environmental degradation: Does financial development matter in GCC countries? Tourism Economics 30 3 680 701

Gaur, M. K., & Squires, V. R. (2018). Geographic extent and characteristics of the world's arid zones and their peoples. In *Climate variability impacts on land use and livelihoods in drylands* (pp. 3–20).

SN Gosling NW Arnell 2016 A global assessment of the impact of climate change on water scarcity Climatic Change 134 371 385

S Hallegatte 2009 Strategies to adapt to an uncertain climate change Global Environmental Change 19 2 240 247

Hassan, Q., Viktor, P., Al-Musawi, T. J., Ali, B. M., Algburi, S., Alzoubi, H. M., Al-Jiboory, A. K., Sameen, A. Z., Salman, H. M.Marek Jaszczur & Jaszczur, M. (2024). The renewable energy role in the global energy transformations. *Renewable Energy Focus*, *48*, 100545.

ME Hereher 2020 Assessment of climate change impacts on sea surface temperatures and sea level rise—The Arabian Gulf Climate 8 4 50

N Howarth N Odnoletkova T Alshehri A Almadani A Lanza T Patzek 2020 Staying cool in A warming climate: Temperature, electricity and air conditioning in Saudi Arabia Climate 8 1 4

H Huang M Lampe von F Tongeren van 2011 Climate change and trade in agriculture Food Policy 36 S9 S13

Hussain, S., Niyazi, B., Elfeki, A., & Masoud, M. (2025). Forecasting 21st century hydrological challenges in arid regions: Climate-induced Shifts in Runoff and Evapotranspiration. *Water Resources Management*, 1–31.

AN Igwe NL Eyo-Udo A Stephen 2024 Strategies for mitigating food pricing volatility: Enhancing cost affordability through sustainable supply chain practices Strategies 13 9 151 163

Koul, S. (2024). Towards a Sustainable Future: Economic Diversification and Growth Sectors in the Middle East. Available at SSRN 5165703

Kuczynski, T., Blanes-Vidal, V., Li, B., Gates, R. S., de Alencar Naas, I., Moura, D. J., Berckmans, D., & Banhazi, T. M. (2011). Impact of global climate change on the health, welfare and productivity of intensively housed livestock. *International Journal of Agricultural and Biological Engineering*, *4*(2), 1–22.

KR Lassey 2008 Livestock methane emission and its perspective in the global methane cycle Australian Journal of Experimental Agriculture 48 2 114 118

Le Mouël, C., Forslund, A., Marty, P., Manceron, S., Marajo-Petitzon, E., Caillaud, M. A., Dumas, P., & Schmitt, B. (2023). Can the Middle East-North Africa region mitigate the rise of its food import dependency under climate change? *Regional Environmental Change*, *23*(2), 52.

MK Linnenluecke J Han Z Pan T Smith 2019 How markets will drive the transition to a low carbon economy Economic Modelling 77 42 54

DB Lobell SM Gourdji 2012 The influence of climate change on global crop productivity Plant Physiology 160 4 1686 1697

A Mazzoni E Heggy G Scabbia 2018 Forecasting water budget deficits and groundwater depletion in the main fossil aquifer systems in North Africa and the Arabian Peninsula Global Environmental Change 53 157 173

MA Mir MW Ashraf 2023 The challenges and potential strategies of Saudi Arabia's water Resources: A review in analytical way Environmental Nanotechnology, Monitoring & Management 20 100855

S Mogren 2015 Saltwater Intrusion in Jizan coastal zone, southwest Saudi Arabia, inferred from geoelectric resistivity survey International Journal of Geosciences 6 3 286 297

Mohamed, H., Elayah, M. A. A., & Schulpen, L. W. M. (2017). *Yemen between the impact of the climate change and the ongoing Saudi-Yemen war: A real tragedy*. Centre for Governance and Peace-building Yemen & Centre for International Development Issues Nijmegen.

F Mushtaq LM Alhems M Farooq SP Tiwari 2025 A comprehensive review of geomatics based coastal zone management in the Realm of Arabian Gulf, Saudi Arabia Discover Applied Sciences 7 2 140

Myers, S. S., Smith, M. R., Guth, S., Golden, C. D., Vaitla, B., Mueller, N. D., Dangour, A. D., & Huybers, P. (2017). Climate change and global food systems: Potential impacts on food security and undernutrition. *Annual Review of Public Health, 38*(1), 259–277.

A Nardone B Ronchi N Lacetera MS Ranieri U Bernabucci 2010 Effects of climate changes on animal production and sustainability of livestock systems Livestock Science 130 1–3 57 69

Naser, H. A. (2014). Marine ecosystem diversity in the Arabian Gulf: threats and conservation. In *Biodiversity—The dynamic balance of the planet*. IntechOpen.

RG Newell BC Prest SE Sexton 2021 The GDP-temperature relationship: Implications for climate change damages Journal of Environmental Economics and Management 108 102445

GO Odhiambo 2017 Water scarcity in the Arabian Peninsula and socio-economic implications Applied Water Science 7 5 2479 2492

GO Odhiambo 2017 Water scarcity in the Arabian Peninsula and socio-economic implications Applied Water Science 7 2479 2492

Patel, S., Dey, A., Singh, S. K., Singh, R., & Singh, H. P. (2021). Socio-economic impacts of climate change. *Climate Impacts on Sustainable Natural Resource Management*, 237–267.

P Patlakas C Stathopoulos H Flocas NS Bartsotas G Kallos 2021 Precipitation climatology for the arid region of the Arabian Peninsula—Variability, trends and extremes Climate 9 7 103

Pelling, M., Patwardhan, A., Hallegatte, S., Maskrey, A., Oki, T., Oswald Spring, U.,... & Viguie, V. (2012). Toward a sustainable and resilient future. In *Managing the risks of extreme events and disasters to advance climate change adaptation: Special report of the intergovernmental panel on climate change* (pp. 437–486). IPCC.

Priyan, K. (2021). Issues and challenges of groundwater and surface water management in semi-arid regions. In *Groundwater resources development and planning in the semi-arid region* (pp. 1–17).

Radford, D. A., Lawler, T. C., Edwards, B. R., Disher, B. R., Maier, H. R., Ostendorf, B., Nairn, J., van Delden, H., & Goodsite, M. (2022). A framework for the mitigation and adaptation from heat-related risks to infrastructure. *Sustainable Cities and Society, 81*, 103820.

Rahman, M. M., Akter, R., Abdul Bari, J. B., Hasan, M. A., Rahman, M. S., Abu Shoaib, S., Shatnawi, Z. N., Alshayeb, A. F., Shalabi, F. I., Rahman, A., Alsnad, M. A., & Rahman, S. M. (2022). Analysis of climate change impacts on the food system security of Saudi Arabia. *Sustainability, 14*(21), 14482.

KK Rao A Al Mandous M Al Ebri N Al Hameli M Rakib S Al Kaabi 2024 Future changes in the precipitation regime over the Arabian Peninsula with special emphasis on UAE: Insights from NEX-GDDP CMIP6 model simulations Scientific Reports 14 1 151

D Renaudeau A Collin S Yahav V Basilio De JL Gourdine RJ Collier 2012 Adaptation to hot climate and strategies to alleviate heat stress in livestock production Animal 6 5 707 728

M Rodell JS Famiglietti DN Wiese JT Reager HK Beaudoing FW Landerer MH Lo 2018 Emerging trends in global freshwater availability Nature 557 7707 651 659

MO Saeed MM Ershath IA Al-Tisan 2019 Perspective on desalination discharges and coastal environments of the Arabian Peninsula Marine Environmental Research 145 1 10

Saleem, A., Anwar, S., Nawaz, T., Fahad, S., Saud, S., Ur Rahman, T., Rasheed J. & Nawaz, T. (2024). Securing a sustainable future: the climate change threat to agriculture, food security, and sustainable development goals. *Journal of Umm Al-Qura University for Applied Sciences*, 1–17.

Samuel, S. A. (2025). Carbon pricing mechanisms for reducing greenhouse gas emissions and encouraging sustainable industrial practices.

Saudi Gazette. (2022, April). *National Foundation for Food Preservation established to tackle SR40bn of food waste in KSA*. Saudi Gazette. https://saudigazette.com.sa/article/619478/SAUDI-ARABIA/National-Foundation-for-Food-Preservation-established-to-tackle-SR40bn-of-food-waste-in-KSA

Sejian, V., Silpa, M. V., Chauhan, S. S., Bagath, M., Devaraj, C., Krishnan, G., Reshma Nair, M. S., Anisha, J. P., Manimaran, A., Koenig, S., Bhatta, R., Dunshea, F. R. (2021). Eco-intensified breeding strategies for improving climate resilience in goats. *Ecological Intensification of Natural Resources for Sustainable Agriculture*, 627–655.

Shahin, M., & Shahin, M. (2007). Water scarcity in the Arab region-major problems and attempts to alleviate their impacts. *Water Resources and Hydrometeorology of the Arab Region*, 519–547.

SM Shahin MA Salem 2015 The challenges of water scarcity and the future of food security in the United Arab Emirates (UAE) Natural Resources and Conservation 3 1 1 6

Sherif, M., Sefelnasr, A., Al Rashed, M., Alshamsi, D., Zaidi, F. K., Alghafli, K., Baig, F., Al-Turbak, A., Alfaifi, H., Loni, O. A., Ahamed, M. B., & Ebraheem, A. A. (2023). A review of managed aquifer recharge potential in the Middle East and North Africa Region with examples from the Kingdom of Saudi Arabia and the United Arab Emirates. *Water*, *15*(4), 742.

A Shirvani SJ Nazemosadat E Kahya 2015 Analyses of the Persian Gulf sea surface temperature: Prediction and detection of climate change signals Arabian Journal of Geosciences 8 2121 2130

S Skendžić M Zovko IP Živković V Lešić D Lemić 2021 The impact of climate change on agricultural insect pests Insects 12 5 440

M Sophocleous 2004 Global and regional water availability and demand: Prospects for the future Natural Resources Research 13 61 75

F Steward 2012 Transformative innovation policy to meet the challenge of climate change: Sociotechnical networks aligned with consumption and end-use as new transition arenas for a low-carbon society or green economy Technology Analysis & Strategic Management 24 4 331 343

F Tahir AA Baloch H Ali 2019 Resilience of desalination plants for sustainable water supply in Middle East Sustainability perspectives: Science, policy and practice: A global view of theories, policies and practice in sustainable development Springer International Publishing 303 329

RK Trivedy 2007 Low cost and energy saving technologies for water and wastewater treatment Journal of Industrial Pollution Control 23 2 403 411

E Vallino L Ridolfi F Laio 2020 Measuring economic water scarcity in agriculture: A cross-country empirical investigation Environmental Science & Policy 114 73 85

MK Aalst Van T Cannon I Burton 2008 Community level adaptation to climate change: The potential role of participatory community risk assessment Global Environmental Change 18 1 165 179

MT Vliet Van D Wiberg S Leduc K Riahi 2016 Power-generation system vulnerability and adaptation to changes in climate and water resources Nature Climate Change 6 4 375 380

D Varisco 2019 Pumping Yemen dry: A history of Yemen's water crisis Human Ecology 47 3 317 329

Wabnitz, C. C., Lam, V. W., Reygondeau, G., Teh, L. C., Al-Abdulrazzak, D., Khalfallah, M., Pauly, D., Deng PaloMares, M. L., Zeller, D., & Cheung, W. W. (2018). Climate change impacts on marine biodiversity, fisheries and society in the Arabian Gulf. *PloS one*, *13*(5), e0194537.

Wagner, D. C. (2020). *The impact of climate variation and change on gender dynamics from a vulnerability and adaptive perspective* [Master's thesis].

Waha, K., Krummenauer, L., Adams, S., Aich, V., Baarsch, F., Coumou, D., Fader, M., Hoff, H., Jobbins, G., Marcus, R., Mengel, M., Otto, I.M., Perrette, M., Rocha, M., Robinson, A., & Schleussner, C. F. (2017). Climate change impacts in the Middle East and Northern

Africa (MENA) region and their implications for vulnerable population groups. *Regional Environmental Change*, *17*, 1623–1638.

LV Weatherdon AK Magnan AD Rogers UR Sumaila WW Cheung 2016 Observed and projected impacts of climate change on marine fisheries, aquaculture, coastal tourism, and human health: An update Frontiers in Marine Science 3 48

Y Wehbe M Temimi 2021 A remote sensing-based assessment of water resources in the Arabian Peninsula Remote Sensing 13 2 247

S Woodruff TK BenDor AL Strong 2018 Fighting the inevitable: Infrastructure investment and coastal community adaptation to sea level rise System Dynamics Review 34 1–2 48 77

Yadav, S. (2025). empowering communities for climate resilience: Bridging knowledge and action to combat climate change. In *Community climate justice and sustainable development* (pp. 83–104). IGI Global Scientific Publishing.

FM Ziadat P Zdruli S Christiansen L Caon M Abdel Monem T Fetsi 2022 An overview of land degradation and sustainable land management in the near East and North Africa Sustainable Agriculture Research 11 1 11 24

Chapter 7
Biodiversity and Ecosystem Dynamics of the Arabian Peninsula: Climate-Driven Changes, Species Vulnerability and Conservation Strategies

Neeta Kumari, Saahil Hembrom, and Soumya Pandey

7.1 Introduction

Climate refers to the long-term weather patterns of a geographical region, encompassing temperature, precipitation, humidity and wind. In contrast, weather represents short-term atmospheric changes over days or weeks. Climate change is a long-term phenomenon driven by both natural processes and human activities, leading to alterations in ecosystems and the organisms that inhabit them (Abbas et al., 2017; Al Solami, 2020; Alghanem et al., 2020; Al-Namazi et al., 2021). It has emerged as a global environmental challenge, disrupting biodiversity and ecological services by altering temperature patterns and water cycles, contributing to issues such as acidification (Prakash, 2021). Changes in temperature, precipitation and other environmental factors influence ecosystems, shaping animal behavior, plant growth and habitat conditions (Lipton et al., 2018). Biodiversity plays a crucial role in maintaining ecosystem balance, promoting adaptability and resilience to environmental changes (May, 2011). The variety of species within an ecosystem enhances resistance to pests and diseases while providing essential resources such as food, raw materials and medicine (Onoh et al., 2024). Beyond ecological benefits, biodiversity also supports economic sectors reliant on natural resources, including oil, gas, freshwater, fiber and fuel, contributing to national economic growth. Additionally, it plays a role in climate regulation, pollution control, soil conservation and ecosystem

N. Kumari (✉) · S. Hembrom
Civil and Environmental Engineering, Birla Institute of Technology, Mesra, Ranchi, Jharkhand, India
e-mail: neetak@bitmesra.ac.in

S. Pandey
Department of Civil Engineering, Kalinga University, Raipur, Chhattisgarh, India

F. Mushtaq et al. (eds.), *Climate Change in the Arabian Peninsula*,
https://doi.org/10.1007/978-3-032-02481-7_7

stability. Biodiversity also holds cultural, aesthetic, medicinal and recreational value, influencing ecotourism and resource conservation efforts (European Commission, 2008).

Spanning approximately 3.2 million km^2, the Arabian Peninsula is the largest peninsula in the world, encompassing Oman, Yemen, the United Arab Emirates (UAE), Kuwait, Qatar, Bahrain, Jordan, Saudi Arabia and Iraq (Masoumi et al., 2024; Almazroui, 2020; Ghazanfar, 1998, 2015). It extends from 12° N in southern Yemen to 32° N in northern Saudi Arabia and from 34° E in northwestern Saudi Arabia to 60° E in eastern Oman. The landscape is characterized by mountains, wadis, plains, gravel expanses and vast sand deserts. Strategically positioned between Africa and Asia, the Peninsula exhibits biogeographical influences from both continents, resulting in a unique ecological composition (White & Léonard, 1991). Despite its arid conditions, the Arabian Peninsula hosts a rich diversity of plant species, many of which are endemic. Some taxa, once widespread, are now confined to mountainous refugia where the climate remains relatively cooler and more humid than in the surrounding plains (Kürschner, 1998). Over the past 50 years, rapid development has led to the degradation of fragile ecosystems, primarily due to overgrazing and urban expansion (Ansari, 2016). However, growing awareness of these challenges has prompted efforts to conserve key plant habitats and restore degraded landscapes to protect regional biodiversity.

The Peninsula's arid to semi-arid climate makes it highly susceptible to climatic variations, characterized by extreme temperatures and minimal precipitation (Alshehri et al., 2021; Almazroui et al., 2020). Temperature variations across the region are significant, with northern areas experiencing seasonal temperatures ranging from 8.57 to 28.32 °C, while southern regions record higher temperatures between 26.68 and 33.97 °C, sometimes exceeding 50 °C in summer (Almazroui et al., 2013, 2020). The monsoon season lasts from May to October, while the dry season extends from June to September (Almazroui et al., 2020). The northwest and western regions experience annual mean temperatures between 18.96 and 31.42 °C. Rainfall remains scarce, with precipitation levels ranging from 25 mm in the northwest to 230 mm in the southwest annually (Almazroui et al., 2013, 2020; Kotwicki & Al Sulaimani, 2009). During the monsoon season (April–May), rainfall varies from 200 to 75 mm, while during the dry period (June–August), it ranges from 20 to 65 mm (Almazroui et al., 2017). These climatic conditions directly influence biodiversity, determining the survival and distribution of plant and animal species. Extreme temperatures and erratic rainfall patterns have led to specialized ecological adaptations, enabling flora and fauna to thrive despite the harsh environment. However, increasing climate variability and rising temperatures present growing threats to ecosystems, underscoring the need for climate monitoring and conservation efforts to ensure ecological stability.

The Arabian Peninsula's diverse geography and biogeography place it at the intersection of three major ecological regions: the Palearctic, Afrotropical and Indomalayan realms (Ghazanfar, 2024) (Fig. 7.1). This convergence has resulted in an exceptionally diverse array of flora and fauna, each adapted to the region's extreme climatic

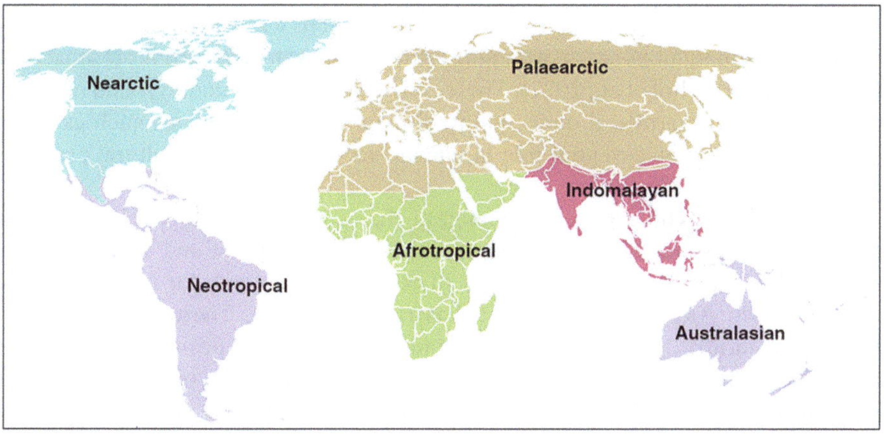

Fig. 7.1 Ecological regions of the world adopted from Bibi and Métais (2016). *Source* Bibi and Métais (2016)

conditions. While all countries within the Peninsula exhibit distinct biological diversity, Yemen, Saudi Arabia and Oman stand out due to their wide range of landforms and ecosystems. The mountainous areas, in particular, provide a cooler and more stable climate than the surrounding lowlands, forming ecological refugia that support high levels of plant endemism. Conversely, the desert ecosystems host lifeforms specifically adapted to severe arid conditions. The region has a long history of human habitation and subsistence agriculture, particularly in the mountains, where crop varieties have evolved to withstand arid conditions. These crops serve as valuable genetic resources, offering potential solutions for climate-resilient agriculture. However, despite its biological richness, the Arabian Peninsula has faced intense environmental pressures over the past five decades, including rapid urbanization, habitat destruction, overgrazing and climate change. These factors continue to threaten the delicate ecological balance of the region, reinforcing the urgent need for sustainable conservation measures.

7.2 Ecological Diversity of the Arabian Peninsula

The diverse ecological landscapes of the Arabian Peninsula have been shaped by its vast and varied terrain, encompassing mountains, deserts, alluvial plains and coastal regions. Covering over 3 million km^2, the peninsula is predominantly characterized by rocky landscapes with sparse vegetation. Positioned at the intersection of Asia and Africa, it serves as a vital biogeographical transition zone, supporting a range of flora and fauna adapted to extreme environmental conditions (Harrison et al., 2015; Ben-Hasan & Christensen, 2019; Ghazanfar, 2024). The region contains four major freshwater ecoregions-the Oman Mountain, Arabian Interior, Lower Tigris and Euphrates and Southwestern Arabian Coast-where water availability is largely confined to mountainous areas sustaining critical freshwater ecosystems (Esmaeili & Eslami, 2023; Freyhof et al., 2021). These ecosystems include underground aquifers, wadi river valleys, artificial dams and traditional irrigation systems such as Aflaj, which have played an essential role in human settlement and agricultural practices (Esmaeili et al., 2022).

The Asir and Hijaz Mountain ranges, characterized by relatively moderate climates, act as biodiversity refugia, harboring endemic plant and animal species distinct from those in the arid deserts of Rub' al Khali and An Nafud (Deil & al Gifri, 1998; Aloufi & Amr, 2015; Ansari et al., 2015). These deserts, shaped by extreme aridity, have led to the evolution of highly specialized organisms adapted to harsh conditions (Enzel et al., 2015). Biological soil crusts, particularly those dominated by cyanobacteria, play a crucial role in desert ecosystems by enhancing soil stability, retaining moisture and supporting seed germination. Studies indicate that cyanobacteria make up nearly 87% of northern Qatar's soil crusts, gradually decreasing toward the south, underscoring their significance in desert resilience (Richer et al., 2012). These microbial communities produce diverse bioactive compounds that enhance stress tolerance and hold untapped potential for biotechnological applications (Metcalf et al., 2015).

The peninsula's coastal and marine ecosystems are equally significant in supporting biodiversity. Extensive mangrove forests, coral reefs and sabkhas create essential habitats for numerous species while providing coastal protection. The coral reefs in the Arabian Gulf, known for their exceptional thermal tolerance, offer critical insights into adaptive resilience under climate change pressures. As environmental shifts continue to influence regional ecosystems, the Arabian Peninsula remains a dynamic landscape where geological evolution and ecological diversity are intricately linked. From towering mountains and vast sand deserts to rich marine ecosystems, the region's natural history underscores its resilience in the face of climatic and geological transformations.

7.3 Phytogeography of Arabian Peninsula

The phytogeography of the Arabian Peninsula is characterized by diverse vegetation types, including grasslands, deciduous and semi-evergreen shrublands and open mountain and deciduous forests. In montane regions, vegetation is dominated by xerophytic plant communities adapted to the steep slopes and harsh climatic conditions. Importantly, three species of *Juniperus J. seravschanica, J. procera and J. phoenicea* are present across the peninsula's mountainous landscapes (Makki et al., 2023a, b; Ghazanfar, 2024). At lower elevations, the vegetation transitions to woodlands dominated by *Vachellia* and *Olea europaea*, while *Vachellia-Commiphora* shrubland mosaics are prominent in Yemen and Oman. In contrast, higher elevations support orophytic grasslands that serve as vital grazing resources, whereas rocky outcrops provide suitable habitats for succulents and other drought-tolerant

Table 7.1 Endemism and species diversity in the Arabian Peninsula Countries

Country	Area (km^2)	Taxa (~)	Endemic Taxa
Yemen	555,000	2580	458 (17.7%)
Saudi Arabia	2.25 million	2250	246 (10.9%)
Oman	309.5000	1440	69 (4.79)
United Arab Emirates	83.6000	598	0
Kuwait	17.82000	404	0
Qatar	11.44000	371	0
Bahrain	786.5	323	0

Source Miller et al. (1996)

shrubs. These montane ecosystems contribute significantly to regional floristic diversity, offering niches for a variety of endemic species. Among the *Juniperus* species, *J. procera* is widespread in the southwestern and western regions of the Arabian Peninsula, particularly in Yemen and Saudi Arabia. J. phoenicea, a Mediterranean element, is predominantly distributed in western Saudi Arabia, where it coexists with J. procera along the escarpment mountains near Taif. Meanwhile, *J. seravschanica*, an Irano-Turanian species, is found in the western Hajar range of northern Oman (Fisher & Gardner, 1995; Hall, 1984). The Arabian Peninsula hosts a rich and diverse flora, with approximately 4000 plant species, of which 20% (about 800 taxa) are endemic, primarily concentrated in Yemen, Saudi Arabia and Oman (Miller et al., 1996; Sulthan, 2016; Chaudhary, 2000) (Table 7.1). These countries represent the region's most significant centers of plant diversity, shaped by a complex interplay of climate, topography and biogeographic history. Several Oman endemics extend their distribution to the eastern mountains of the UAE.

Wadis, or seasonal riverbeds, form important freshwater ecoregions in the Arabian Peninsula, supporting distinct riparian vegetation. While typically dry, they become ephemeral rivers during seasonal rainfall, creating oases and agricultural zones. The vegetation within wadis is influenced by drainage, groundwater levels and sediment composition, leading to the formation of unique plant communities. Tamarix species are commonly found, along with vegetation types that reflect Mediterranean, Sub-Saharan and North Arabian pseudo-savanna influences. These wadis serve as important ecological corridors, linking different plant communities across the arid landscape and sustaining localized biodiversity. The Arabian Peninsula is largely dominated by desert ecosystems, encompassing the vast expanses of the Rub' al Khali (Empty Quarter), An Nafud and Ad Dhana deserts. These arid landscapes are characterized by extensive sand dunes, gravel plains and rocky plateaus, each supporting distinct desert-adapted flora. Sand dune deserts, which account for over one-third of the Peninsula's land area, are a defining feature of the region. Compared to silty or clay-rich soils, sand has high permeability and porosity, allowing for deep moisture penetration while remaining structurally unstable and nutrient-deficient.

Despite these harsh conditions, certain plant species have adapted to thrive in these environments. *Calligonum comosum* is one such species, capable of surviving in sand dunes due to its extensive horizontal root system. This adaptation allows it to efficiently absorb moisture, as the uppermost layers of sand lose water rapidly through evaporation, whereas the deeper layers retain moisture for a longer duration (Danin, 2012; Ghazanfar, 2024). Additionally, on the soft sands of slip faces, sedges such as *Cyperus aucheri* and tussock grasses like *Centropodia* are able to establish themselves. In sandy soils, rhizosheaths play a crucial role in water absorption and nitrogen fixation, particularly in desert perennial grasses such as *Stipagrostis plumosa.* These rhizosheaths consist of sand grains and matted root hairs, bound together by mucilage secretions, creating a microenvironment that supports nitrogen-fixing bacteria. These bacteria, in turn, enhance soil fertility by supplying essential nutrients to the plant, thereby enabling survival in nutrient-poor desert conditions (Danin, 1996; Mandaville, 1998).

The coastal ecoregions of the Arabian Peninsula, shaped by geomorphological variability, host specialized plant species adapted to high salinity and water scarcity. Sabkhas (coastal salt flats) are among the most extreme environments, with low plant diversity due to high evaporation rates and saline soils. Mangroves, particularly Avicennia marina, are distributed unevenly along the Red Sea, Arabian Gulf and Arabian Sea, forming fragmented stands in estuaries and sheltered coastlines. These coastal zones provide critical habitats for salt-tolerant vegetation, contributing to the region's floristic uniqueness. The Arabian Peninsula's plant diversity reflects its biogeographic connections with Africa, Iran and the Mediterranean region. The southwestern Peninsula shares species with northeastern and eastern Africa, indicating historical migration patterns during wetter climatic periods. Similarly, northern Oman, the UAE, Baluchistan and southwest Iran exhibit strong floristic similarities, highlighting past ecological linkages. Surveys of northwestern Saudi Arabia have identified it as a refuge for Mediterranean and Irano-Turanian plant species, further emphasizing the complex evolutionary history of the Peninsula's flora. The diverse ecoregions-ranging (Fig. 7.2) from montane forests and wadis to vast deserts and coastal habitats support a highly distinctive and specialized flora, shaped by a combination of climatic conditions, historical migrations and geographic isolation.

7.4 Floristic Diversity and Biogeography of the Arabian Peninsula

The Arabian Peninsula's distinct geographical position and diverse biogeographic zones have shaped its plant diversity through complex migration patterns and ecological adaptations. The region's flora exhibits strong historical connections to the Mediterranean, Eurasian and African regions, forming small, isolated plant populations that are crucial for genetic conservation and climate adaptation. Despite the region's limited crop diversity, the Peninsula is home to wild relatives of cultivated

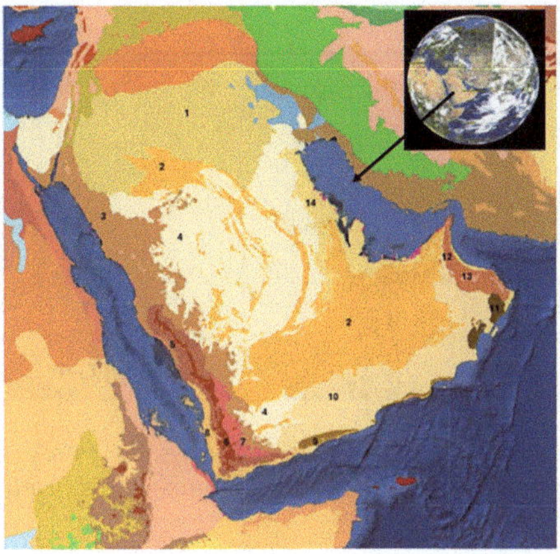

Map Label	Ecoregion Name
1	North Arabian Desert
2	Arabian Desert
3	Red Sea–Arabian Desert Shrubland
4	Arabian Sand Desert
5	Southwest Arabian Escarpment Shrublands and Woodlands
6	Southwest Arabian Montane Woodlands and Grassland
7	Southwest Arabian Highland Xeric Scrub
8	Southwest Arabian Coastal Xeric Shrublands
9	South Arabian Fog Woodlands, Shrublands, and Dunes
10	South Arabian Plains and Plateau Desert
11	East Arabian Fog Shrublands and Sand Desert
12	Al-Hajar Xeric Woodland and Shrubland
13	Al-Hajar Montane Woodland and Shrubland
14	Arabian–Persian Gulf Coastal Plain Desert

Fig. 7.2 Biogeographic zones of the Arabian Peninsula adopted from Ghazanfar (2024). *Source* Ghazanfar (2024)

species, such as wild olive *(Olea europaea)* and *Pistacia khinjuk*, which possess valuable genetic traits for drought and salinity tolerance (Gros-Balthazard & Flowers, 2021; FAO, 2004; Ghazanfar, 2024) (Fig. 7.3). A key example is Abelmoschus esculentus, a drought-resistant species found in Oman, believed to be an ancestor of cultivated okra. The floristic connections of the Arabian Peninsula highlight its historical ties to Africa, Iran and the southwestern Mediterranean. The southwestern region, including parts of Saudi Arabia, Yemen and Dhofar (Oman), shares strong geological and biogeographic links with Africa. Many plant species below 1500 m in southern Arabia are also found in northeastern and eastern Africa, including various species of *Vachellia, Commiphora, Gymnosporia, Lannea, Turraea* and *Boswellia sacra* (White & Léonard, 1991). These plants likely migrated from Africa during wetter climatic periods, when desert barriers were reduced, allowing for easier plant dispersal. Additionally, Afromontane genera such as *Anagyris, Ceratonia* and *Juniperus* extend from Africa into southwest Saudi Arabia, Yemen and southern Oman, further supporting the hypothesis of past ecological connectivity between these regions.

Many species in the Arabian Peninsula have evolved specialized physiological, morphological and symbiotic adaptations to survive in extreme desert environments. Key traits such as salt excretion, succulence, specialized photosynthesis and resilient proteins enable these plants to endure high temperatures, aridity and saline soils (Ali-Thani & Yasseen, 2018). Halophytic species, in particular, demonstrate exceptional resilience to saline conditions by hosting salt-tolerant bacteria in their root systems, enhancing their ability to thrive under harsh conditions. These microbial associations have significant implications for sustainable agriculture, as they can improve salt tolerance in non-halophytic crops, helping to mitigate the effects of increasing soil salinization (Ali-Thani & Yasseen, 2018; Danin, 1996; Etesami & Beattie, 2018). In addition, nitrogen-fixing bacteria linked to rhizosheaths mats of root hairs bound with

Fig. 7.3 Photo of wild olive (*Olea europaea*) and *Pistacia khinjuk*. *Source* By Forest & Kim Starr, https://commons.wikimedia.org/w/index.php?curid=71836928. *Source* https://earthpedia.earth.com/

sand grains and mucilage aid in improving soil fertility and enhancing plant resilience to arid conditions, particularly in desert-adapted grasses such as *Stipagrostis plumosa* (Danin, 1996; Mandaville, 1998).

The Arabian Peninsula's mountains serve as critical biodiversity refuges, supporting a rich array of endemic and relict plant species. Mountain ecosystems exhibit stratified vegetation, where different plant communities occupy distinct altitudinal zones. The Sarawat Mountains, located in the Asir range of southwestern Saudi Arabia, represent one of the most significant centers of plant diversity in the Arabian Peninsula, accounting for approximately 60% of Saudi Arabia's recorded plant species (Seraj et al., 2014). Jandaf Mountain, within the Asir region, is another ecologically significant site characterized by steep slopes and deep valleys, functioning as a biogeographic and bioclimatic refuge. Moisture inputs from the Red Sea, hydrology and air currents play a crucial role in shaping vegetation patterns, creating diverse microhabitats that support a wide range of plant species (Al-Namazi et al., 2022). Geological heterogeneity, variations in soil types and diverse rock formations significantly influence vegetation distribution, with higher plant diversity observed in the mountains compared to the surrounding lowlands (Abulfatih, 1984; Elvidge & Lyon, 1985). Additionally, mountain plant communities are strongly shaped by climatic factors such as temperature fluctuations, moisture availability and seasonal weather variability (Oldfather et al., 2016).

The montane regions of northern Oman, the UAE, Baluchistan and southwest Iran exhibit strong floristic similarities, reflecting historical plant migrations during humid climatic periods. Species such as *Juniperus seravschanica* and *Prunus arabica* once had a widespread distribution but are now restricted to higher elevations due to increasing aridity. Some species, such as *Ceratonia oreothauma*, have become relictual, persisting in isolated moist areas that serve as remnants of a wetter past (Kürschner, 1998). Recent botanical surveys in northwestern Saudi Arabia have identified mountain summits as bioclimatic refuges for Mediterranean and Irano-Turanian plant species. These high-altitude zones maintain cool and humid microclimates, fostering evergreen woodlands and dense ground vegetation. Rare species like

Pistacia khinjuk are predominantly found at higher elevations, suggesting historical migrations from the Kurdo-Zagrosian mountains before the Pleistocene. Similarly, in central Oman, research has revealed a Pleistocene refugium with a high concentration of endemic species in monsoon-influenced regions, further underscoring the Arabian Peninsula's role as a biodiversity hotspot shaped by past climatic fluctuations.

Beyond the montane regions, the Arabian Peninsula's broader flora exhibits connections to the Sahara-Arabian desert belt. Desert-adapted plants such as *Anabasis* and *Atriplex* demonstrate affinities with the Sahara, while endemic species such as *Euryops jaberiana*, unique to northwestern Saudi Arabia and *Euryops arabicus*, found in northern Oman, Yemen, Djibouti and Somalia, highlight the floristic overlap between Arabia, Africa and surrounding regions. Some species also extend to Socotra and Ethiopia, reinforcing the biogeographic links between the Arabian Peninsula and the Horn of Africa. These floristic connections, shaped by historical climatic shifts, geological evolution and ecological corridors, provide critical insights into the evolution and distribution of plant life across the Arabian Peninsula and its neighboring regions.

7.5 Climate Change and Its Effects on the Ecosystem

The Arabian Peninsula is experiencing growing environmental pressures, making it one of the more climate-vulnerable regions in the Middle East. Saudi Arabia, in particular, experiences extreme heat and intense solar radiation, with some of the highest recorded temperatures globally (Al-Ghamdi et al., 2023; Al-Namazi et al., 2021; Ansari et al., 2022). While humidity levels generally remain low, coastal areas along the Red Sea and the Arabian Gulf face unique climatic conditions. These water bodies, due to their relatively narrow and enclosed geography, retain significant heat during the summer, limiting nighttime cooling. As a result, temperatures in these coastal regions often remain as high as 29–30 °C during the night (Al-Ghamdi et al., 2023). Temperature variations across the peninsula are largely influenced by altitude and proximity to the sea. The western mountain ranges provide relatively moderate temperatures, whereas the adjacent Red Sea coastal plain and the high plateau to the east experience more extreme climatic conditions. Interior regions frequently record daytime temperatures between 38 and 43 °C, while the northwestern mountainous areas can witness sharp nighttime temperature drops, sometimes leading to frost and even snowfall during winter (Al-Ghamdi et al., 2023). Beyond extreme heat, the Arabian Peninsula is also highly susceptible to sand and dust storms, localized flooding and coastal humidity surges. The Shamal wind, occurring in late spring and early summer, significantly exacerbates dust storms, particularly in eastern Arabia, reducing visibility and posing serious respiratory health risks (Al-Ghamdi et al., 2023).

In recent decades, a clear warming trend has been observed across the region. Between 1891 and 1990, temperatures increased by 0.36 °C, with a more pronounced warming rate of 0.16 °C per decade recorded in Bahrain between 1947 and 2005.

Seasonal temperature analyses from 1986 to 2015 further illustrate these changes (Amin et al., 2016). During winter, northern regions experience lower temperatures around 10 °C, increasing to 20 °C in the south, largely influenced by cold Siberian winds, Mediterranean cyclones and the strong northern Shamal winds (Patlakas et al., 2019). Coastal areas along the Red Sea generally remain warmer than the interior, with temperatures ranging between 15 °C and 20 °C depending on local topography and wind patterns. In contrast, high-altitude regions such as Makkah and Asir experience cooler temperatures, averaging around 15 °C. By the spring months (March to May), temperatures rise across the Peninsula, with the northern regions warming under Mediterranean influences and reaching up to 35 °C in the south and 20 °C in the north by April (Amin et al., 2016).

Rising temperatures and declining rainfall across the Arabian Peninsula are expected to intensify in the coming years, exacerbating water scarcity and habitat degradation (Quamar, 2024; Rao et al., 2024; Ansari and Ghanem, 2017, 2019; Odhiambo, 2017). Freshwater ecosystems are particularly vulnerable to these climatic shifts. For instance, in Oman, species distribution modeling has identified montane areas in the northern part of the country, as well as in northeastern UAE, as critical habitats for the Cyprinion muscatense cyprinid fish species (Howells et al., 2016, 2021). The projected changes in temperature and precipitation patterns pose significant threats to these ecosystems, highlighting the growing risks to freshwater biodiversity across the region (Gholamhosseini et al., 2024). The broader impacts of climate change on different ecosystems in the Arabian Peninsula are further detailed in Table 7.2.

Impact on Flora

The flora of the Arabian Peninsula is undergoing significant transformations due to rising temperatures, shifting precipitation patterns and increased evapotranspiration rates. These climatic changes threaten plant survival and ecosystem stability, particularly in desert regions that cover nearly one-third of the Peninsula. The sandy soils in these areas, while highly permeable and porous, are nutrient-deficient. Consequently, native plant species such as *Calligonum comosum*, *Centropodia* spp. and *Cyperus aucheri* have evolved extensive root systems to efficiently utilize the limited moisture available (Mandaville, 1998). One of the most concerning impacts of climate change on the region's vegetation is the decline in species distribution and habitat range. A study on *Rubus asirensis*, an endemic shrub of the Arabian Peninsula, found that its population size and habitat extent have significantly decreased over the past century due to increasing temperatures and reduced precipitation. These changes not only threaten plant species but also disrupt ecosystem stability, as these plants provide critical habitats for various animal species (Bedair et al., 2024).

Additionally, intensified evapotranspiration losses have accelerated soil moisture depletion, further limiting vegetation growth and reducing water availability in an already arid landscape (Ajjur et al., 2022). Mountain summits in northwest Saudi Arabia act as refuges for relict plant species originating from Irano-Turanian and Mediterranean regions. Species such as *Globularia arabica*, *Juniperus phoenicea*, *Retama raetam*, *Pistacia khinjuk* and *Moringa peregrina* are increasingly shifting

Table 7.2 Climate change impacts on different ecosystems in the Arabian Peninsula

Ecosystem	Flora and fauna species	Climate change impacts	References
Marine ecosystem	**Coral reefs**: Coral, fish and invertebrates **Seagrass beds**: Dugongs and Sea Turtles **Mangrove forest**: *Avicennia marina*	Sea temperature rise Sea-level Rise Increased salinity Ocean acidification Extreme weather events Human-Induced Stress	Ghazanfar (2024); Yadav et al. (2024); Wabnitz et al. (2018)
Desert ecosystem	**Flora**: Acacia trees, palms, succulents, spiny shrubs and grasses **Fauna:** Nubian ibex, Fennec fox, Rüppell's fox, *Uromastyx aegyptia*, Sinai agama, Arabian Oryx	Increased temperatures Desertification and land degradation Altered precipitation patterns Dust storms and aerosols	Al-Blooshi et al., (2020); Ivory et al., (2021)
Mountain ecosystem	**Flora**: Endemic species-*Juniperus phoenicea, Olea europaea, Tamarix aphylla, Pistacia vera, Pistacia lentiscus, Ficus carica* **Fauna**: Mountain animals-Arabitragus jayakari, *Panthera pardus nimr*	Temperature increase Alteration pattern of precipitation Habitat fragmentation Soil degradation	Ghazanfar, (2024); Kotwicki and Al Sulaimani (2009)

to higher elevations in response to rising temperatures and habitat fragmentation (Danin, 1996). Similarly, Irano-Turanian flora including *Astracantha echinus* subsp. *arabica*, *Artemisia sieberi* and *Ephedra pachyclada* are facing habitat degradation as climatic conditions continue to shift (Danin, 1996). Land degradation across the Arabian Peninsula is accelerating due to the combined effects of climate change and human activities, particularly in overgrazed and deforested regions. A recent study by Bhatt et al., 2023 underscores the potential role of halophytes (salt-tolerant plants) in land restoration and mitigating freshwater scarcity. Researchers identified 107 halophyte species with significant ecological and economic benefits, with species such as *Alhagi graecorum*, *Phragmites australis*, *Arthrocnemum macrostachyum* and *Bassia muricata* being particularly valuable for restoration projects (Bhatt et al., 2023) (Fig. 7.4).

Although large-scale flowering events are rare in the Arabian Peninsula, extreme rainfall in March 2024 triggered an unusual bloom of *Picris asplenioides*, *Zygophyllum simplex* and *Linaria haelava* in northwestern Saudi Arabia (Ghazanfar, 2024). Similarly, in February 2023, a super bloom of *Horwoodia dicksoniae* was recorded in Rafha, attributed to an unusually strong winter monsoon (Ghazanfar, 1997; Ghazanfar et al., 2019). While these events are striking, they are becoming increasingly infrequent due to long-term climatic shifts. The increasing frequency of extreme heat, prolonged droughts and soil degradation poses a serious threat to

Fig. 7.4 Few important halophyte species important for restoration of degraded lands in Arabian Peninsula: **a** *Alhagi graecorum*; **b** *Phragmites australis*; **c** *Arthrocnemum macrostachyum* adopted from Bhatt et al. (2023). *Source* Bhatt et al. (2023)

the Arabian Peninsula's endemic and indigenous plant species. For instance, *Stipagrostis plumosa*, a crucial desert species, plays an essential role in nitrogen fixation and water absorption in sandy soils through its rhizoshealths (Mandaville, 1998). The continued decline of such species could have cascading effects on the region's fragile ecosystems, potentially disrupting entire ecological networks. The Peninsula's coastal vegetation, especially in sabkha (salt-flat) environments, is also under threat. Dominated by halophytes such as *Avicennia marina*, these ecosystems are highly sensitive to rising sea levels, coastal erosion and increasing salinity. Inland sabkhas, which are already largely barren except for a few salt-tolerant species, are experiencing further biodiversity decline as aridity intensifies (Mandaville, 1998).

Impact on Fauna

The rising temperatures in the Arabian Peninsula are posing severe threats to animal species also, particularly those in arid environments where extreme heat is already a limiting factor. Many desert-adapted species are nearing their physiological limits for thermoregulation, which significantly increases their vulnerability to climate change (Williams et al., 2012).

Impact on Terrestrial Fauna

Desert fauna, including the Arabian oryx, sand gazelles and reptiles such as the spiny-tailed lizard (*Uromastyx aegyptia*), are highly susceptible to prolonged heatwaves (Fig. 7.5). Rising temperatures reduce access to critical resources such as food and water, forcing species to shift their activity patterns or migrate to cooler areas.

Fig. 7.5 Photo of Arabian oryx and sand gazelles by Charles J. Sharp and spiny-tailed lizard by Eitan Ferman. *Source* Charles J. Sharp, Own work from Sharp Photography (https://commons.wikimedia.org/w/index.php?curid=31333868, https://commons.wikimedia.org/w/index.php?curid=126862712), & Eitan Ferman, Own work (https://commons.wikimedia.org/w/index.php?curid=26226179)

However, habitat fragmentation limits their ability to relocate, increasing the risk of population decline or even extinction (Williams et al., 2012). The mountain regions of the Arabian Peninsula serve as a refuge for species that depend on cooler climates, such as the Arabian leopard (*Panthera pardus nimr*). However, as global temperatures continue to rise, these species face habitat loss and increasing competition for limited resources (Soultan et al., 2019). The fragmentation of high-altitude forests and grasslands further exacerbates biodiversity loss.

Impact on Aquatic Fauna

Freshwater ecosystems across the Arabian Peninsula are facing severe stress due to a combination of climate change and human-induced factors. Rising temperatures, declining precipitation and excessive groundwater extraction are accelerating habitat degradation, leading to significant declines in freshwater biodiversity. Oman, in particular, is experiencing acute water scarcity, with increasing groundwater salinity and heavy reliance on desalinated water disrupting the natural balance of freshwater ecosystems (Al Charaabi & Al-Yahyai, 2013). The impacts of climate-driven water scarcity are further compounded by habitat degradation, pollution, invasive species and land-use changes (Dudgeon et al., 2006; Stewart et al., 2022). Many freshwater fish species, including *Cyprinion muscatense, Cyprinion acinaces* and *Arabibarbus arabicus*, are particularly vulnerable, as shifts in habitat distribution are closely linked to both temperature and precipitation patterns (Gholamhosseini et al., 2024; Weiskopf et al., 2022). In semi-arid and arid regions such as Saudi Arabia, increasing soil temperatures often exceeding 60 °C during summer exacerbate the vulnerability of freshwater species (Tieleman et al., 2003). Climate models project a temperature rise of 3–5 °C in Saudi Arabia by the end of the twenty-first century, further intensifying pressure on freshwater habitats (Williams et al., 2012). These changes underscore need for sustainable water management and conservation strategies to mitigate biodiversity loss and ensure the resilience of aquatic ecosystems in the face of climate change.

Fig. 7.6 Arabian tahr by Wajhuallah. *Source* Wajhuallah; Own work, https://commons.wikimedia.org/w/index.php?curid=68527625)

Impact on Endemic and Indigenous Species

Approximately 17% of endemic mammals in the Afro-Arabian region (including the Arabian Peninsula) are at risk of extinction due to climate change (Soultan et al., 2019). Many of these species have limited dispersal ability, making them particularly vulnerable to habitat loss. The Arabian tahr (*Arabitragus jayakari*), which is restricted to the mountainous regions of Oman and the UAE, is one such species facing increased pressure from changing climatic conditions (Fig. 7.6). Additionally, the Arabian Peninsula's extreme climate places stress on migratory species, including birds that rely on stopover sites along the Red Sea and Arabian Gulf. Habitat degradation in these regions disrupts their migratory routes, potentially leading to declines in population numbers (Hume et al., 2015).

Impacts on Coastal Ecosystems and Marine Biodiversity, Including Mangroves

Coastal ecosystems in the Arabian Peninsula are increasingly vulnerable to the impacts of climate change, driven by rising sea levels, coastal erosion and increasing salinity (Barth et al., 2002; Ibraheem et al., 2014; Lüning and Vahrenholt, 2019). Mangrove forests, seagrass beds and coral reefs key coastal habitats are experiencing significant stress due to higher sea surface temperatures and ocean acidification. *Avicennia marina*, the dominant mangrove species in the region, is at risk as saltwater intrusion alters coastal hydrology, reducing its growth and survival. Similarly, seagrass meadows, which serve as critical feeding grounds for marine species like dugongs and sea turtles, are declining due to sedimentation and temperature-induced stress (Borrell et al., 2019). Coral reefs, particularly in the Red Sea and the Arabian Gulf, are facing frequent bleaching events, with extreme temperatures causing widespread coral mortality (Jennings et al., 2015; Chatziefthimiou et al., 2018). These environmental shifts not only threaten biodiversity but also undermine the resilience of coastal communities that rely on these ecosystems for fisheries, tourism and coastal protection (Sherman et al., 2007). The impact of climate change on various mangrove species is listed in Table 7.3.

Table 7.3 Climate change impact on mangroves species

Middle east mangroves	Name of mangrove species	Factors affecting the mangroves	References
Oman Qurm, Muhut Island, Wadi Dayqah, Flamingo Lake, Shinas Mangrove Park, Bandar AL khairan	*Avicennia marina*	Collapse of climate shifts has led to increased soil salinity and reduced vegetation cover, causing erosion and silting of lagoons in the region	Al-Nadabi & Sulaiman (2018)
Bahrain Ras Hayyan, Mangrove Forest, Arad Bay, Bahrain Mangrove Forest, Tubli Bay Coast, Ras Sanad Mangrove Forest	*Avicennia marina*	The underground springs and nearby farms discharge low-salinity water and minimal motion of seawater affects the habitat distribution and mangrove's health	Abou Seedo et al. (2017)
Saudi Arabia Zifaf Island, Qandal Forest, Solain Island, Shibara Island, Darin Mangrove Forest, Ras, Tarut Bay, Mangrove Forest North and South Sanabis, Mangrove Eco Park, Umm Al-Qandal Island, Ras Hatiba, Farasan Islands, Umm Al-Rubais, etc.	*Rhizophora Mucronate Avicennia marina*	The Tarut Bay region experiences high temperatures and very low precipitation	Saifullah et al. (1996)
United Arab Emirates Marawah Island, Abu-Dhabi Mangroves, Saadiyat Island, Ras Ghurab–Abu Dhabi, Khor Al Beidah Wetlands, Khor Kalba Mangrove Center, Sir Bani Yas Island etc.	*Avicennia marina*	The impact of climate change includes surface temperature, rise of sea level, salinity and coastal inundation	Raihan et al. (2024); Saenger et al. (2002)
Qatar Al Thakira mangrove forest, Umm Al Hul, Al Ruwais, All Dhakira Mangroves, Al Wakra, Purple Island, Simaisma, Fuwairit	*Avicennia marina*	Mangroves in Oatar face climate changes, increasing salinity, altering hydrodynamics, threats from urbanization, industrialization, recreational activities and pollution	Abdel-Razik et al. (1991); Pernot et al. (2015)

(continued)

Table 7.3 (continued)

Middle east mangroves	Name of mangrove species	Factors affecting the mangroves	References
Yemen North and South Al-Luhayah, Kamran Island, Midi, Mangrove Garden in Yemen, Al-Habl, Al-Buhays, El-Ghurairah at Bab al-Mandab	*Rhizophora Mucronate Avicennia marina*	The ecosystem faces significant climate change alteration, which includes sea water salinity, temperature rise and also contamination of microplastics and heavy metals, which degrade the health of the mangroves	Waleed et al., (2024)
Sudan Agig, Arakiyai, Mersa Atta, Haydob, Ashat, Mukawwar Island, Mohammed Qol, Klanieb, Halut, Fagum-Lagagengeeb	*Avicennia marina*	The Mangrove along Sudan's southeast Red Sea coast faces climate change, the rise of sea level and urban development; overgrazing, excessive deforestation and improper management have led to decline and fluctuations in mangrove area	Osman et al., (2019)

Adopted and Modified from Waleed et al., (2024)

7.6 Conservation Strategies for Biodiversity Preservation

The Arabian Peninsula is home to a unique and diverse range of flora and fauna, making their conservation and sustainable management imperative. Given the high rate of endemism, particularly in freshwater biodiversity, an integrated conservation approach is essential. This can be achieved through a combination of in-situ and ex-situ conservation strategies, including the establishment of botanical and zoological gardens, national parks and protected areas. Strengthening environmental legislation, raising awareness through education and promoting eco-tourism are also crucial components. Eco-tourism, in particular, can serve as a sustainable economic driver by generating local employment, stimulating markets and improving regional transportation infrastructure (Masoumi et al., 2024). Scientific collaboration and research play a vital role in biodiversity conservation. The systematic assessment of species data can help to identify optimal conservation strategies, while biodiversity surveys can contribute to the creation of a regional expert network for continuous species monitoring and data updates. The Red List of Threatened Species (IUCN, 2012) remains a critical global resource for tracking the conservation status of flora and fauna. Additionally, the implementation of robust policy frameworks is essential to promote sustainability and mitigate habitat degradation (Masoumi et al., 2024).

Recognizing these challenges, Saudi Arabia has already taken significant initiatives to protect its biodiversity. As part of its commitment to global conservation efforts, the country has ratified the Convention on Biological Diversity (CBD) and actively participates in the United Nations Framework Convention on Climate

Change (UNFCCC) meetings. A national target has been set to bring 17% of the country's land under protection (Thomas et al., 2017; WWF, 2022; PERSGA/GEF, 2004). Recent assessments of 797 taxa identified 232 species (29%) as threatened, leading to the IUCN's release of a red list specifically for plant conservation in the Arabian Peninsula (Forrest & Neal, 2023). Biodiversity hotspots in the region, including the southwestern Yemen Mountain range (part of the Afromontane Biodiversity Hotspot), Dhofar in Oman, Harrat Al-Harrah in southwestern Arabia, Socotra Island, Jebel Areys and Hadramaut, host a significant concentration of endemic species. The Socotra Archipelago, a UNESCO World Heritage Site, is recognized for its exceptional biodiversity, while Ramsar-protected wetlands, such as Bahramand Island and Mubarak Al Kabeer in Kuwait, further underscore the region's ecological significance. In response, regional governments have implemented conservation strategies, including ecological surveys and habitat mapping, to support biodiversity protection (Allen et al., 2021; Patezelt, 2009; Thomas, 2024). Beyond habitat protection, preserving Plant Genetic Resources (PGRs) particularly Crop Wild Relatives (CWRs) is vital for agricultural sustainability and climate resilience. Saudi Arabia and other nations in the region actively participate in international agreements on PGRs, which aim to safeguard genetic diversity for future food security (Gros-Balthazard et al., 2017; FAO, 2004).

Despite the region's extreme aridity, over 400 wild relative species, including approximately 70 food and feed crops, demonstrate remarkable adaptability. These genetic resources not only provide valuable traits for breeding climate-resilient crops but also contribute to conservation-driven agricultural initiatives (Almarri et al., 2024; Kameswara Rao, 2013). To ensure the long-term conservation of key crop species, efforts have also focused on seed banking. For example, high-value crop seeds from the Jazan province in Saudi Arabia have been collected and preserved for future cultivation (Al-Turki et al., 2019). Additionally, advancements in ecological modeling, such as species distribution models, spatial analysis and climate-environmental variable modeling, are being leveraged to develop sustainable biodiversity management plans. These models offer critical insights into habitat suitability and climate change impacts, guiding conservation actions at both national and regional levels (Bedair et al., 2024; Gholamhosseini et al., 2024; Soultan et al., 2019).

A comprehensive and multi-faceted approach is essential for ensuring the long-term preservation of biodiversity in the Arabian Peninsula. Strengthening policy enforcement, restoring degraded habitats, advancing scientific research and fostering community engagement will be critical in mitigating biodiversity loss. Expanding cross-border collaborations, increasing the extent of protected areas and leveraging cutting-edge conservation technologies will further enhance efforts to safeguard ecological resources. Moreover, integrating traditional ecological knowledge with modern conservation strategies can provide a holistic approach to biodiversity management. By adopting sustainable practices, enhancing conservation planning and prioritizing climate resilience, the region can secure the sustainability of its unique ecological heritage while addressing the ongoing challenges posed by climate change and human activities.

7.7 Conclusion

The ecosystems and biodiversity of the Arabian Peninsula are undergoing important transitions, influenced by a dynamic interplay between natural adaptation processes and emerging environmental challenges. Climate change, alongside habitat degradation and resource exploitation, is accelerating ecosystem shifts, threatening species survival and altering ecological interactions. While the region has long been home to species uniquely adapted to extreme conditions, the rapid pace of environmental change now outstrips their natural resilience, necessitating urgent and sustained intervention. Beyond the immediate ecological concerns, biodiversity conservation in the Arabian Peninsula is deeply linked to broader sustainability goals, including food security, climate resilience and economic stability. The loss of key species and habitats not only weakens natural ecosystems but also affects human well-being, particularly in communities reliant on biodiversity for livelihoods, agriculture and traditional practices. The decline of freshwater systems, coastal mangroves and desert flora indicates the broader consequences of environmental degradation, extending beyond conservation to fundamental aspects of survival in an arid landscape. While significant steps have been taken including the expansion of protected areas, participation in global biodiversity agreements and policy reforms there remains a need for continuous adaptation and innovation. Conservation must evolve from reactive protection to proactive ecosystem management, leveraging new scientific tools, predictive modeling and technology-driven monitoring. Future efforts should focus not only on preserving species but on maintaining ecological processes that sustain biodiversity in the long term. Strengthening governance frameworks, ensuring sustainable land use and fostering cross-border collaboration will be key to addressing these complex environmental challenges. Ultimately, securing the region's biodiversity is not just about conservation it is about resilience, sustainability and the ability of both ecosystems and human societies to adapt to a changing world. The Arabian Peninsula has historically demonstrated remarkable ecological endurance and with strategic, science-backed and community-driven efforts, it can continue to do so in the face of future challenges.

References

Abbas, Z. K., Saggu, S., Rehman, H., Al Thbiani, A., & Ansari, A. A. (2017). Ecological variations and role of heat shock protein in Artemisia judaica L. in response to temperature regimes of Tabuk, Saudi Arabia. *Saudi Journal of Biological Sciences, 24*(6), 1268–1273.

Abdel-Razik, M. S. (1991). Population structure and ecological performance of the mangrove *Avicennia marina* (Forssk.) Vierh on the Arabian Gulf coast of Qatar. *Journal of Arid Environments, 20*(3), 331–338.

Abou Seedo, K., Abido, M. S., Salih, A., & Abahussain, A. (2017). Structure and composition of mangrove associations in Tubli Bay of Bahrain as affected by municipal wastewater discharge and anthropogenic sedimentation. *International Journal of Biodiversity, 2017*(1), 2084256.

Abulfatih, H. A. (1984). Elevationally restricted floral elements of the Asir Mountains, Saudi Arabia. *Journal of Arid Environments, 7*(1), 35–41.

Ajjur, S. B., & Al-Ghamdi, S. G. (2022, January). Assessment of the Climate Change Impact on the Arabian Peninsula Evapotranspiration Losses. In *World Environmental and Water Resources Congress 2022* (pp. 492–498).

Al Charaabi, Y., & Al-Yahyai, S. (2013). Projection of future changes in rainfall and temperature patterns in Oman. *Journal of Earth Science and Climate Change, 4*, 154. https://doi.org/10.4172/2157-7617.1000154

Al-Ghamdi, G., Alzahrani, A., Al-Ghamdi, S., Alghamdi, S., Al-Ghamdi, A., Alzahrani, W., & Zinner, D. (2023). Potential hotspots of hamadryas baboon-human conflict in Al-Baha region, Saudi Arabia. *Diversity, 15*(11), 1107.

Al Solami, M. A. (2020). Comparative response of red and green algae to the quality of coastal water of Red Sea, Haql, Saudi Arabia. *Journal of Environmental Protection, 11*(10), 793–806.

Al-Blooshi, L. S., Issa, S., & Ksiksi, T. (2020). Assessing the environmental impact of climate change on desert ecosystems: A review. *Advances in Ecological and Environmental Research, 5*(2), 27–52.

Alghanem, S. M., Al-Atwi, H. Q., Al-Saiari, M. O., Al-Balawi, A. M., Al-Zahrani, S. A., & Al-Sayed, A. M. (2020). Floristic diversity and perennial vegetation analysis of Al-Wadi Al-akhder, Tabuk Region, Saudi Arabia. *Int. J. Plant Sci. Ecol, 6*, 31–38.

Allen, D. J., Westrip, J. R. S., Puttick, A., Harding, K. A., Hilton-Taylor, C., & Ali, H. (2021). *UAE National Red List of Vascular Plants* (Technical Report). Ministry of Climate Change and Environment. Retrieved July 17, 2024, from https://www.nationalredlist.org/publication/uae-national-red-list-vascular-plants

Almarri, N. B., Ahmad, S., & Elshal, M. H. (2024). Diversity and conservation of plant genetic resources in Saudi Arabia. In *Sustainable utilization and conservation of plant genetic diversity* (Vol. 35, pp. 1009–1031).

Almazroui, M. (2020). Changes in temperature trends and extremes over Saudi Arabia for the period 1978–2019. *Advances in Meteorology, 2020*(1), 8828421.

Almazroui, M., Islam, M. N., & Jones, P. D. (2013). Urbanization effects on the air temperature rise in Saudi Arabia. *Climatic Change, 120*, 109–122.

Almazroui, M., Islam, M. N., Saeed, S., Saeed, F., & Ismail, M. (2020). Future changes in climate over the Arabian Peninsula based on CMIP6 multimodel simulations. *Earth Systems and Environment, 4*, 611–630.

Almazroui, M., Nazrul Islam, M., Saeed, S., Alkhalaf, A. K., & Dambul, R. (2017). Assessment of uncertainties in projected temperature and precipitation over the Arabian Peninsula using three categories of Cmip5 multimodel ensembles. *Earth Systems and Environment, 1*, 1–20.

Al-Nadabi, A., & Sulaiman, H. (2018, May). Carbon sink potential of Avicennia marina in the Al-Qurm nature reserve, Muscat, Oman. In *Iop conference series: earth and environmental science* (Vol. 151, No. 1, p. 012003). IOP Publishing.

Al-Namazi, A. A., Algarni, S. M., Wan, J. S., Al Mosallam, M. S., & Alotaibi, F. (2022). Floristic composition of Jandaf Mountain as biodiversity hotspot area in southwestern Saudi Arabia. *Saudi Journal of Biological Sciences, 29*(5), 3654–3660.

Al-Namazi, A. A., Al-Khulaidi, A. W. A., Algarni, S., & Al-Sagheer, N. A. (2021). Natural plant species inventory of hotspot areas in Arabian Peninsula: Southwest Al-Baha region, Saudi Arabia. *Saudi Journal of Biological Sciences, 28*(6), 3309–3324.

Aloufi, A. A., & Amr, Z. S. (2015). On the herpetofauna of the Province of Tabuk, northwest Saudi Arabia. *Herpetozoa, 27*(3/4), 147–158.

Alshehri, M. A., Aziz, A. T., Alzahrani, O., Osman, G., & Alasmari, A. (2021). Molecular characterization of some algae by protein banding pattern and ISSR markers collected from the Gulf of Aqaba, Saudi Arabia. *Pakistan Journal of Botany, 53*(2), 707–713.

Al-Thani, R. F., & Yasseen, B. T. (2018). Biological soil crusts and extremophiles adjacent to native plants at Sabkhas and Rawdahs, Qatar: The possible roles. *Frontiers in Environmental Microbiology, 4*(2), 55–70.

Al-Turki, T. A., Al-Namazi, A. A., & Masrahi, Y. S. (2019). Conservation of genetic resources for five traditional crops from Jazan, SW Saudi Arabia, at the KACST Gene-Bank. *Saudi Journal of Biological Sciences, 26*(7), 1626–1632.

Amin, M. T., Mahmoud, S. H., & Alazba, A. A. (2016). Observations, projections and impacts of climate change on water resources in Arabian Peninsula: Current and future scenarios. *Environmental Earth Sciences, 75*, 864. https://doi.org/10.1007/s12665-016-5684-4

Ansari, A. A. (2016). Seasonal dynamics in the relative density of aquatic flora along some coastal areas of the Red Sea, Tabuk, Saudi Arabia. *The Egyptian Journal of Aquatic Research, 42*(3), 273–280.

Ansari, A. A., & Ghanem, S. M. (2017). Seasonal variation in the growth responses of some chlorophytic algal flora of the Red Sea. *The Egyptian Journal of Aquatic Research, 43*(2), 129–134.

Ansari, A. A., & Ghanem, S. M. (2019). Growth attributes and biochemical composition of *Padina pavonica* (L.) from the Red Sea, in response to seasonal alterations of Tabuk, Saudi Arabia. *The Egyptian Journal of Aquatic Research, 45*(2), 139–144.

Ansari, A. A., Abbas, Z. K., Saggu, S., Rehman, H., & Moawed, M. M. (2014). Growth responses of *Lavandula pubescens* to temperature regimes of Tabuk, Saudi Arabia. *Journal of Medicinal Plants Studies, 2*(6), 38–41.

Ansari, A. A., Alnashib, A. H. A., Albalwi, A. A., Alenazi, W. A., & Aldahash, K. S. K. (2020). Quantitative analysis on the diversity of algal flora along some selected coastal sites of the Red Sea at Haql, Tabuk, Saudi Arabia. *International Journal of Botany Studies, 5*, 97–101.

Ansari, A. A., Ghanim, S. A., Trivedi, S., Rehman, H., Abbas, Z. K., & Saggu, S. (2015). Seasonal dynamics in the trophic status of water, floral and faunal density along some selected coastal areas of the Red Sea, Tabuk, Saudi Arabia. *International Aquatic Research, 7*, 337–348.

Ansari, A. A., Siddiqui, Z. H., Alatawi, F. A., Alharbi, B. M., & Alotaibi, A. S. (2022). An assessment of biodiversity in tabuk region of Saudi Arabia: A comprehensive review. *Sustainability, 14*(17), 10564.

Barth, H. J., & Böer, B. (Eds.). (2002). *Sabkha ecosystems: volume I: the Arabian Peninsula and adjacent countries* (Vol. 1). Springer Science & Business Media.

Bedair, H., Badawy, N. K., Morsy, A., Rashad, H., & Dakhil, M. A. (2024). Impact of climate change on the spatial distribution of the endemic shrub *Rubus asirensis* in the Arabian Peninsula. In: Plant Ecology (pp. 1–10).

Ben-Hasan, A., & Christensen, V. (2019). Vulnerability of the marine ecosystem to climate change impacts in the Arabian Gulf—An urgent need for more research. *Global Ecology and Conservation, 17*, Article e00556.

Bhatt, A., Bhat, N. R., Suleiman, M. K., & Al-Mansour, H. (2023). Prioritization of potential native plants from Arabian peninsula based on economic and ecological values: implication for restoration. *Sustainability, 15*(7), 6139.

Bibi, F., & Métais, G. (2016). Evolutionary history of the large herbivores of south and Southeast Asia (Indomalayan Realm). In *The ecology of large herbivores in south and southeast Asia* (pp. 15–88).

Borrell, J. S., Al Issaey, G., Lupton, D. A., Starnes, T., Al Hinai, A., Al Hatmi, S., Senior, R. A., Wilkinson, T., Milborrow, J. L. H., & Patzelt, A. (2019). Islands in the desert: Environmental distribution modelling of endemic flora reveals the extent of Pleistocene tropical relict vegetation in southern Arabia. *Annals of Botany, 124*(3), 411–422.

Chatziefthimiou, A. D., Deitch, E. J., Glover, W. B., Powell, J. T., Banack, S. A., Richer, R. A., Cox, P. A., & Metcalf, J. S. (2018). Analysis of neurotoxic amino acids from marine waters, microbial mats, and seafood destined for human consumption in the Arabian Gulf. *Neurotoxicity Research, 33*, 143–152.

Chaudhary, S. A. (2000). *Flora of the Kingdom of Saudi Arabia* (Vol. 2, Part 3, pp. 191–192). Ministry of Agriculture & Water Research Center, National Herbarium.

Danin, A. (2012). *Plants of desert dunes*. Springer Science & Business Media.

Danin, A., and Danin, A. (1996). Plant adaptations to environmental stresses in desert dunes. *Plants of Desert Dunes*, 133–152.

Deil, U., &, al Gifri, A. N. (1998). Montane and Wadi vegetation. In *Vegetation of the Arabian Peninsula* (pp. 125–174). Springer Netherlands.

Dudgeon, D., Arthington, A. H., Gessner, M. O., Kawabata, Z. I., Knowler, D. J., Lévêque, C., Naiman, R. J., Prieur-Richard, A. H., Soto, D., Stiassny, M. L., & Sullivan, C. A. (2006). Freshwater biodiversity: importance, threats, status and conservation challenges. *Biological Reviews, 81*(2), 163–182.

Elvidge, C. D., & Lyon, R. J. (1985). Influence of rock-soil spectral variation on the assessment of green biomass. *Remote Sensing of Environment, 17*(3), 265–279.

Enzel, Y., Kushnir, Y., & Quade, J. (2015). The middle Holocene climatic records from Arabia: reassessing lacustrine environments, shift of ITCZ in Arabian Sea, and impacts of the southwest Indian and African monsoons. *Global and Planetary Change, 129*, 69–91.

Esmaeili, H. R., & Eslami Barzoki, Z. (2023). Climate change may impact Nile Tilapia, *Oreochromis niloticus* (Linnaeus, 1758) Distribution in the Southeastern Arabian Peninsula through Range Contraction under Various Climate Scenarios. *Fishes, 8*(10), 481.

Esmaeili, H. R., Jufaili, S. A., Masoumi, A. H., & Zarei, F. (2022). Ichthyodiversity in southeastern Arabian Peninsula: Annotated checklist, taxonomy, short description and distribution of Inland fishes of Oman. *Zootaxa, 5134*(4), 451–503.

Etesami, H., & Beattie, G. A. (2018). Mining halophytes for plant growth-promoting halotolerant bacteria to enhance the salinity tolerance of non-halophytic crops. *Frontiers in Microbiology, 9*, 148.

European Commission. (2008). *The European Union's biodiversity action plan: Halting the loss of biodiversity by 2010 – and beyond*. European Commission. Retrieved November 27, 2024, from http://www.ab.gov.tr/files/ardb/evt/1_avrupa_birligi/1_6_raporlar/1_3_diger/environment/eu_biodiversity_action_plan___2010_and_beyond.pdf

Fisher, M., & Gardner, A. S. (1995). The status and ecology of a *Juniperus excelsa* subsp. *polycarpos* woodland in the northern mountains of Oman. *Vegetatio, 119*, 33–51.

Food and Agriculture Organization of the United Nations (FAO). (2004). *The International Treaty for Plant Genetic Resources for Food and Agriculture*. FAO. Retrieved July 17, 2024, from http://www.planttreaty.org/training/annex1_en.htm

Forrest, A., & Neale, S. (2023). *The conservation status of the plants of the Arabian Peninsula: Endemic taxa, trees, and aloes* (p. 48). Environment and Protected Areas Authority.

Freyhof, J., Els, J., Feulner, G. R., Hamidan, N. A., & Krupp, F. (2021). Freshwater fishes of the Arabian Peninsula. *Taxa, 20309*.

Ghazanfar, S. A. (1997). The phenology of desert plants: A 3-year study in a gravel desert Wadi in northern Oman. *Journal of Arid Environments, 35*(3), 407–417. https://doi.org/10.1006/jare.1996.0132

Ghazanfar, S. A. (1998). Status of the flora and plant conservation in the Sultanate of Oman. *Biological Conservation, 85*(3), 287–295. https://doi.org/10.1016/S0006-3207(97)00122-3

Ghazanfar, S. A. (2015). *Flora of the Sultanate of Oman; Loganiaceae-Asteraceae* (Vol. 3, pp. 236–237). Scripta Botanica Belgica Series 55. National Botanic Garden of Belgium.

Ghazanfar, S. A. (2024). Biogeography and conservation in the Arabian Peninsula: A present perspective. *Plants, 13*(15), 2091.

Ghazanfar, S. A., Böer, B., Al Khulaidi, A. W., El-Keblawy, A., & Alateeqi, S. (2019). Plants of sabkha ecosystems of the Arabian Peninsula. In *Sabkha ecosystems: Volume VI: Asia/Pacific* (pp. 55–80). Springer. https://doi.org/10.1007/978-3-030-04261-5_3

Gholamhosseini, A., Yousefi, M., & Esmaeili, H. R. (2024). Predicting climate change impacts on the distribution of endemic fish Cyprinion muscatense in the Arabian Peninsula. *Ecology and Evolution, 14*(7), Article e11720.

Gros-Balthazard, M., & Flowers, J. M. (2021). A brief history of the origin of domesticated date palms. In *The Date Palm Genome, Vol. 1: Phylogeny, Biodiversity and Mapping* (pp. 55–74). Springer International Publishing.

Gros-Balthazard, M., Galimberti, M., Kousathanas, A., Newton, C., Ivorra, S., Paradis, L., Vigouroux, Y., Carter, R., Tengberg, M., Battesti, V., Santoni, S., Falquet, L., Pintaud, J.-C., Terral, J.-F., & Wegmann, D. (2017). The discovery of wild date palms in Oman reveals a complex domestication history involving centers in the Middle East and Africa. *Current Biology, 27*(14), 2211–2218.
Hall, J. B. (1984). Juniperus excelsa in Africa: a biogeographical study of an Afromontane tree. *Journal of biogeography*, 47–61.
Harrison, I., Cox, N., & Tognelli, M. F. (2015). *The status and distribution of freshwater biodiversity in the Arabian Peninsula.* IUCN.
Howells, E. J., Abrego, D., Liew, Y. J., Burt, J. A., Meyer, E., & Aranda, M. (2021). Enhancing the heat tolerance of reef-building corals to future warming. *Science Advances, 7*(34), eabg6070.
Howells, E. J., Abrego, D., Meyer, E., Kirk, N. L., & Burt, J. A. (2016). Host adaptation and unexpected symbiont partners enable reef-building corals to tolerate extreme temperatures. *Global Change Biology, 22*(8), 2702–2714.
Hume, B. C., D'Angelo, C., Smith, E. G., Stevens, J. R., Burt, J., & Wiedenmann, J. (2015). *Symbiodinium thermophilum* sp. nov., a thermotolerant symbiotic alga prevalent in corals of the world's hottest sea, the Persian/Arabian Gulf. *Scientific Reports, 5*(1), 8562.
Ibraheem, I. B. M., Alharbi, R. M., Abdel-Raouf, N., & Al-Enazi, N. M. (2014). Contributions to the study of the marine algae inhabiting Umluj Seashore, Red Sea. *Beni-Suef University Journal of Basic and Applied Sciences, 3*(4), 278–285.
IUCN. (2012). *IUCN Red List categories and criteria: Version 3.1* (2nd ed.). Gland, Switzerland, and Cambridge, UK: International Union for Conservation of Nature and Natural Resources.
Ivory, S. J., Dinies, M., & Lézine, A. M. (2021). Ecosystem change and human-environment interactions of Arabia. In *Quaternary vegetation dynamics* (pp. 157–182). CRC Press.
Jennings, R. P., Singarayer, J., Stone, E. J., Krebs-Kanzow, U., Khon, V., Nisancioglu, K. H., Pfeiffer, M., Zhang, X., Parker, A., Parton, A., Groucutt, H. S., White, T. S., Drake, N. A., & Petraglia, M. D. (2015). The greening of Arabia: Multiple opportunities for human occupation of the Arabian Peninsula during the Late Pleistocene inferred from an ensemble of climate model simulations. *Quaternary International, 382*, 181–199.
Kameswara Rao, N. (2013). Crop wild relatives from the Arabian Peninsula. *Genetic Resources and Crop Evolution, 60*, 1709–1725.
Kotwicki, V., & Al Sulaimani, Z. (2009). Climates of the Arabian Peninsula–past, present, future. *International Journal of Climate Change Strategies and Management, 1*(3), 297–310.
Kürschner, H. (1998). Biogeography and introduction to vegetation. *Vegetation of the Arabian Peninsula* (pp. 63–98). Springer.
Lipton, D., Rubenstein, M. A., Weiskopf, S. R., Carter, S., Peterson, J., Crozier, L., Fogarty, M., Gaichas, S., Hyde, K. J. W., Morelli, T. L., Morisette, J., Moustahfid, H., Muñoz, R., Poudel, R., Staudinger, M. D., Stock, C., Thompson, L., Waples, R., & Weltzin, J. F. (2018). Ecosystems, ecosystem services, and biodiversity. In D. R. Reidmiller, C. W. Avery, D. R. Easterling, K. E. Kunkel, K. L. M. Lewis, T. K. Maycock, & B. C. Stewart (Eds.), *Impacts, risks, and adaptation in the United States: Fourth National Climate Assessment, Volume II* (pp. 268–321). U.S. Global Change Research Program. https://doi.org/10.7930/NCA4.2018.CH7
Lüning, S., & Vahrenholt, F. (2019). Holocene climate development of North Africa and the Arabian Peninsula. *The Geology of the Arab World—An Overview*, 507–546.
Makki, T., Mostafavi, H., Matkan, A. A., Aghighi, H., Valavi, R., Chee, Y. E., & Teimori, A. (2023a). Impacts of climate change on the distribution of riverine endemic fish species in Iran, a biodiversity hotspot region. *Freshwater Biology, 68*(6), 1007–1019.
Makki, T., Mostafavi, H., Matkan, A. A., Valavi, R., Hughes, R. M., Shadloo, S., Aghighi, H., Abdoli, A., Teimori, A., Eagderi, S., & Coad, B. W. (2023b). Predicting climate heating impacts on riverine fish species diversity in a biodiversity hotspot region. *Scientific Reports, 13*(1), 14347.
Mandaville, J. P. (1998). Vegetation of the sands. In *Vegetation of the Arabian Peninsula* (pp. 191–208). Springer Netherlands.

Masoumi, A. H., Esmaeili, H. R., Khosravi, R., Gholamhosseini, A., Korkmaz, M., & Jeppesen, E. (2024). Species on the move: Impacts of climate change on the spatial range of endemic fishes of the eco-sensitive semi-arid area of the Arabian Peninsula. *Science of the Total Environment, 947*, Article 174095.

May, W. (2011). The sensitivity of the Indian summer monsoon to a global warming of 2 C with respect to pre-industrial times. *Climate Dynamics, 37*(9), 1843–1868.

Metcalf, J. S., Banack, S. A., Richer, R., & Cox, P. A. (2015). Neurotoxic amino acids and their isomers in desert environments. *Journal of Arid Environments, 112*, 140–144.

Miller, A. G., Cope, T. A., & Nyberg, J. A. (1996). Flora of the Arabian Peninsula and Socotra.

Odhiambo, G. O. (2017). Water scarcity in the Arabian Peninsula and socio-economic implications. *Applied Water Science, 7*(5), 2479–2492.

Oldfather, M. F., Britton, M. N., Papper, P. D., Koontz, M. J., Halbur, M. M., Dodge, C., Flint, A. L., Flint, L. E., & Ackerly, D. D. (2016). Effects of topoclimatic complexity on the composition of woody plant communities. *AoB Plants, 8*, plw049.

Onoh, U. C., Ogunade, J., Owoeye, E., Awakessien, S., & Asomah, J. K. (2024). Impact of climate change on biodiversity and ecosystem services. *International Journal of Geography and Environmental Management, 10*(1), 77–93.

Patlakas, P., Stathopoulos, C., Flocas, H., Kalogeri, C., & Kallos, G. (2019). Regional climatic features of the Arabian Peninsula. *Atmosphere, 10*(4), 220.

Patzelt, A. (2009). *Oman Red Data Book* (p. 310). Diwan of Royal Court, Office for Conservation of the Environment.

Pernot, O., Simon, L., Al-Khyatt, J., Al-Ghouti, M., & Abu-Dieyeh, M. H. (2015). Ecosystem services and mangroves in Qatar: Preservation issues. *Qscience Proceedings, 2015*(5), 37.

PERSGA/GEF. (2004). *Status of mangroves in the Red Sea and Gulf of Eden.* Retrieved from http://www.persga.org

Prakash, S. (2021). Impact of climate change on aquatic ecosystem and its biodiversity: An overview. *International Journal of Biological Innovations, 3*(2).

Quamar, M. F. (2024). Palynological Perspective on Understanding Climate Change in India Over the Pre-industrial Common Era (CE; Past Ca. 2000 Years): a Comprehensive Review and a Critical Evaluation. *Applications of Palynology in Stratigraphy and Climate Studies*, 205–232.

Raihan, A., Ali, T., & Mortula, M. (2024). Unveiling coastal ecosystem dependencies: Multifactorial analysis of climate change influencing mangrove system in the United Arab Emirates over the period 1990–2020. *Discover Sustainability, 5*(1), 498.

Rao, K. K., Al Mandous, A., Al Ebri, M., Al Hameli, N., Rakib, M., & Al Kaabi, S. (2024). Future changes in the precipitation regime over the Arabian Peninsula with special emphasis on UAE: Insights from NEX-GDDP CMIP6 model simulations. *Science and Reports, 14*, 151.

Richer, R., Anchassi, D., El-Assaad, I., El-Matbouly, M., Ali, F., Makki, I., & Metcalf, J. S. (2012). Variation in the coverage of biological soil crusts in the State of Qatar. *Journal of Arid Environments, 78*, 187–190.

Saenger, P., Blasco, F., Youssef, A. M., Loughland, R. A., & Wrydani, S. (2002). The mangrove vegetation of the United Arab Emirates, with particular emphasis on those of the Abu-Dhabi Emirate. In *Proceedings of the 2nd International Symposium and Workshop on Arid Zone Environments: Research and Management Options for Mangrove and Saltmarsh Ecosystems* (pp. 36–49). ERWDA.

Saifullah, S. S. (1996). Mangrove ecosystem of Saudi Arabian Red Sea coast-an overview. *Marine Scienes-Ceased lssuerg, 17*(1), 1–2.

Seraj, S. S., Jrais, R. N., & Ayyad, S. K. (2014). Floristic composition, life form and chorology of plant life at Al-Saoda, Asir Region, South-Western Saudi Arabia. *Journal of Biology, Agriculture and Healthcare, 4*(26), 60–67.

Sherman, R., Mullen, R., Li, H., Fang, Z., & Wang, Y. (2007). Alpine ecosystems of northwest Yunnan, China: An initial assessment for conservation. *Journal of Mountain Science, 4*, 181–192.

Soultan, A., Wikelski, M., & Safi, K. (2019). Risk of biodiversity collapse under climate change in the Afro-Arabian region. *Scientific Reports, 9*(1), 955.

Stewart, P. S., Voskamp, A., Santini, L., Biber, M. F., Devenish, A. J., Hof, C., Willis, S. G., & Tobias, J. A. (2022). Global impacts of climate change on avian functional diversity. *Ecology Letters, 25*(3), 673–685.
Sulthan, S. M. I. (2016). Climate change in Saudi Arabia due to industrial revolution. *IJCEM, 3*(3).
Thomas, J. (n.d.). *Current status of plant conservation in Saudi Arabia.* Retrieved July 17, 2024, from https://www.plantdiversityofsaudiarabia.info/Biodiversity-Saudi-Arabia/Conservation/Conservation.htm
Thomas, J., El-Sheikh, M. A., & Alatar, A. A. (2017). Endemics and endangered species in the biodiversity hotspot of the Shada Mountains, Saudi Arabia. *Journal of Arid Land, 9*, 109–121.
Tieleman, B. I., Williams, J. B., Buschur, M. E., & Brown, C. R. (2003). Phenotypic variation among and within larks along an aridity gradient: Are desert birds more flexible? *Ecology, 84*(7), 1800–1815. https://doi.org/10.1890/0012-9658(2003)084[1800:PVAAWL]2.0.CO;2
Wabnitz, C. C., Lam, V. W., Reygondeau, G., Teh, L. C., Al-Abdulrazzak, D., Khalfallah, M., Pauly, D., Deng Palomares, M. L., Zeller, D., & Cheung, W. W. (2018). Climate change impacts on marine biodiversity, fisheries and society in the Arabian Gulf. *PLoS ONE, 13*(5), Article e0194537.
Waleed, T. A., Abdel-Maksoud, Y. K., Kanwar, R. S., & Sewilam, H. (2024). Mangroves in Egypt and the Middle East: current status, threats, and opportunities. *International Journal of Environmental Science and Technology*, 1–38.
Weiskopf, S. R., Harma.kova, Z. V., Johnson, C. G., Londono-Murcia, M. C., Miller, B. W., Myers, B. J., Pereira, L., Arce-Plata, M. L., Blanchard, J. L., Ferrier, S., Fulton, E. A., & Rosa, I. M. (2022). Increasing the uptake of ecological model results in policy decisions to improve biodiversity outcomes. *Environmental Modelling & Software, 149*, 105318.
White, F., & Léonard, J. (1991). Phytogeographical links between Africa and southwest Asia.
Williams, J. B., Muñoz-Garcia, A., & Champagne, A. (2012). Climate change and cutaneous water loss of birds. *Journal of Experimental Biology, 215*(7), 1053–1060.
WWF. (2022). *Living planet report 2022—Building a nature-positive society*. In: R. E. A. Almond, M. Grooten, D. Juffe Bignoli, & T. Petersen (Eds.). WWF.
Yadav, A. R., Yadav, M. R., & Mohite, V. T. (2024). Effect of climate change on Arabian Sea: a review. *Journal of Experimental Zoology India, 27*(1).

Chapter 8
Climate Change and Public Health in the Arabian Peninsula: Epidemiological Threats, Vulnerabilities and Adaptive Pathways

Mohammad Muslim and Hepa Susami

8.1 Introduction

Climate change poses an unprecedented and systemic threat to human health, exerting widespread effects on the biophysical environment, socio-economic structures and public health systems (Gkouliaveras et al., 2025). It functions as a threat multiplier, compounding existing health challenges and deteriorating the foundations of health security. The intensification and increased frequency of extreme weather events such as heatwaves, droughts, floods, storms and wildfires are directly associated with elevated morbidity and mortality, while also contributing to indirect health outcomes, including the rising burden of non-communicable diseases, the resurgence and geographical spread of vector- and water-borne diseases and a growing incidence of climate-related health emergencies (Ma et al., 2022; Patz et al., 2014). The impacts of climate change extend to health service delivery, with adverse effects on infrastructure, supply chains and the health workforce, thereby impeding efforts to achieve universal health coverage (UHC) (WHO, 2021). In parallel, long-term climate stressors such as rising temperatures, shifting precipitation patterns and sea-level rise degrade critical environmental and social determinants of health, including air and water quality, food systems, housing and livelihood security (Ebi & Hess, 2020). These systemic disruptions affect both physical and mental health outcomes across populations, with disproportionate impacts on vulnerable and marginalized

M. Muslim (✉)
Department of Environmental Science, University of Kashmir, Hazratbal, Srinagar, India
e-mail: mmuslim@uok.edu.in

H. Susami
Abhipraya Foundation, Jati Karya, Jati Sampurna, Bekasi, Indonesia
e-mail: eva.susami@gmail.com

F. Mushtaq et al. (eds.), *Climate Change in the Arabian Peninsula*,
https://doi.org/10.1007/978-3-032-02481-7_8

groups (Cianconi et al., 2020). Without urgent and integrated mitigation and adaptation measures, the health consequences of climate change are projected to intensify, reversing decades of global health progress and compromising international commitments to uphold the right to health as established in international human rights frameworks. The Intergovernmental Panel on Climate Change (IPCC) Sixth Assessment Report (AR6) affirms that climate-related hazards are manifesting more rapidly and with greater intensity than previously estimated, thereby reducing the temporal and institutional capacity for effective adaptation (IPCC, 2023). The report highlights that ~3.6 billion individuals inhabit regions characterized by high vulnerability to climate-induced risks. Despite contributing minimally to cumulative greenhouse gas emissions, low-income countries and small island developing states (SIDS) disproportionately experience the adverse health consequences of climate change. Epidemiological evidence indicates that, over the past decade, mortality rates from climate-exacerbated extreme events have been up to 15 times higher in highly vulnerable regions compared to more resilient counterparts.

The health impacts of climate change are multifaceted, encompassing both direct and indirect pathways (Fig. 8.1). These include increased mortality and morbidity associated with extreme weather events (e.g., heatwaves, floods, cyclones), disruptions to agricultural production and food security, expansion of climate-sensitive infectious diseases (including zoonotic, vector-borne, food- and water-borne pathogens) and a growing burden of mental health disorders linked to environmental stress and displacement. Additionally, climate change undermines essential social determinants of health, such as economic stability, healthcare access, housing and social protection systems. These climate-sensitive health burdens are inequitably distributed, with increased vulnerability observed among socioeconomically and structurally marginalized populations including women, children, the elderly, persons with chronic illnesses, displaced populations and ethnic minorities due to intersecting exposures and constrained adaptive capacities (Anjum & Aziz, 2025). Although the relationship between climate change and adverse health outcomes is well established, accurately quantifying the magnitude, distribution and attribution of climate-sensitive health risks remains methodologically challenging. Nevertheless, recent advancements in epidemiological modeling and attribution science increasingly enable the isolation of climate-related contributions to morbidity and mortality (Lu et al., 2021). These developments enhance the ability to assess the scope and intensity of health threats driven by anthropogenic climate change. According to the World Health Organization (WHO), ~2 billion individuals currently lack access to safely managed drinking water, while an estimated 600 million suffer from foodborne illnesses annually of which children under five account for nearly 30% of associated fatalities. Climatic stressors such as rising temperatures, extreme precipitation variability and deteriorating water quality intensify the transmission dynamics of waterborne and foodborne diseases. In 2020, ~770 million individuals experienced hunger, with the burden predominantly concentrated in Africa and Asia. Climate change adversely impacts food availability, nutritional quality and dietary diversity, thereby exacerbating both acute and chronic forms of food insecurity and malnutrition.

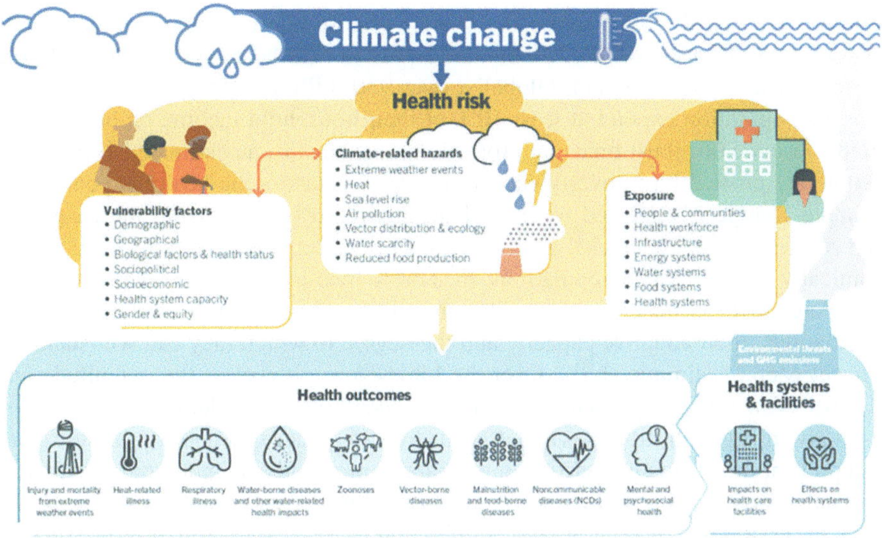

Fig. 8.1 Pathways of climate-sensitive health risks, illustrating exposure mechanisms and key vulnerability determinants. Climate change influences health through both direct and indirect pathways, with outcomes shaped by interacting environmental, socio-economic and public health system factors. *Source* World Health Organization (WHO, 2021)

Furthermore, altered temperature and precipitation regimes are facilitating the expansion and seasonality of vector-borne diseases (VBD), including malaria, dengue and chikungunya (Pandey et al., 2021). In addition to physical health effects, climate change is contributing to the growing prevalence of mental health disorders. Acute events such as heatwaves and natural disasters are associated with increased incidence of anxiety, depression and post-traumatic stress disorder (PTSD), while chronic stressors such as forced displacement, livelihood disruption and ecological degradation contribute to long-term psychological distress. Emerging evidence reinforces the scale of these health threats. A recent global attribution study estimates that 37% of heat-related deaths can be directly attributed to anthropogenic climate change. Between 2000 and 2019, the rate of heat-related mortality among adults aged 65 and older increased by over 70%. In 2020 alone, 98 million more people experienced food insecurity compared to the 1981–2010 baseline average. WHO projects that climate change will contribute to ~250,000 additional deaths annually between 2030 and 2050, due to malaria, diarrheal diseases, heat stress and coastal flooding. However, these projections likely underestimate the true burden, as current models face limitations in capturing the complex interactions between climate drivers and indirect health outcomes such as drought-induced food shortages, climate migration and disruptions to health service delivery.

The climate change threatens to reverse decades of progress in global health, sustainable development and poverty reduction. It amplifies existing health inequities within and between populations and impedes progress toward achieving UHC.

Climate-induced shocks increase the overall disease burden while simultaneously undermining the accessibility, quality and equity of healthcare systems often during periods of increased demand. Globally, over 930 million individuals ~12% of the world's population expend at least 10% of their household income on health care. In the absence of robust financial protection mechanisms, an estimated 100 million people are pushed into poverty each year due to catastrophic health expenditures. Climate change is expected to exacerbate this trend by increasing the frequency and severity of health shocks, disproportionately affecting uninsured and socioeconomically disadvantaged populations. In short to medium term, the health outcomes associated with climate change will primarily be shaped by the vulnerability of populations, their adaptive capacity and the degree to which they can withstand the current pace of climatic shifts. Over the longer term, however, the magnitude and severity of health impacts will increasingly depend on the implementation of transformative mitigation strategies undertaken today to reduce greenhouse gas emissions and prevent the crossing of critical temperature thresholds and potential irreversible climate system tipping points. The multifaceted nature of climate-sensitive health risks necessitates a more granular examination of key exposure pathways such as air pollution, vector- and water-borne diseases and food insecurity through which climate change exerts its most profound and measurable impacts on human health.

8.2 The Spread of Vector-Borne and Waterborne Diseases

Vector-borne diseases (VBDs) and water-borne diseases (WBDs) are major public health challenges with growing concern under changing climate regimes. VBD and WBD form two epidemiologically distinct, yet climatologically intertwined, classes of infection. VBDs are infections transmitted to humans (and frequently other vertebrates) via the bite of infected haematophagous arthropods such as mosquitoes, ticks, sandflies, blackflies and tsetse flies. WHO lists malaria, dengue, chikungunya, yellow fever, Zika, lymphatic filariasis, schistosomiasis, onchocerciasis, Chagas disease, leishmaniasis and Japanese encephalitis among the most serious VBDs. Collectively, they account for *>17%* of all infectious diseases and at least *700,000* deaths each year; Malaria alone is responsible for ~619,000 deaths annually (Shin et al., 2024), while dengue generates >4 million clinically apparent cases with the only VBD disability-adjusted life-years (DALYs) still rising globally (Zhang et al., 2024). WBDs are infections acquired through ingestion, inhalation or dermal contact with water contaminated by pathogenic bacteria, viruses or parasites. Classical examples include cholera (*Vibrio cholerae*), typhoid fever (*Salmonella typhi*), hepatitis A virus, rotavirus diarrhoea and protozoan and helminthic infections such as cryptosporidiosis, giardiasis and schistosomiasis. The WHO attributes *≈485,000 diarrhoeal deaths* annually to unsafe water, sanitation and hygiene, with total water-related mortality exceeding *1.2 million* in 2016, including 300,000 children <5 years. Economic losses are estimated at US $12 billion per year through healthcare costs, lost productivity and impacts on tourism (Das et al., 2025). VBD and WBD are profoundly sensitive to environmental conditions, particularly temperature, humidity and precipitation patterns. Over the past half-century, a convergent body of entomological, hydrological and epidemiological research has demonstrated that anthropogenic warming, shifting precipitation regimes and the growing frequency of hydrometeorological extremes are reshaping the ecology, seasonality and geography of both VBDs and WBDs (IPCC, 2021; Mora et al., 2017; Patz et al., 2005). Temperature governs vector competence and pathogen replication; rainfall and drought modulate larval habitats and water quality; extreme events intensify multiple risks by displacing populations and disrupting control systems.

Vectorial capacity, the composite of vector abundance, daily survival, biting frequency and the pathogen's extrinsic incubation period (EIP) exhibits a strong, often nonlinear, temperature dependence. Laboratory and field data show that each 1 °C rise within the 23–34 °C window can shorten the EIP of *Plasmodium falciparum* in *Anopheles* mosquitoes by 8–10%, allowing infected females to become infectious several days earlier and markedly raising the basic reproduction number (Paaijmans et al., 2009). Comparable thermal responses have been documented for dengue, chikungunya and Zika viruses in *Aedes aegypti* and *Aedes albopictus* and for West Nile virus in *Culex* vectors (Delrieu et al., 2023). As these physiological accelerations accumulate across populations, they manifest as geographic and seasonal expansion. A multispecies model published in *The Lancet Planetary Health* estimated that climatic suitability for dengue transmission has risen from 58 to 132 countries since 1960, now placing more than half of the world's people at seasonal risk. WHO surveillance mirrors that projection: the global incidence of dengue has climbed 30-fold in five decades, with 50–100 million symptomatic infections and >4 million officially reported cases from 79 countries in 2023 (Jing & Wang, 2023). Thermal thresholds underpin similar poleward shifts for Lyme disease (northwards into southern Canada and high-latitude Europe) and for *Anopheles stephensi*, an urban malaria vector now detected well outside its South-Asian origin (Ryan et al., 2023).

Precipitation influences VBDs primarily through the creation and destruction of aquatic breeding sites. Heavy rains leave transient pools ideal for culicine mosquitoes; conversely, droughts trigger domestic water storage that inadvertently breeds *Aedes* larvae. A longitudinal study in Dhaka found that rainfall >200 mm month^{-1} was associated with a 15% rise in the combined incidence of dengue, malaria and chikungunya, with vector abundance peaking 2–4 weeks after rain events. In contrast, anomalously dry pre-monsoon months in Bangladesh drove households to store water, sustaining *Aedes* populations and contributing to the country's record (>1000 deaths) dengue epidemic in 2023 (Islam et al., 2021). Hydrological links are even more direct for WBDs. Floods, storm surges and intense rainfall events mobilise faecal pathogens from latrines, livestock yards and overwhelmed sewers into drinking-water networks; droughts concentrate contaminants and reduce residual chlorine in piped systems. Classic UK and US outbreak investigations document spikes of cryptosporidiosis, giardiasis and campylobacteriosis after heavy rainfall compromised surface or groundwater sources (Gillingham et al., 2023). A systematic review of 27 studies across Western Asia similarly found a consistently positive association between temperature spikes and WBD incidence, yet heterogeneous responses to rainfall and humidity, reflecting local sanitation capacity and pathogen ecology (Teymouri & Dehghanzadeh, 2022).

Heatwaves, tropical cyclones and large-scale climate oscillations (El Niño/La Niña) synchronise multiple risk factors. During the 2015–2016 super-El Niño, unusually warm, wet conditions across equatorial America coincided with social disruption from flooding, producing explosive co-epidemics of Zika, dengue and chikungunya; Brazil alone recorded 214,000 Zika cases and an unprecedented surge in microcephaly (Metelmann, 2019). Similar compound hazards were evident when Cyclone Idai's floods (Muleia et al., 2024) and Pakistan's monsoon deluge 2022 precipitated large cholera waves by destroying water and sanitation infrastructure (Noureen et al., 2022). The IPCC Sixth Assessment Report (WG II) warns that, under a high-emissions pathway (SSP5-8.5), WBD burdens could increase 15–40% by 2100 as flood frequency and severity escalate (IPCC, 2023). Climate change also acts directly on pathogens. *Vibrio cholerae* and *V. vulnificus* replicate faster in warmer coastal waters (>25 °C), lengthening the season for wound and seafood infections in temperate latitudes; fatal cases have occurred in the Baltic region in 2018 and the north-east United States in 2023. Freshwater warming fosters cyanobacterial blooms that release hepatotoxins, creating secondary gastrointestinal risks. For arboviruses, higher temperatures raise salivary gland infection rates and shorten viral EIP, compounding the vector-level accelerations discussed earlier.

Although the impacts of climate change are increasingly observable, they often interact with other environmental and social factors. Rapid urban growth supplies dense human hosts, increases impermeable surfaces that collect water and generates heat-island amplification; global trade moves vectors (e.g., *A. albopictus* eggs in used tyres) and pathogens across continents; and lapses in vector control, vaccination and WaSH infrastructure open ecological niches. A decomposition of 2007–2017 Disability-Adjusted Life Year (DALY) trends revealed that malaria, yellow fever and leishmaniasis declined >20% despite warming due to sustained control investments whereas dengue DALYs rose 26%, reflecting the combined pressures of urbanisation, inadequate mosquito control and climate suitability. Thus, climate change modulates, but does not single-handedly dictate, global disease trajectories.

8.3 Vector- and Waterborne Diseases in the Arabian Peninsula

Stretching from the southern shores of the Gulf to the Arabian Sea, the countries of the Arabian Peninsula share a hyper-arid climate, where annual rainfall rarely exceeds 100 mm, except in the southwestern highlands. The Arabian Peninsula is undergoing profound climatic transformations, making it one of the most climate-vulnerable regions in the world (Almazroui et al., 2021). Characterized by rising temperatures, prolonged heatwaves, changing precipitation patterns and an increase in extreme weather events, these shifts have severe and far-reaching consequences for human health (Mashat et al., 2018). The region, already recognized for its harsh desert environment, is witnessing an acceleration of these changes due to anthropogenic activities, including fossil fuel emissions, urbanization and land-use changes (Khan et al., 2023). The complex interplay between climate variability and public health underscores the urgent need to examine the growing health hazards in the Arabian Peninsula.

Regional climate-model ensembles project that mean surface temperatures are rising at roughly twice the global average and that the frequency of heatwaves surpassing 50 °C and short-duration cloud-burst rains will continue to rise throughout the twenty-first century (Melebari et al., 2021). These climatic shifts intersect with exceptionally rapid urbanisation, desalination-dependent water supply, high migrant-labour turnover together creates environmental niches and social conditions favourable to the emergence and redistribution of climate-sensitive infectious diseases. Mass gatherings such as the Hajj and Umrah, which draw millions of pilgrims each year and large refugee flows from neighbouring conflict zones add further layers of vulnerability by intensifying population mixing and straining water-sanitation infrastructure during peak climatic stress. VBD burdens vary markedly across the Peninsula, with dengue highest in Yemen and Saudi Arabia, malaria concentrated in Oman and Yemen and cutaneous leishmaniasis more evenly spread (Fig. 8.2).

Vector-Borne Diseases

Dengue: Although the Arabian Peninsula is predominantly arid, vector-borne infections have gained epidemiological significance over the last two decades as rising night-time minima and irregular precipitation create micro-habitats for mosquitoes, sandflies and ticks. Dengue fever, a mosquito-borne viral infection transmitted mainly by Aedes aegypti, was virtually unknown in Saudi Arabia until the early 1990s. The first confirmed case appeared in 1993 in Jeddah, likely introduced through international travellers, given the country's role as a major religious and economic hub (Fakeeh & Zaki, 2001). A year later, the Disease Control Division was established and the first outbreak occurred with 289 reported cases. Infections remained sporadic

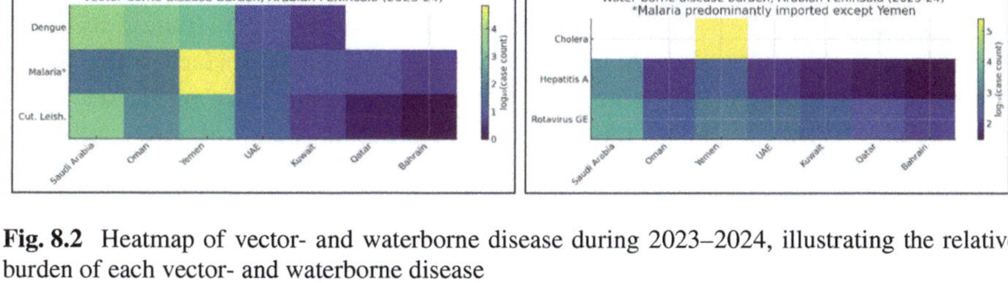

Fig. 8.2 Heatmap of vector- and waterborne disease during 2023–2024, illustrating the relative burden of each vector- and waterborne disease

through the late 1990s, with nearly no activity between 1998 and 2003. A resurgence occurred in 2004, marking a significant outbreak involving multiple serotypes (DENV-1, DENV-2 and DENV-3) and establishing endemic status in the western regions. Since then, the number of cases has continued to rise, particularly in Jeddah, Makkah and Jizan, fuelled by urbanization, water storage practices and favourable climatic conditions. Major surges occurred in 2013 and 2016, with over 4400 cases each year. In Makkah, dengue first appeared in 2004 and outbreaks among construction workers between 2008 and 2012 resulted in over 4000 cases. Madinah saw its first confirmed case in 2008, with a gradual rise to 27 confirmed cases in 2023. Riyadh, previously unaffected, reported its first-ever cases in 2022, indicating a geographical expansion. In Jizan, outbreaks have been recurrent since 2005, with over 4600 cases reported by 2020, predominantly involving DENV-2. Seasonal patterns align with rainfall, increasing mosquito breeding (Alyahya, 2023).

Malaria: Malaria by contrast to other VBDs, illustrates how determined control can offset climatic risk. Bahrain, Kuwait, Qatar and the UAE have reported zero indigenous cases since the mid-1990s; Oman and Saudi Arabia remain in the "pre-elimination" phase, with most infections imported by migrant workers from the Horn of Africa and South Asia. Nonetheless, ecological niche mapping indicates that the southwestern Asir and Jazan highlands regions that receive summer monsoon spillover from the Indian Ocean now satisfy temperature precipitation thresholds for seasonal *Anopheles* proliferation for up to 5 months of the year, a full month longer than in the 1980s. The recent detection of *Anopheles stephensi* in displaced-persons camps in Aden underscores the risk of urban malaria resurgence if vector surveillance lapses (Al-Awadhi et al., 2021).

Cutaneous leishmaniasis: This remains the most widely distributed sandfly-borne infection in the Peninsula. Saudi Arabia recorded 5687 cases in 1999, falling to 2549 by 2009 after intensified rodent control, yet ecological data reveal 25 resident sandfly species and persistent zoonotic foci in Riyadh, Eastern Province and the mountainous south-west (Abass et al., 2020). In Oman, incidence peaks have coincided with anomalously wet winters in 2010 and 2018, suggesting that pulse rainfall events expand rodent and sandfly habitats and trigger outbreak cycles several months later (Majidnia et al., 2023).

Water-Borne Diseases

Water-borne disease burdens also vary markedly across the Peninsula (Fig. 8.2). Cholera peaks in Yemen, while hepatitis A and rotavirus gastroenteritis are most common in Saudi Arabia and Oman, with much lower rates elsewhere. Water-borne infections in the rich Gulf monarchies are generally constrained by high-technology desalination and chlorination, yet vulnerabilities surface when floods compromise drainage or when migrant labour communities rely on tanker deliveries. Saudi Ministry of Health reports recurrent hepatitis A and rotavirus outbreaks among children in peri-urban settlements lacking piped water, with incidence surges following flash-flood events in 2009, 2008 in Jeddah and 2013 in Riyadh (Abid et al., 2023). Yemen presents the most critical WBD picture. Between 2016 and 2022 the country endured the largest recorded cholera epidemic in modern history, with more than 2.5 million suspected cases and over 4000 associated deaths. Epidemiological deconstruction shows case accelerations after each monsoon-driven flood season and sharp declines during prolonged dry spells, confirming a precipitation-amplified contamination cycle in a nation where only 50% of health facilities remain functional. Although 2021–2022 surveillance suggested a downward trend partly attributable to mass oral-cholera vaccination and Water, Sanitation and Hygiene (WaSH) interventions the risk of resurgence persists as climate projections foresee further intensification of extreme rainfall on Yemen's Red Sea escarpment (UNICEF, 2024). Coastal warming is also lengthening the season for non-cholera *Vibrio* infections in the western Gulf, with surveillance hospitals in Kuwait recording their first locally acquired *Vibrio vulnificus* septicemias during the exceptionally hot summer of 2020; sea-surface temperatures exceeded 31 °C for nine consecutive weeks, 2 weeks longer than the 1982–2012 climatology (Rafei et al., 2024).

Temporal Patterns Over the Last 15 Years

Synthesising surveillance and climatological datasets reveals four consistent tendencies since 2010. First, dengue incidence in Saudi Arabia has grown both numerically and spatially, with outbreaks in 2008, 2011, 2019 and 2022 progressively encroaching into inland cities and extending the transmission season to late autumn. Second, Oman's post-cyclone leishmaniasis spikes illustrate how episodic rainfall transforms arid ecologies into transient disease niches, a pattern similarly observed in Riyadh, where anomalous rainfall events have been associated with localized outbreaks of cutaneous leishmaniasis (Lotfy & Alsaqabi, 2016). Third, Yemen's cholera peaks (2017, 2020) align with the dual stressors of El Niño-enhanced rainfall and conflict-related water-system collapse. Fourth, extreme-heat and pluvial-flood events in the UAE during 2020 and again in the abnormally wet summer of 2023 temporarily overwhelmed drainage in low-income labour districts, prompting the Dubai Health Authority to report spikes in acute gastroenteritis and heat-related illness within days of the storms, though comprehensive published numbers remain limited.

8.4 Heatwaves and Public Health Burden

Heat waves defined as prolonged periods of excessively high temperatures have become more frequent and intense due to climate change. The human body's thermoregulation mechanisms become overwhelmed during extreme heat events, leading to multi-organ failure and increased mortality rates. Vulnerable groups, including the elderly, children, pregnant women and individuals with pre-existing health conditions, face the highest risks (Balmain et al., 2018; Khan & Lawrence, 2022). The phenomenon of rising night-time temperatures has added another layer of complexity to heat-related illnesses. This condition, commonly referred to as the urban heat island effect, occurs when metropolitan areas experience significantly higher temperatures than surrounding rural areas due to heat absorption by buildings, concrete and asphalt. In cities like Riyadh, Abu Dhabi and Kuwait City, night-time temperatures often remain above 35 °C, preventing the human body from recovering from daytime heat exposure (Elhacham & Alpert, 2021; Harris, 2019). Heat stress arises when the combined metabolic and environmental heat load exceeds the capacity of physiological and behavioural thermoregulation to maintain core temperature close to 37 °C. The primary autonomic defences cutaneous vasodilation and sweating progressively lose effectiveness as skin blood flow approaches its maximum and sweat output is constrained by dehydration and glandular fatigue. Laboratory studies conducted over the last decade confirm that the critical environmental limit for uncompensated heat strain in healthy adults is a humid-gauge wet-bulb temperature (T_{wb}) of roughly 30–31 °C, lower than the 35 °C formerly assumed in global risk assessments, with even tighter margins for older adults, children, pregnant women and people with chronic cardiovascular disease (Abdul-Nabi et al., 2025).

Analyses collated in the *Annual Review of Public Health* underscore a steep rise in heat-attributable morbidity and mortality. Extreme heat is now estimated to contribute to ~4,89,000 premature deaths per year worldwide and the number of acute health emergencies triggered by compound events (heat plus wildfire smoke, tropical cyclones or power outages) is increasing faster than demographic growth (Ebi et al., 2021). A bibliometric synthesis of 463 peer-reviewed articles on heat and health reveals that the research base itself mirrors this surge, with publications growing from <20 per year in the early 2000s to >110 per year by 2020; nevertheless, key exposure metrics such as night-time heat load, hydration status and individual clothing insulation remain under-documented in many world regions (Ncongwane et al., 2021). Workforce productivity has been significantly affected as well. The International Labour Organization projects that by 2030 the global economy will forfeit 2.2% of total working hours because ambient conditions exceed safety thresholds, with agriculture and construction accounting for almost 80% of the loss. Independent estimates place the 2023 penalty at roughly 512 billion potential labour-hours 49% above the 1990s baseline (Moreno, 2023). These global aggregates, however, conceal pronounced regional asymmetries that are widening as climate change intensifies.

Long-term perturbations to Earth's radiative balance are shifting the distribution of temperature extremes. The frequency of days above local 99th-percentile thresholds has already doubled in many subtropical latitudes and the likelihood of multi-day compound heat events is growing disproportionately fast (Ebi et al., 2021). Concurrent rises in atmospheric moisture amplify physiological risk because high specific humidity raises the evaporative resistance of air, impairing sweat evaporation. Recent global reanalyses show that since 2000 the average person has experienced ~50 additional days each year in which combinations of temperature and humidity crossed widely accepted danger limits for outdoor work (Matthews, 2018).

8.5 Human Health Impacts and Mortality from Extreme Heat in the Arabian Peninsula

The Arabian Peninsula already records some of the highest sensible heat loads on Earth, a product of intense solar radiation, subsidence under the descending branch of the Hadley cell and sea-to-land advection of humid air from the Arabian Gulf. A machine-learning analysis of long-term records from Dammam, Saudi Arabia, finds a statistically significant upward trend in specific humidity of 0.025 g kg^{-1} $year^{-1}$ (Sen's Slope), with the steepest seasonal increases in July–August (0.054–0.059 g kg^{-1} $year^{-1}$) and mean mid-summer maxima of 44 °C (Aldosary et al., 2025). Because evaporative cooling efficacy declines exponentially with humidity, these moisture trends convert directly into higher physiological heat stress. Analyses based on ERA5 (ECMWF Reanalysis version 5) a global climate reanalysis dataset produced by the European Centre for Medium-Range Weather Forecasts (ECMWF) indicate spatially coherent warming trends in extreme-temperature indices, with increases exceeding 1 °C per decade across much of northern Saudi Arabia, Kuwait and interior Oman. In some areas, local temperature anomalies have already reached up to 3 °C above the 1951–1980 baseline. Modelled population-exposure studies suggest that, even under the intermediate SSP2-4.5 scenario, more than 60% of residents in North Africa and the Arabian Peninsula will experience at least one episode each year in the "very strong" to "extreme" heat-stress classes of the Universal Thermal Climate Index (UTCI > 38 °C) by mid-century. Because evaporative cooling efficiency declines exponentially with humidity, the documented moisture trend over Gulf coastal zones is expected to push wet-bulb temperatures towards critical human limits on an ever-growing fraction of days (Hamed et al., 2024). The Arab countries in the Arabian Gulf will have the greatest relative annual increases, with particularly large changes projected for the latter half of the century under the SSP3-7.0 and SSP5-8.5 scenarios. By 2061–2080, Saudi Arabia could see a 13-fold increase in heat-related death rates even under a low emissions scenario (SPP1-2.6); however, this could rise to a 63-fold increase under the SSP5-8.5 scenario (Hajat et al., 2023) (Fig. 8.3).

An illustrative case is the 2015 heatwave in the Middle East, where temperatures soared to 54 °C in Mitribah, Kuwait and Ahvaz, Iran one of the highest recorded temperatures on Earth. The extreme conditions led to thousands of hospitalizations and hundreds of reported deaths, highlighting the devastating impact of prolonged heat exposure on human health. Two recent Mecca-centred investigations illuminate real-world consequences. A distributed-lag nonlinear study of 37,178 non-accidental deaths (2006–2015) documented a U-shaped temperature mortality curve with a minimum-mortality temperature (MMT) of 31.8 °C and showed that days exceeding 38 °C raised all-cause mortality significantly; altogether, 6.9% of deaths were attributable to temperature deviations, almost entirely on the heat side of the spectrum (Yezli et al., 2023). A systematic review of heat-exhaustion cases during successive Hajj pilgrimages synthesised 1194 clinically confirmed presentations. Patients arrived with core temperatures up to 40 °C, most frequently older

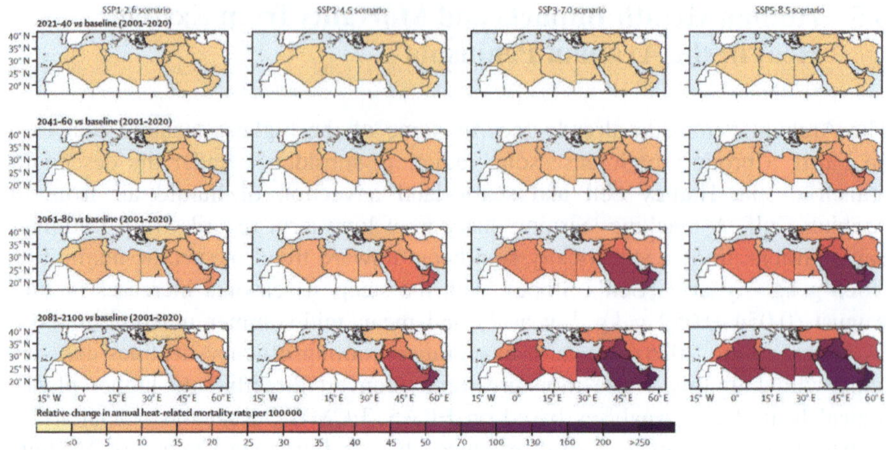

Fig. 8.3 Projected relative change in annual heat-related mortality rate (per 100,000 population) across the Middle East and North Africa under four CMIP6 SSP scenarios for four future periods relative to the 2001–2020 baseline as studied by Hajat et al. (2023). The color scale indicates percentage increases, with darker shades representing larger rises in heat-related deaths. *Source* Hajat et al. (2023)

overweight males from the Middle East and North Africa. Despite rapid cooling and hydration lowering the pooled case-fatality rate to just 0.11%, the review highlights substantial cardiac strain, electrolyte imbalance and acid–base disturbance even in non-fatal cases, reinforcing the need for early detection and treatment (Yezli et al., 2024). Extreme heat has also been linked to an increase in mental health issues, including anxiety, aggression and depression. A recent study showed that heat stress affects cognitive function and decision-making abilities, leading to higher risks of workplace and traffic accidents (Mazlomi et al., 2016). As a result, governments in the region have started implementing heat stress mitigation policies, such as mandatory work breaks during peak heat hours, though enforcement remains inconsistent. Outdoor and manual workers, particularly in construction, agriculture and oil industries and also athletes and civil protection employees have a higher risk of getting more exposed to heat stress. The long hours spent under direct sun exposure led to severe dehydration, chronic kidney disease and even fatalities in extreme cases. For instance, in Saudi Arabia and the UAE, many migrant laborers experience heat-related illnesses due to strenuous working conditions with limited access to cooling breaks and hydration (Ahmed et al., 2020; Al Hurini et al., 2024; Al-Bouwarthan et al., 2019).

The conjunction of escalating sea-surface temperatures and low-level moisture transport means that dangerous heat in the peninsula is increasingly humidity-driven. Studies across Gulf cities reveal that apparent temperatures now routinely exceed 50 °C during afternoon peaks and can remain above 31 °C wet-bulb well into the night owing to sea-breeze surges a phenomenon that prevents physiological recovery between work shifts. Construction-site monitoring in the UAE indicates that "no-work" periods prescribed by the Thermal Work Limit index have increased by 38% when comparing the baseline 2005–2009 period with 2020–2024 values. Meanwhile, agriculture and logistics workers along Kuwait's coast already endure ~20 days per summer with heat-index values $\geq$54 °C and the record air temperature of 54 °C at Mitribah Kuwait in 2016 remains the third-highest reliably measured worldwide (Abbas, 2024; Elhacham & Alpert, 2021; Raymond et al., 2024; Safieddine et al., 2022). Mass gatherings also compound exposure. During the June 2024 Hajj, wet-bulb temperatures in the Grand Mosque reached 33 °C and air temperatures topped 51.8 °C, contributing to more than 1300 excess deaths. Attribution studies suggest anthropogenic warming added ~2.5 °C to ambient conditions relative to a counterfactual climate (Saeed, 2024). Furthermore, climate models project that the frequency and intensity of extreme heat events during Hajj will increase in the coming decades (Pal & Eltahir, 2016; Ullah et al., 2024). The projected heat-related deaths in the Middle East and North Africa (MENA) region could increase by over 80% if global warming is not mitigated. In the absence of adequate adaptation strategies, heat-related deaths could surpass those from all other climate-related disasters combined. The impact of extreme heat on human mortality is already evident in Saudi Arabia and the UAE, where heat-related emergency room visits surge during summer (Schmidt, 2023). Reports indicate that a significant proportion of Kuwait's heat-related fatalities occur among migrant workers, particularly from South Asia, who often perform physically demanding jobs outdoors (Hajat et al., 2023). Many face poor working and housing conditions, with limited access to cooling, hydration, or healthcare services. Studies have also found a link between heat exposure and decreased cognitive performance, raising occupational safety concerns (Alahmad et al., 2022).

The Fig. 8.4 reveals a steep gradient in heat-related mortality across the Middle East and North Africa. Iran stands out with an annual rate of about 11 deaths per 100,000 population nearly three times higher than the next-ranked countries, Israel and Cyprus suggesting that a combination of high exposure and limited protective infrastructure magnifies risk. Mortality then declines roughly logarithmically through Türkiye, Iraq, Lebanon and Egypt, settling into a cluster of moderate values ($\approx$1–2 per 100,000) among Syria, Palestine and several North African states. Wealthier Gulf nations Qatar, UAE, Kuwait and Oman occupy the lower tail (<1 per 100,000), implying that extensive cooling access and health-system capacity can offset extreme meteorological conditions. This has led to surge in energy demand, primarily due to the widespread use of air conditioning to combat extreme heat like in UAE. The government has introduced initiatives like district cooling systems and green building standards (e.g., Estidama and the Dubai Green Building Regulations) to improve energy efficiency. However, rising urban temperatures continue to drive electricity consumption, raising long-term concerns about energy sustainability and carbon emissions (Thornton et al., 2021; Zhang et al., 2022). Urban heat island effects exacerbate cooling needs, especially in high-density areas with limited green infrastructure (Middel et al., 2014; Ziter et al., 2019). The integration of renewable energy sources, particularly solar, can offset the increasing energy burden.

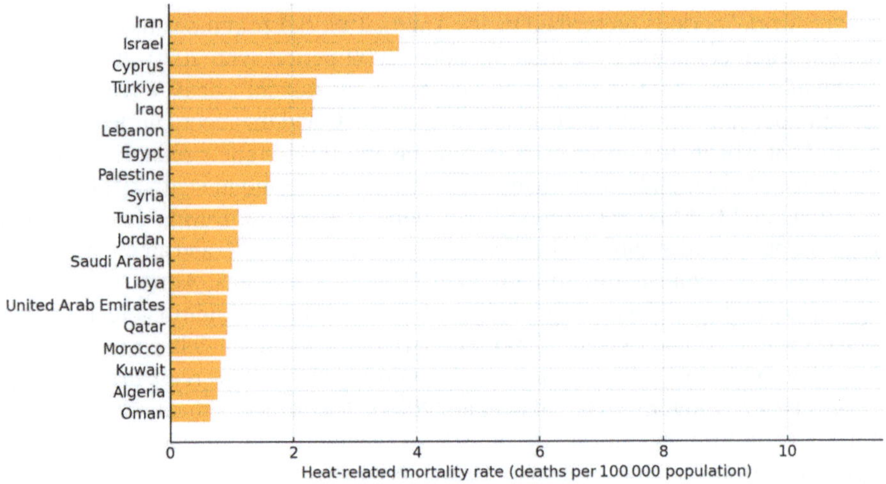

Fig. 8.4 Current annual heat related mortality rates in MENA. (Derived using data from Hajat et al., 2023)

8.6 Respiratory Diseases Due to Impacts of Air Quality on Climate Change

The U.S. Environmental Protection Agency (EPA, 2023) defines air pollution as the presence of substances or agents that degrade air quality and harm the atmosphere. Key pollutants such as particulate matter (PM), ground-level ozone (O_3), nitrogen dioxide (NO_2) and sulfur dioxide (SO_2) originate from both human and natural sources. Human activities are the primary contributors, with energy and agriculture each accounting for 25%, industry 21%, transportation 14% and other sources 16%. Urban transportation is a major emitter of NO_2, PM, SO_2 and organic aerosols. Natural sources like wildfires, volcanic eruptions and dust storms also release significant amounts of PM and SO_2 (Koengkan et al., 2021). Climate change and air pollution are closely interconnected, often intensifying each other's impacts. Global warming, driven primarily by greenhouse gas emissions such as carbon dioxide (CO_2), methane (CH_4) and nitrous oxide (N_2O), stems largely from fossil fuel use, deforestation and agriculture (IPCC, 2021). Atmospheric inversions, where warm air overlays cooler air, trap pollutants near the surface, especially during high-pressure systems, worsening air quality. Extreme weather events like heatwaves, storms and wildfires further exacerbate air pollution. Heatwaves accelerate photochemical reactions, increasing ground-level ozone (O_3), a known respiratory irritant. Rising temperatures also enhance O_3 formation by increasing photochemical activity and volatile organic compound (VOC) emissions (Hui et al., 2023). Storms contribute to dust and particulate emissions through disturbed surfaces and damaged infrastructure, while wildfires emit large quantities of PM, black carbon and other pollutants that can be transported over long distances (Hassan et al., 2023). Black carbon, in particular, degrades air quality and accelerates ice melt, reinforcing warming trends. While secondary organic aerosols (SOAs) formed in the atmosphere can reflect sunlight and provide some cooling, this effect is insufficient to offset the warming driven by greenhouse gases. These interactions form feedback loops that magnify both air quality degradation and associated health risks (Fig. 8.5).

The combined effects of climate change and air pollution pose serious global health threats, particularly in relation to respiratory diseases. Climate change exacerbates air pollution, collectively intensifying respiratory health risks. Studies highlight that air pollution significantly aggravates conditions like asthma, COPD, lung cancer and respiratory infections (Maung et al., 2022; Song et al., 2023). Climate-induced factors such as heatwaves heighten pollutant effects, increasing asthma incidence, particularly among youth (Eguiluz-Gracia et al., 2020). COPD-related mortality also rises with higher $PM_{2.5}$ and NO_2 levels, both influenced by climatic shifts (Chen et al., 2022; Yuan et al., 2023). Furthermore, climate-driven air pollution correlates with increased respiratory infections, including tuberculosis, influenced by variables like temperature, humidity and wind speed (Sun et al., 2023; Ward et al., 2021; Ziou et al., 2022).

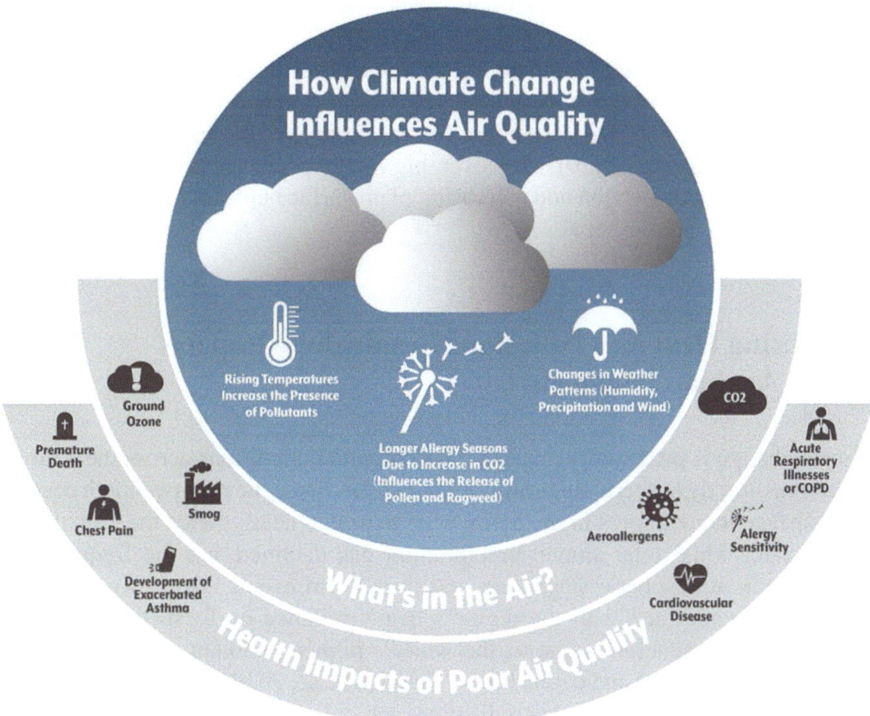

Fig. 8.5 Pathways through which climate change influences air quality and exacerbates health risks. *Source* https://www.aqi.in/blog/us/air-pollution-and-climate-change/

In addition, to these chronic impacts, climate change contributes to acute health crises. It leads to increased mortality from extreme weather events due to shifts in temperature and relative humidity (Ebi et al., 2021). It also exacerbates air pollution, which is linked to ~11.8% of global deaths, accounting for 6.7 million premature fatalities annually (WHO, 2022). Individuals with pre-existing respiratory conditions are especially vulnerable to these changing environmental stressors (WHO, 2021). Variations in meteorological conditions can intensify the effects of climate change by affecting the formation, transport and dispersion of pollutants (Ramanathan, 2020). A study by Joshi et al. (2020) determined that heatwaves and temperature inversions contributed to increases in ozone (O_3) concentrations by 28 and 14%, respectively. Zhang et al. (2021) reported that a 10 $\mu g/m^3$ rise in O_3 levels was linked to a 0.55% increase in mortality from respiratory diseases in China. Additionally, climate change-driven events such as heatwaves, storms and wildfires can worsen air pollution, increasing the risk of respiratory complications and hospital admissions (Kinney & Pinkerton, 2022; Walsh et al., 2020).

Changes in climate variables, particularly temperature and relative humidity (RH), can directly affect respiratory health through inflammation, oxidative stress and altered immune responses (Grigorieva & Lukyanets, 2021). High temperatures promote the formation of reactive oxygen species (ROS), triggering oxidative damage and inflammation in lung tissue (Jacobs et al., 2020). These processes can worsen respiratory conditions and modulate immune responses to infections (Grigorieva & Lukyanets, 2021). Rising temperatures and RH also enhance pollutant reactivity and ground-level ozone (O_3) formation, increasing respiratory irritation and asthma severity (d'Amato et al., 2020). Moreover, high RH can elevate pollutant concentrations by hindering dispersion and increasing plant surface uptake, further amplifying exposure risks (Guarnieri et al., 2023).

8.7 Rising Heat and Pollution: Respiratory Impacts in Arabian Peninsula

Climate change is profoundly altering environmental conditions across the Arabian Peninsula, directly exacerbating respiratory disease. Since the mid-twentieth century, mean surface temperatures in the MENA have risen at nearly twice the global land-average rate, while total annual precipitation has declined and the frequency of extreme-heat, atmospheric stagnation and dust-storm events has grown markedly. These shifts, combined with the region's heavy reliance on fossil-fuel extraction and combustion, generate some of the world's highest ambient concentrations of particulate matter (PM) and ozone (O_3) (Alahmad et al., 2020; Aldosary et al., 2025; Pasin & Öner, 2018). Rapid industrialization, vehicular emissions and dust mobilization further elevate PM, nitrogen dioxide (NO_2) and sulfur dioxide (SO_2). During intense summer anticyclonic heat events, accelerated photochemical reactions boost ground-level O_3, while stagnant air traps NO_x and volatile organic compounds from traffic, power plants and petrochemical complexes. Drought-driven land degradation in Iraq, Syria and the Rub' al-Khali now mobilizes vast quantities of sand and silt: annual dust-storm days in Kuwait, eastern Saudi Arabia and central Iraq have risen by roughly 30% since the early 2000s and are projected to double by mid-century under moderate-emission scenarios (Bolleter et al., 2021; Ozdemir et al., 2018). Natural and anthropogenic aerosols combine to create a complex air-quality profile with severe respiratory implications. Farahat's (2016) six-country study demonstrated significant variability in emission patterns across the Peninsula. Qatar led in per-capita CO_2 emissions (40.3 t $year^{-1}$ in 2010), whereas Saudi Arabia dominated cement-industry CO_2 output. AERONET observations reveal a pervasive mix of desert dust and anthropogenic aerosol, driving daily PM_{10} peaks above 280 $\mu g\ m^{-3}$ in major cities (Goudie, 2014; WHO, 2016). In Kuwait and the UAE, PM_{10} regularly exceeds WHO guidelines by five- to ten-fold and Riyadh and Doha frequently record 24-h PM_{10} above 500 $\mu g\ m^{-3}$ during dust events. Secondary organic aerosols from Gulf algal-bloom precursors further amplify O_3 and fine PM in coastal zones (Farahat, 2016).

Teather et al. (2013) conducted one of the region's few studies assessing combined health impacts of climate change and air pollution, focusing on Qatar. They reported that rising ambient temperatures significantly raised ground-level O_3 and PM, pollutants strongly linked to higher cardiovascular and respiratory disease prevalence. As in other Peninsula countries, Qatar's declining air quality from both dust events and anthropogenic emissions poses growing health risks. Spatially, these emission sources cluster around urban and industrial hubs (Fig. 8.6). Carbon monoxide and nitrogen oxides peak over major cities and petrochemical districts; sulfur dioxide concentrates at Gulf-coast refineries and power plants; volatile organic compounds mirror these industrial signatures into surrounding zones; and both black carbon and organic carbon are elevated near combustion sources and downwind of dust-storm regions. This pattern underscores the dual influence of human activities and natural dust mobilization on regional pollutant burdens.

Climate change also increases temperature inversions, trapping pollutants near the surface and further degrading air quality. Such combined exposures raise rates of asthma, COPD and cardiovascular conditions (Zhu et al., 2024). Time-series data from Riyadh show that each 10 $\mu g\ m^{-3}$ rise in daily $PM_{2.5}$ corresponds to a 4% increase in asthma emergency-department visits, while dust-storm days elevate COPD admissions by 15–25% in the Eastern Province (Tyrovolas et al., 2020). In Doha, dust episodes account for nearly one-third of pediatric asthma hospitalizations and Kuwait City records PM_{10} exceedances on nearly 20% of days, driving spikes in acute lower-respiratory infections and bronchial asthma (Al-Hemoud et al., 2018; Khaniabadi et al., 2017). Nationally, the UAE's mean annual $PM_{2.5}$ of ~90 $\mu g\ m^{-3}$ three times the interim WHO target correlates with a 13% adult asthma prevalence and elevated premature mortality from ischemic heart disease and stroke (MacDonald Gibson et al., 2013). Coastal cities in Morocco and Tunisia see summertime O_3 peaks ($\geq$80 ppb) that raise asthma admissions by about 7% and Lebanon's wildfire-smoke episodes trigger 34% surges in COPD exacerbations within 72 h of fires (Dalibalta et al., 2018; Khaniabadi et al., 2019).

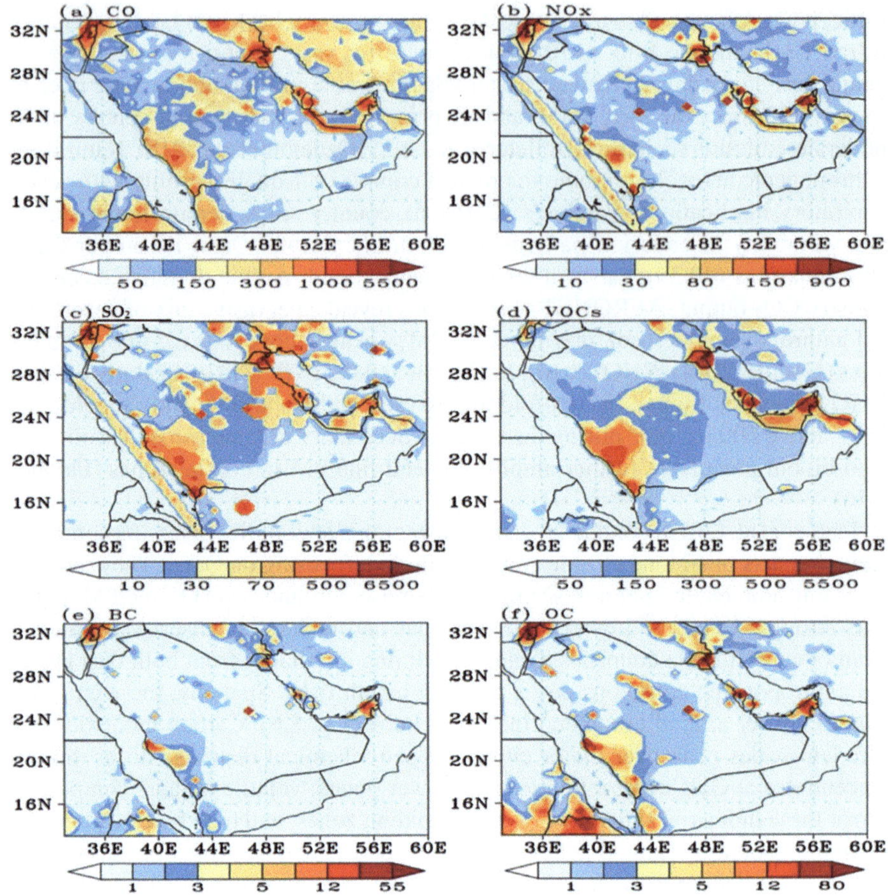

Fig. 8.6 The annual pollutant emissions (kg·m^{-2} × 106) during the period 1960–2020 over KSA studied by Al-Mutairi et al. (2023). *Source* Al-Mutairi et al. (2023)

From a physiological standpoint, heat and pollutants exert a dual assault on the respiratory system. Elevated temperatures dehydrate airway surfaces, thicken mucus and impair ciliary clearance, increasing epithelial permeability to inhaled particles and pathogens. Coarse PM_{10} laden with silica abrades bronchial tissues, while fine $PM_{2.5}$ carries soluble metals, endotoxins and allergens into the alveoli, provoking oxidative stress and inflammation. In vitro studies of Gulf dust show that exposure doubles interleukin-8 secretion in bronchial epithelial cells within hours and clinical data indicate that each 5 °C rise above 32 °C reduces FEV_1 by 3–6% in COPD patients (Tyrovolas et al., 2020). Recurrent exposure drives airway remodeling, bronchial hyperresponsiveness and progressive lung-function decline. Vulnerable subpopulations bear a disproportionate share of this burden. Children account for 60% of dust-related emergency visits in Qatari hospitals and exhibit long-term lung-growth deficits following prenatal $PM_{2.5}$ exposure in the UAE (Barakat-Haddad et al., 2015). Outdoor laborers including migrant construction workers may inhale annual PM_{10} loads exceeding 1300 $\mu g\ m^{-3}$, correlating with increased bronchitis and occupational silicosis. Hajj pilgrims, exposed to wet-bulb temperatures above 29 °C and pervasive dust, report respiratory symptoms in up to 40% of cases (Yeatts et al., 2012). Indoor exposures in North African oases where winter biomass burning and sand infiltration yield indoor PM comparable to outdoor levels drive unexpectedly high female COPD prevalence despite low smoking rates (Dastoorpoor et al., 2018).

Looking ahead, models under the SSP2-4.5 scenario project 2–4 °C warming and a 10% rainfall decline by 2050 across the Peninsula, conditions expected to double dust-storm days and raise summertime O_3 by 20%. Health-impact projections suggest climate-attributable cardiorespiratory mortality could climb from 92 to 137 per 100,000 population by mid-century (Khaniabadi et al., 2017). Yet targeted interventions have demonstrated significant benefits: Morocco's renewable-energy expansion cut urban $PM_{2.5}$ by 12 $\mu g\ m^{-3}$; Qatar's satellite-based dust early-warning system reduced emergency visits by 12%; and Abu Dhabi's blue-green corridors achieved 6 ppb O_3 reductions and 18% fewer self-reported asthma symptoms (Alahmad et al., 2020). Enforcement of Kuwait's 2018 Clean-Air Law and Saudi Arabia's draft emissions standards could avert one-sixth of projected COPD deaths by 2030, while emerging health-sector adaptation plans aim to lower asthma burdens by 10% through enhanced surveillance, hospital cooling upgrades and community alert systems (Aldosary et al., 2025; Dalibalta et al., 2018).

8.8 Socioeconomic Disparities and Health Risks

Socioeconomic disparities play a critical role in determining population's vulnerability to climate-related health risks. Low-income communities, migrant workers and displaced populations face disproportionate exposure to extreme heat, infectious diseases and inadequate living conditions (Rocklöv & Dubrow, 2020; Watts et al., 2015). These groups often reside in poorly insulated housing, lack access to air conditioning and have limited healthcare services, increasing their susceptibility to heat-related illnesses, respiratory diseases and vector-borne infections (Mazloumi et al., 2014; Mojahed et al., 2022). The individuals in the lowest income category usually reported the lowest rates of periodic preventive health check-ups. In contrast, a significantly higher proportion of respondents in the highest income category (≥30,000 SR) reported undergoing such check-ups (Al-Hanawi & Chirwa, 2021) (Fig. 8.7).

In the Gulf Cooperation Council (GCC) countries, migrant workers form a substantial part of the labour force, particularly in the construction, domestic and industrial sectors. These workers are often employed in high-risk outdoor jobs, making them more vulnerable to extreme heat. For example, reports from Qatar and the UAE indicate a significant rise in heatstroke and dehydration cases among migrant labourers working in open spaces without adequate heat protection (Pal &

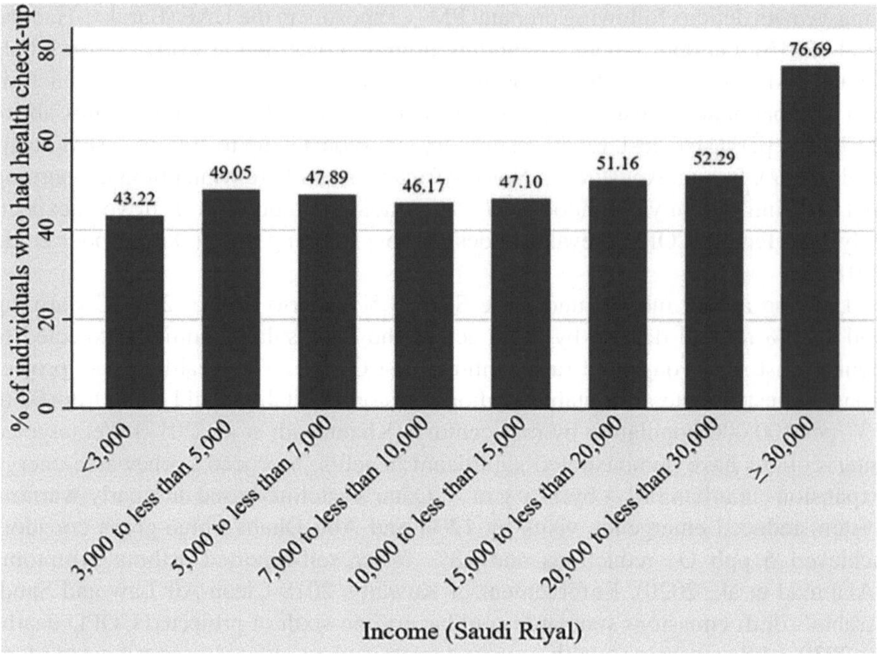

Fig. 8.7 Distribution of periodic preventive health check-ups across income. Adopted from Al-Hanawi and Chirwa (2021)

Eltahir, 2016; Ullah et al., 2024). Additionally, low-income communities frequently inhabit urban heat islands densely populated areas where concrete, asphalt and industrial emissions contribute to elevated temperatures. Studies in Riyadh and Dubai have shown that temperatures in poorer neighbourhoods are 5–7 °C higher than in greener, wealthier districts due to the lack of tree cover, reflective surfaces and green infrastructure (Mostafa & Alshahrani, 2024). This temperature disparity directly affects health outcomes, leading to increased emergency hospital visits for heat exhaustion and heatstroke (Rizmie et al., 2022). Beyond heat stress, access to clean drinking water and sanitation remains a pressing issue for vulnerable populations. In Yemen, where ongoing conflict has devastated public infrastructure, millions lack access to potable water, increasing their risk of waterborne diseases such as cholera and dysentery (Al-Hmani et al., 2024; Paz, 2019). Malnutrition, poor sanitation and inadequate medical care exacerbate these health challenges, placing immense strain on humanitarian aid efforts (Al-Muhannadi & Buheji, 2024; Smith & Haddad, 2015).

Impact on Elderly Population

Certain demographic groups are particularly at risk due to physiological and developmental factors. Elderly individuals, for instance, have reduced ability to regulate body temperature, making them more susceptible to heat exhaustion, dehydration and cardiovascular stress. The combination of extreme heat and pre-existing conditions such as hypertension, diabetes and respiratory diseases places older adults at an increased risk of fatal heat-related illnesses. During the 2015 heat wave in the Middle East, a significant rise in elderly mortality rate was recorded, particularly in Saudi Arabia and Iraq, where daytime temperatures exceeded 50 °C (Neira et al., 2023). Many elderly individuals, especially those in rural or lower-income urban areas, lacked access to cooling systems, making heat adaptation nearly impossible. Even in wealthier nations such as Kuwait and the UAE, emergency hospital visits for heat stress among seniors have surged during extreme weather events.

Impact on Children

Children are another highly vulnerable group. Infants and young children have an underdeveloped thermoregulation system, meaning their bodies absorb heat more quickly and struggle to cool down efficiently. This makes them highly prone to dehydration, heat rashes and febrile seizures. Furthermore, air pollution exacerbated by climate change increases the prevalence of pediatric asthma and respiratory infections. In highly urbanized cities like Doha and Abu Dhabi, rising levels of fine particulate matter ($PM_{2.5}$) and ozone have been linked to increased cases of childhood asthma hospitalizations (Isaifan, 2023).

Impact on Pregnant Women

Pregnant women also face increased risks due to extreme heat exposure, which has been linked to preterm births, low birth weight and pregnancy complications such as hypertension and gestational diabetes. Studies indicated that prolonged exposure to high temperatures during pregnancy can cause fetal stress and developmental delays. In Saudi Arabia, a study found that pregnant women exposed to excessive heat in their third trimester had a 30% higher risk of preterm labour compared to those in cooler environments (Konkel, 2019). Furthermore, dehydration in expectant mothers can negatively impact placental function, leading to fetal growth restrictions and increased risks of stillbirth.

In conclusion, growing evidence indicates that climate change is amplifying respiratory illness across the Arab and MENA region by intensifying heat stress, dust exposure and photochemical smog. However, significant knowledge gaps—particularly regarding long-term, combined heat–pollution impacts, indoor exposures and multi-pollutant interactions—continue to limit precise risk estimation and the effective targeting of vulnerable populations. Nonetheless, the region already possesses effective measures to protect lung health, deep cuts in greenhouse-gas emissions, stricter air-quality standards, expansion of renewable energy, urban greening and climate-informed health-system planning each yield tangible reductions in asthma, COPD and related hospitalisations. Investing in robust long-term population studies, enhanced indoor-air studies and integrated exposure models will improve the precision of cost benefit analyses, but even now, decisive policy action and coordinated adaptation can and must flatten the projected rise in respiratory disease as the climate continues to warm. Research exploring the combined impact of air pollution and climate change on human health remains limited compared to studies addressing these factors separately. Nevertheless, several international investigations have underscored the critical need to examine their synergistic interactions to fully understand their compounded effects on public health (Dionisio et al., 2017; Fiore et al., 2015; Sá et al., 2016).

8.9 Adaptation and Mitigation Strategies

The growing link between climate change and the spread of infectious diseases in the Arabian Peninsula calls for a twofold strategy one that quickly addresses the current risks and another that focuses on long-term efforts to reduce the root causes of future threats. Both global guidance and local studies highlight that any effective response must involve multiple sectors, ensure equity across communities and be built on strong monitoring systems. At the same time, it's important to aim for added benefits by aligning public health goals with actions in areas like clean energy, water security and city planning. This section outlines the main pillars of such an approach, using insights from international reviews, WHO recommendations and local case examples. According to the WHO, a climate-resilient health system is one that can anticipate, respond to and recover from climate-related shocks and pressures, while also adapting over time. To apply this idea to the Arabian Peninsula, the focus should be on three key areas: (i) collecting detailed data on both diseases and climate to guide decision-making, (ii) building strong systems for clean water, sanitation and controlling disease-carrying insects and (iii) ensuring good coordination between health, weather, environmental and municipal services. Since adaptation and mitigation often overlap, it's also essential to weigh their impacts carefully. For instance, air conditioning helps people cope with extreme heat but may also contribute to more emissions. Therefore, any policy decisions must consider both the trade-offs and the potential for joint benefits.

Climate-Informed Surveillance and Early-Warning Systems

Improving both symptom-based and lab-based disease tracking and combining this information with weather forecasts and satellite data is often mentioned as one of the most effective early steps to deal with climate-related health risks. In Saudi Arabia, the Health Electronic Surveillance Network already collects real-time updates on diseases like dengue, leishmaniasis and cholera (Almayahi et al., 2021). If this system were connected with local weather forecasts for temperature and humidity, it could help send out location-specific health warnings 3–4 weeks before the time when mosquitoes and other carriers of disease are most active. In Yemen, a similar approach has been used for cholera. By combining rainfall forecasts with mobile health data, health officials were able to cut the time between detecting a threat and responding to it by 40% during the most recent outbreak. Sharing data across countries through the GCC Public Health Committee could also help countries like Kuwait, Bahrain and Qatar where less health information is currently available benefit from Saudi Arabia's efforts to map disease-carrying insects.

Integrated Vector Management and Personal Protection

WHO's Global Vector Control Response 2017–2030 emphasises integrated vector management, community engagement and judicious insecticide use. For the Peninsula, the priority bundle includes indoor residual spraying and larviciding in residual malaria foci of Jazan and Asir; Wolbachia-based *Aedes* suppression currently piloted in Jeddah; and cross-border tick surveillance to curb Crimean Congo haemorrhagic fever emergence along livestock corridors. Personal protection repellents, insecticide-treated bednets for migrant workers, window screening remains cost-effective, yet uptake among low-income expatriate communities is uneven and requires behaviour-change communication tailored to linguistic diversity.

Water, Sanitation and Hygiene (WaSH) Resilience

Systematic reviews show that improved WaSH infrastructure and household-level water treatment are among the most proven interventions for reducing water-borne disease under climate stress. Gulf cities benefit from near-universal piped supply, but peri-urban settlements and labour camps reliant on tanker deliveries are vulnerable when flash floods contaminate surface reservoirs or overwhelm storm drains, as documented in Jeddah's 2009 and 2018 outbreaks of hepatitis A. Adaptation priorities include flood-proofing desalination intakes, expanding dual-reticulation networks for potable and non-potable water and enforcing minimum residual-chlorine standards during peak-temperature months when bacterial growth accelerates.

Urban Planning, Drainage and Heat-Island Mitigation

Fast-growing cities like Riyadh, Dubai and Doha have expanded so quickly that their drainage systems have been outpaced by the speed of urbanization. This has led to the creation of water-filled areas that make it easier for mosquitoes like Aedes to breed, while also causing pollution to build up in stormwater. Studies from Dhaka and other large cities show that when rainfall goes above 200 mm in a month, the number of diseases spread by mosquitoes can increase by around 15% especially if stormwater is not properly managed. To deal with these challenges, city planners are focusing on using paving materials that let water soak into the ground, building green spaces and water systems to handle drainage better and using cooling designs that help reduce the extra heat trapped in cities. These steps not only help control flooding and pollution but also lower temperatures, making it harder for mosquitoes to grow quickly and spread disease.

Climate-Resilient Health-System Capacity

Regional hospitals already rely heavily on air conditioning to keep their facilities cool. However, a review by Teymouri and Dehghanzadeh (2022) points out that much more needs to be done. This includes training healthcare staff to better understand how climate affects disease patterns, preparing for sudden increases in patients during heatwaves or floods and keeping enough vaccines in stock for high-risk areas such as oral cholera vaccines for Yemen or yellow fever vaccines for pilgrims coming from regions where the disease is common. In addition, models that explore possible future scenarios show that quick diagnosis and treatment, along with targeted mosquito control, can stop the spread of diseases like dengue in the early stages. Acting early while the virus's ability to spread is still moderate can help avoid large outbreaks, even if future climate conditions become more extreme.

Policy Integration and Financing

Progress in protecting public health from climate risks will slow down without steady funding and strong coordination across different sectors. During the first-ever Health Day at Conference of the Parties, 28th session (COP28), governments and their partners committed around US$1 billion to support climate-related health efforts. These include building local capacity, improving disease tracking and encouraging innovation. This funding presents a chance for the GCC to jointly support pilot adaptation projects in lower-income countries like Yemen and Oman. To make the most of such efforts, national climate plans should include health benefits in all major sectors, such as energy, transport and housing. For example, the energy used for desalination should be matched with increased use of renewable sources. Similarly, building rules that help keep cities cooler should also support efforts to control mosquito populations and reduce the spread of diseases.

Community Engagement and Education

Community based programmes, such as water safety lessons in schools and clean up campaigns led by mosques before the rainy season, help improve participation and give support to vulnerable groups. Studies show that raising public awareness encourages people to remove mosquito breeding spots at home and avoid unsafe water storage, especially when messages are shared in the native languages of migrant communities. In Yemen, combining hygiene education with the distribution of chlorine tablets led to a 27% drop in diarrhoea cases in displacement camps within 6 months, showing that simple, low cost measures can still be effective during times of conflict.

Mitigation Co-benefits

While adaptation dominates near-term planning, cutting greenhouse-gas emissions will attenuate the long-run expansion of climate-suitable niches for vectors and water pathogens. The Saudi Green Initiative's targets to plant ten billion trees and reach net-zero by 2060 and the UAE's aim for 44% renewables by 2050, can modestly reduce regional warming and indirectly slow VBD spread. Decarbonising municipal fleets and enhancing public transport concurrently shrink diesel storage pools that serve as mosquito habitats, providing an immediate ancillary benefit.

8.10 Conclusion

The Arabian Peninsula stands is confronting a period of escalating climate threats with serious implications for public health, economic resilience and regional stability. Rising temperatures, water scarcity and the proliferation of climate-sensitive diseases disproportionately affect the most vulnerable populations, emphasizing the urgency of climate-adaptive health strategies. A coordinated, multi-sectoral approach is essential one that integrates climate risk assessments into urban planning, enforces workplace protections, strengthens healthcare infrastructure and fosters regional cooperation. Effective adaptation depends on both top-down policy initiatives and bottom-up community engagement, ensuring that responses are inclusive, locally relevant and sustainable. By investing in climate-smart public health systems, promoting sustainable cooling technologies and prioritizing cross-border collaboration, the region can not only mitigate immediate health risks but also pave the way for long-term resilience. The path forward is clear proactive, inclusive and scientifically informed action must be taken today to protect the health and future of generations to come.

References

Abass, E., Al-Hashem, Z., & Yamani, L. Z. (2020). Leishmaniasis in Saudi Arabia: Current situation and future perspectives. *Pakistan Journal of Medical Sciences, 36*(4), 836.

Abbas, W. (2024). Urban climate of the Middle East and North Africa region: A comprehensive anatomization. In *Hydroclimatic extremes in the middle east and North Africa* (pp. 177–228). Elsevier.

Abdul-Nabi, S. S., Al Karaki, V., Khalil, A., & El Zahran, T. (2025). Climate change and its environmental and health effects from 2015 to 2022: A scoping review. *Heliyon, 11*, 533.

Abid, I., Blanco, A., Al-Otaibi, N., Guix, S., Costafreda, M. I., Pintó, R. M., & Bosch, A. (2023). Dynamic and seasonal distribution of enteric viruses in surface and well water in Riyadh (Saudi Arabia). *Pathogens, 12*(12), 1405. https://doi.org/10.3390/pathogens12121405

Ahmed, H. O., Bindekhain, J. A., Alshuweihi, M. I., Yunis, M. A., & Matar, N. R. (2020). Assessment of thermal exposure level among construction workers in UAE using WBGt, HIS and TWL indices. *Industrial Health, 58*(2), 170–181. https://doi.org/10.2486/indhealth.2019-0149

Al Hurini, A., Nemmar, A., Moawad, K., et al. (2024). The impact of increased heat on the physical, mental, and social health domains of adults in the United Arab Emirates in 2024. *Cureus, 16*(12), e74951. https://doi.org/10.7759/cureus.74951

Alahmad, B., Shakarchi, A. F., Khraishah, H., Alseaidan, M., Gasana, J., Al-Hemoud, A., et al. (2020). Extreme temperatures and mortality in Kuwait: Who is vulnerable? *Science of the Total Environment, 732*, 139289. https://doi.org/10.1016/j.scitotenv.2020.139289

Alahmad, B., Vicedo-Cabrera, A. M., Chen, K., Garshick, E., Bernstein, A., & Koutrakis, P. (2022). Climate change and health in Kuwait: Temperature and mortality projections under different climatic scenarios. *Environmental Research Letters, 17*(7), 074012. https://doi.org/10.1088/1748-9326/ac70e3

Al-Awadhi, M., Ahmad, S., & Iqbal, J. (2021). Current status and the epidemiology of malaria in the middle east region and beyond. *Microorganisms, 9*(2), 338. https://doi.org/10.3390/microorganisms9020338

Al-Bouwarthan, M., Quinn, M. M., Kriebel, D., & Wegman, D. H. (2019). Assessment of heat stress exposure among construction workers in the hot desert climate of Saudi Arabia. *Annals of Work Exposures and Health, 63*(5), 505–520. https://doi.org/10.1093/annweh/wxz019

Aldosary, A. S., Al-Ramadan, B., Kafy, A. A., Altuwaijri, H. A., & Rahaman, Z. A. (2025). Forecasting climate risk and heat stress hazards in arid ecosystems: Machine learning and ensemble models for specific humidity prediction in Dammam, Saudi Arabia. *Natural Hazards, 15*, 1–29. https://doi.org/10.1007/s11069-025-06540-7

Al-Hanawi, M. K., & Chirwa, G. C. (2021). Economic analysis of inequality in preventive health check-ups uptake in Saudi Arabia. *Frontiers in Public Health, 9*, 745356. https://doi.org/10.3389/fpubh.2021.745356

Al-Hemoud, A., Al-Dousari, A., Al-Shatti, A., Al-Khayat, A., Behbehani, W., & Malak, M. (2018). Health impact assessment associated with exposure to PM_{10} and dust storms in Kuwait. *Atmosphere, 9*(1), 6. https://doi.org/10.3390/atmos9010006

Al-Hmani, A., Ben-Jamaa, N., Kharroubi, A., Agoubi, B., & Alwabr, M. A. G. (2024). Case-control study of drinking water quality in Yemen. *Eastern Mediterranean Health Journal, 30*(3), 212–220.

Almayahi, Z. K., Alswaidi, F., & Alzahrani, A. (2021). Perception of the health surveillance users on the health electronic surveillance network (HESN), Saudi Arabia, 2016. *Journal of the Egyptian Public Health Association, 96*(1), 17.

Almazroui, M., Saeed, F., Saeed, S., Ismail, M., Ehsan, M. A., Islam, M. N., et al. (2021). Projected changes in climate extremes using CMIP6 simulations over SREX regions. *Earth Systems and Environment, 5*(3), 481–497. https://doi.org/10.1007/s41748-021-00250-5

Al-Muhannadi, K., & Buheji, M. (2024). Keeping the rhythm of employing social problem-solving to sustain resilience: Cases from Gaza during genocide (2023–2024). *Community Practitioner, 21*(12), 1–2.

Al-Mutairi, M., Al-Otaibi, N., Saber, A., Abdel Basset, H., & Morsy, M. (2023). Climatological study of air pollutant emissions in Saudi Arabia. *Atmosphere, 14*(4), 729.

Alyahya, H. S. (2023). Prevalence of dengue fever in Saudi Arabia: Jeddah as a case study. *Entomological Research, 53*(12), 539–553. https://doi.org/10.1111/1748-5967.12726

Anjum, G., & Aziz, M. (2025). Climate change and gendered vulnerability: A systematic review of women's health. *Women's Health, 21*, 17455057251323644. https://doi.org/10.1177/17455057251323645

Balmain, B. N., Jay, O., Morris, N. R., Shiino, K., Stewart, G. M., Jayasinghe, R., Chan, J., & Sabapathy, S. (2018). Thermoeffector responses at a fixed rate of heat production in heart failure patients. *Medicine and Science in Sports and Exercise, 50*(3), 417–426. https://doi.org/10.1249/MSS.0000000000001455

Barakat-Haddad, C., Zhang, S., Siddiqua, A., & Dghaim, R. (2015). Air quality and respiratory health among adolescents from the United Arab Emirates. *Journal of Environmental and Public Health, 2015*, 284595. https://doi.org/10.1155/2015/284595

Bolleter, J., Grace, B., Hooper, P., & Foster, S. (2021). Wet-bulb temperature and sea-level rise in the United Arab Emirates: Planning responses. *Planning Practice and Research, 36*(4), 408–429. https://doi.org/10.1080/02697459.2021.1879731

Chen, Z., Liu, N., Tang, H., Gao, X., Zhang, Y., Kan, H., et al. (2022). Health effects of exposure to sulfur dioxide, nitrogen dioxide, ozone, and carbon monoxide between 1980 and 2019: A systematic review and meta-analysis. *Indoor Air, 32*(11), e13170. https://doi.org/10.1111/ina.13170

Cianconi, P., Betrò, S., & Janiri, L. (2020). The impact of climate change on mental health: A systematic descriptive review. *Frontiers in Psychiatry, 11*, 490206. https://doi.org/10.3389/fpsyt.2020.490206

d'Amato, G., Chong-Neto, H. J., Monge Ortega, O. P., Vitale, C., Ansotegui, I., Rosario, N., et al. (2020). The effects of climate change on respiratory allergy and asthma induced by pollen and mold allergens. *Allergy, 75*(9), 2219–2228. https://doi.org/10.1111/all.14476

Dalibalta, S., Samara, F., Qadri, H., & Adouchana, H. (2018). Potential causes of asthma in the United Arab Emirates: Drawing insights from the Arabian Gulf. *Reviews on Environmental Health, 33*(2), 205–212. https://doi.org/10.1515/reveh-2017-0016

Das, S., Malhari, M., Starkl, M., Brunner, N., Mishra, P., & Sahoo, S. S. (2025). Health and economic consequences due to inadequate water quality: A case study from eastern India. *Journal of Water and Health, 23*(3), 397–412. https://doi.org/10.2166/wh.2025.093

Dastoorpoor, M., Idani, E., Goudarzi, G., & Khanjani, N. (2018). Acute effects of air pollution on spontaneous abortion, premature delivery, and stillbirth in Ahvaz, Iran: A time-series study. *Environmental Science and Pollution Research, 25*, 5447–5458. https://doi.org/10.1007/s11356-017-0743-3

Delrieu, M., Martinet, J. P., O'Connor, O., Viennet, E., Menkes, C., Burtet-Sarramegna, V., et al. (2023). Temperature and transmission of chikungunya, dengue, and Zika viruses: A systematic review of experimental studies on *Aedes aegypti* and *Aedes albopictus*. *Current Research in Parasitology and Vector-Borne Diseases, 4*, 100139. https://doi.org/10.1016/j.crpvbd.2023.100139

Dionisio, K. L., Nolte, C. G., Spero, T. L., Graham, S., Caraway, N., Foley, K. M., & Isaacs, K. K. (2017). Characterizing the impact of projected changes in climate and air quality on human exposures to ozone. *Journal of Exposure Science and Environmental Epidemiology, 27*(3), 260–270. https://doi.org/10.1038/jes.2016.69

Ebi, K. L., & Hess, J. J. (2020). Health risks due to climate change: Inequity in causes and consequences. *Health Affairs, 39*(12), 2056–2062. https://doi.org/10.1377/hlthaff.2020.01125

Ebi, K. L., Vanos, J., Baldwin, J. W., Bell, J. E., Hondula, D. M., Errett, N. A., et al. (2021). Extreme weather and climate change: Population health and health system implications. *Annual Review of Public Health, 42*(1), 293–315. https://doi.org/10.1146/annurev-publhealth-012420-105026

Eguiluz-Gracia, I., Mathioudakis, A. G., Bartel, S., Vijverberg, S. J., Fuertes, E., Comberiati, P., et al. (2020). The need for clean air: The way air pollution and climate change affect allergic rhinitis and asthma. *Allergy, 75*(9), 2170–2184. https://doi.org/10.1111/all.14177

Elhacham, E., & Alpert, P. (2021). Temperature patterns along an arid coastline experiencing extreme and rapid urbanization: Case study: Dubai. *Science of the Total Environment, 784*, 147168. https://doi.org/10.1016/j.scitotenv.2021.147168

Environmental Protection Agency (EPA). (2023). *Focus on local authority environmental enforcement: Performance report 2022*. Environmental Protection Agency, Ireland. https://www.epa.ie/publications/compliance--enforcement/public-authorities/Focuson-Local-Authority-Environmental-Enforcement---Performance-Report-2022.pdf

Fakeeh, M., & Zaki, A. M. (2001). Virologic and serologic surveillance for dengue fever in Jeddah, Saudi Arabia, 1994–1999. *The American Journal of Tropical Medicine and Hygiene, 65*(6), 764–767.

Farahat, A. (2016). Air pollution in the Arabian Peninsula (Saudi Arabia, the United Arab Emirates, Kuwait, Qatar, Bahrain, and Oman): Causes, effects, and aerosol categorization. *Arabian Journal of Geosciences, 9*, 1–17. https://doi.org/10.1007/s12517-015-2103-1

Fiore, A. M., Naik, V., & Leibensperger, E. M. (2015). Air quality and climate connections. *Journal of the Air and Waste Management Association, 65*(6), 645–685. https://doi.org/10.1080/10962247.2015.1040526

Gillingham, E. L., Lake, I., Iacono, G. L., & Nichols, G. (2023). Effect of climate change on infectious diseases in the UK. In *Health effects of climate change (HECC) in the UK: 2023 report*. https://www.gov.uk/government/publications/climate-change-infectious-disease-risk-assessment-in-the-uk

Gkouliaveras, V., Kalogiannidis, S., Kalfas, D., & Kontsas, S. (2025). Effects of climate change on health and health systems: A systematic review of preparedness, resilience, and challenges. *International Journal of Environmental Research and Public Health, 22*(2), 232. https://doi.org/10.3390/ijerph22020232

Goudie, A. S. (2014). Dust storms and human health. *Clean Air Journal, 24*(2), 7.

Grigorieva, E., & Lukyanets, A. (2021). Combined effect of hot weather and outdoor air pollution on respiratory health: Literature review. *Atmosphere, 12*(6), 790. https://doi.org/10.3390/atmos12060790

Guarnieri, G., Olivieri, B., Senna, G., & Vianello, A. (2023). Relative humidity and its impact on the immune system and infections. *International Journal of Molecular Sciences, 24*(11), 9456. https://doi.org/10.3390/ijms24119456

Hajat, S., et al. (2023). Current and future trends in heat-related mortality in the MENA region: A health impact assessment. *The Lancet Planetary Health, 7*(4), e282–e290. https://doi.org/10.1016/S2542-5196(23)00013-1

Hamed, M. M., Al-Hasani, A. A. J., Nashwan, M. S., Saadi, Z., & Shahid, S. (2024). Assessing the growing threat of heat stress in the North Africa and Arabian Peninsula region connected to climate change. *Journal of Cleaner Production, 447*, 141639. https://doi.org/10.1016/j.jclepro.2023.141639

Harris, P. G. (Ed.). (2019). *Climate change and ocean governance: Politics and policy for threatened seas*. Cambridge University Press. https://doi.org/10.1017/9781108502238

Hassan, M. A., Mehmood, T., Liu, J., Luo, X., Li, X., Tanveer, M., et al. (2023). A review of particulate pollution over Himalaya region: Characteristics and salient factors contributing ambient PM pollution. *Atmospheric Environment, 294*, 119472. https://doi.org/10.1016/j.atmosenv.2022.119472

Hui, K., Yuan, Y., Xi, B., & Tan, W. (2023). A review of the factors affecting the emission of the ozone chemical precursors VOCs and NO_x from the soil. *Environment International, 172*, 107799. https://doi.org/10.1016/j.envint.2023.107799

IPCC. (2021). Climate change 2021: The physical science basis. In V. Masson-Delmotte, P. Zhai, A. Pirani, et al. (Eds.), *Contribution of working group I to the sixth assessment report of the intergovernmental panel on climate change*. Cambridge University Press. https://doi.org/10.1017/9781009157896

IPCC. (2023). Summary for policymakers. In Core Writing Team, H. Lee, & J. Romero (Eds.), *Climate change 2023: Synthesis report. Contribution of working groups I, II and III to the sixth assessment report of the intergovernmental panel on climate change*. IPCC. https://doi.org/10.59327/IPCC/AR6-9789291691647.001

Isaifan, R. J. (2023). Air pollution burden of disease over highly populated states in the Middle East. *Frontiers in Public Health, 10*, 1002707. https://doi.org/10.3389/fpubh.2022.1002707

Islam, S., Haque, C. E., Hossain, S., & Hanesiak, J. (2021). Climate variability, dengue vector abundance and dengue fever cases in Dhaka, Bangladesh: A time-series study. *Atmosphere, 12*(7), 905. https://doi.org/10.3390/atmos12070905

Jacobs, P. J., Oosthuizen, M. K., Mitchell, C., Blount, J. D., & Bennett, N. C. (2020). Heat and dehydration-induced oxidative damage and antioxidant defenses following incubator heat stress and a simulated heat wave in wild-caught four-striped field mice (*Rhabdomys dilectus*). *PLoS ONE, 15*, e0242279. https://doi.org/10.1371/journal.pone.0242279

Jing, Q., & Wang, M. (2023). Progress in dengue epidemiology. In *Progress in China epidemiology: Volume 1* (pp. 121–137). Springer. https://doi.org/10.1007/978-981-19-6026-6_9

Joshi, M., Goraya, H., Joshi, A., & Bartter, T. (2020). Climate change and respiratory diseases: A 2020 perspective. *Current Opinion in Pulmonary Medicine, 26*(2), 119–127.

Khan, T., & Lawrence, A. J. (2022). Health risk assessment associated with air pollution through technological interventions: A futuristic approach. In *Integrating IoT and AI for indoor air quality assessment* (pp. 149–167). Springer. https://doi.org/10.1007/978-3-030-96495-5_7

Khan, N., Ma, J., Zhang, H., & Zhang, S. (2023). Climate change impact on sustainable agricultural growth: Insights from rural areas. *Atmosphere, 14*(8), 1194. https://doi.org/10.3390/atmos14081194

Khaniabadi, Y. O., Fanelli, R., De Marco, A., Daryanoosh, S. M., Kloog, I., Hopke, P. K., et al. (2017). Hospital admissions in Iran for cardiovascular and respiratory diseases attributed to the Middle Eastern dust storms. *Environmental Science and Pollution Research, 24*, 16860–16868. https://doi.org/10.1007/s11356-017-9269-8

Khaniabadi, Y. O., Sicard, P., Takdastan, A., Hopke, P. K., Taiwo, A. M., Khaniabadi, F. O., et al. (2019). Mortality and morbidity due to ambient air pollution in Iran. *Clinical Epidemiology and Global Health, 7*(2), 222–227. https://doi.org/10.1016/j.cegh.2018.07.003

Kinney, P. L., & Pinkerton, K. E. (2022). Heatwaves and air pollution: A deadly combination. *American Journal of Respiratory and Critical Care Medicine, 206*(9), 1060–1062. https://doi.org/10.1164/rccm.202208-1463ED

Koengkan, M., Fuinhas, J. A., & Silva, N. (2021). Exploring the capacity of renewable energy consumption to reduce outdoor air pollution death rate in Latin America and the Caribbean region. *Environmental Science and Pollution Research, 28*, 1656–1674. https://doi.org/10.1007/s11356-020-10501-6

Konkel, L. (2019). Taking the heat: Potential fetal health effects of hot temperatures. *Environmental Health Perspectives, 127*(10), 102002. https://doi.org/10.1289/EHP6221

Lotfy, W. M., & Alsaqabi, S. M. (2016). Climate change and epidemiology of human parasitoses in Saudi Arabia: A review. *Journal of Coastal Life Medicine, 4*(7), 580–588.

Lu, T. S., Olsen, J., & Kinney, P. L. (2021). Climate change and temperature-related mortality: Implications for health-related climate policy. *Biomedical and Environmental Sciences, 34*(5), 379–386. https://doi.org/10.3967/bes2021.048

Ma, J., Guo, Y., Gao, J., Tang, H., Xu, K., Liu, Q., & Xu, L. (2022). Climate change drives the transmission and spread of vector-borne diseases: An ecological perspective. *Biology, 11*(11), 1628. https://doi.org/10.3390/biology11111628

MacDonald Gibson, J., Thomsen, J., Launay, F., Harder, E., & DeFelice, N. (2013). Deaths and medical visits attributable to environmental pollution in the United Arab Emirates. *PLoS ONE, 8*(3), e57536. https://doi.org/10.1371/journal.pone.0057536

Majidnia, M., Ahmadabadi, Z., Zolfaghari, P., & Khosravi, A. (2023). Time series analysis of cutaneous leishmaniasis incidence in Shahroud based on ARIMA model. *BMC Public Health, 23*(1), 1190. https://doi.org/10.1186/s12889-023-15819-6

Mashat, A. W. S., Alamoudi, A. O., Awad, A. M., & Assiri, M. E. (2018). Seasonal variability and synoptic characteristics of dust cases over southwestern Saudi Arabia. *International Journal of Climatology, 38*(1), 105–124. https://doi.org/10.1002/joc.5164

Matthews, T. (2018). Humid heat and climate change. *Progress in Physical Geography: Earth and Environment, 42*(3), 391–405. https://doi.org/10.1177/0309133318776490

Maung, T. Z., Bishop, J. E., Holt, E., Turner, A. M., & Pfrang, C. (2022). Indoor air pollution and the health of vulnerable groups: A systematic review focused on particulate matter (PM), volatile organic compounds (VOCs) and their effects on children and people with pre-existing lung disease. *International Journal of Environmental Research and Public Health, 19*(14), 8752. https://doi.org/10.3390/ijerph19148752

Mazloumi, A., Golbabaei, F., Mahmood Khani, S., et al. (2014). Evaluating effects of heat stress on cognitive function among workers in a hot industry. *Health Promotion Perspectives, 4*(2), 240–246. https://doi.org/10.5681/hpp.2014.032

Mazlomi, A., Golbabaei, F., Mahmood Khani, S., et al. (2016). The influence of occupational heat exposure on cognitive performance and blood level of stress hormones: A field study report. *International Journal of Occupational Safety and Ergonomics, 23*(3), 431–439. https://doi.org/10.1080/10803548.2016.1213115

Melebari, S., Bakri, R., Hafiz, A., Qabbani, F., Khogeer, A., Alharthi, I., et al. (2021). The epidemiology and incidence of dengue in Makkah, Saudi Arabia, during 2017–2019. *Saudi Medical Journal, 42*(11), 1173–1181.

Metelmann, S. (2019). *Development of climate-driven models for mosquito-borne disease risk in the UK* (Unpublished doctoral dissertation). University of Liverpool.

Middel, A., Häb, K., Brazel, A. J., Martin, C. A., & Guhathakurta, S. (2014). Impact of urban form and design on mid-afternoon microclimate in Phoenix local climate zones. *Landscape and Urban Planning, 122*, 16–28. https://doi.org/10.1016/j.landurbplan.2013.11.004

Mojahed, N., Mohammadkhani, M. A., & Mohamadkhani, A. (2022). Climate crises and developing vector-borne diseases: A narrative review. *Iranian Journal of Public Health, 51*, 2664.

Mora, C., Dousset, B., Caldwell, I., Powell, F., Geronimo, R., Bielecki, C., Counsell, C., Dietrich, B., Johnston, E., Louis, L., et al. (2017). Global risk of deadly heat. *Nature Climate Change, 7*(7), 501–506. https://doi.org/10.1038/nclimate3322

Moreno, M. I. R. (2023). Climate change and working time: A complex challenge. *Modernising European Legal Education (MELE), 183*, 11545.

Mostafa, A. M., & Alshahrani, A. (2024). Humanizing sustainable development through green spaces: A case study of Saudi cities. *Frontiers in Sustainable Cities, 6*, 1416983. https://doi.org/10.3389/frsc.2024.1416983

Muleia, R., Maúre, G., José, A., Maholela, P., Adjei, I. A., Karim, M. R., & Marrufo, T. (2024). Assessing the vulnerability and adaptation needs of Mozambique's health sector to climate: A comprehensive study. *International Journal of Environmental Research and Public Health, 21*(5), 532.

Ncongwane, K. P., Botai, J. O., Sivakumar, V., & Botai, C. M. (2021). A literature review of the impacts of heat stress on human health across Africa. *Sustainability, 13*(9), 5312. https://doi.org/10.3390/su13095312

Neira, M., Erguler, K., Ahmady-Birgani, H., Al-Hmoud, N. D., Fears, R., Gogos, C., Hobbhahn, N., Koliou, M., Kostrikis, L. G., Lelieveld, J., Majeed, A., Paz, S., Rudich, Y., Saad-Hussein, A., Shaheen, M., Tobias, A., & Christophides, G. (2023). Climate change and human health in the Eastern Mediterranean and Middle East: Literature review, research priorities and policy suggestions. *Environmental Research, 216*, 114537. https://doi.org/10.1016/j.envres.2022.114537

Noureen, A., Aziz, R., Ismail, A., & Trzcinski, A. P. (2022). The impact of climate change on waterborne diseases in Pakistan. *Sustainability and Climate Change, 15*(2), 138–152.

Ozdemir, E. T., Korkmaz, F. M., & Yavuz, V. (2018). Synoptic analysis of dust storm over the Arabian Peninsula: A case study on February 28, 2009. *Natural Hazards, 92*, 805–827. https://doi.org/10.1007/s11069-018-3239-0

Paaijmans, K. P., Read, A. F., & Thomas, M. B. (2009). Understanding the link between malaria risk and climate. *Proceedings of the National Academy of Sciences of the United States of America, 106*(33), 13844–13849. https://doi.org/10.1073/pnas.0903423106

Pal, J., & Eltahir, E. (2016). Future temperature in the Arabian Peninsula: A deadly heat threshold. *Nature Climate Change, 6*, 197–200. https://doi.org/10.1038/nclimate2833

Pandey, V., Ranjan, M. R., & Tripathi, A. (2021). Climate change and its impact on the outbreak of vector-borne diseases. In *Recent technologies for disaster management and risk reduction: Sustainable community resilience and responses* (pp. 203–228). Springer. https://doi.org/10.1007/978-3-030-70151-0_10

Pasin, B., & Öner, A. C. (2018). Mega-urban developments on the Arabian Peninsula for a post-oil future. In *Handbook of emerging 21st-century cities* (p. 173). https://doi.org/10.1007/978-3-319-96535-2_9

Patz, J. A., Campbell-Lendrum, D., Holloway, T., & Foley, J. A. (2005). Impact of regional climate change on human health. *Nature, 438*(7066), 310–317. https://doi.org/10.1038/nature04188

Patz, J. A., Grabow, M. L., & Limaye, V. S. (2014). When it rains, it pours: Future climate extremes and health. *Annals of Global Health, 80*(4), 332–344. https://doi.org/10.1016/j.aogh.2014.09.007

Paz, S. (2019). The cholera epidemic in Yemen—How did it start? The role of El Niño conditions followed by regional winds. *Environmental Research, 176*, 108571. https://doi.org/10.1016/j.envres.2019.108571

Rafei, R., Osman, M., Kassem, I. I., Dabboussi, F., Weill, F. X., & Hamze, M. (2024). Spotlight on the epidemiology and antimicrobial susceptibility profiles of Vibrio species in the MENA region, 2000–2023. *Future Microbiology, 19*(15), 1333–1353. https://doi.org/10.2217/fmb-2023-0209

Ramanathan, V. (2020). Climate change, air pollution, and health: Common sources, similar impacts, and common solutions. In W. Al-Delaimy, V. Ramanathan, & M. Sanchez Sorondo (Eds.), *Health of people, health of planet and our responsibility: Climate change, air pollution and health* (pp. 49–59). Springer Open. https://doi.org/10.1007/978-3-030-31125-2_5

Raymond, C., Matthews, T., & Tuholske, C. (2024). Evening humid-heat maxima near the southern Persian/Arabian Gulf. *Communications Earth and Environment, 5*(1), 591. https://doi.org/10.1038/s43247-024-01417-3

Rizmie, D., de Preux, L., Miraldo, M., & Atun, R. (2022). Impact of extreme temperatures on emergency hospital admissions by age and socio-economic deprivation in England. *Social Science and Medicine, 308*, 115193. https://doi.org/10.1016/j.socscimed.2022.115193

Ryan, S. J., Lippi, C. A., Villena, O. C., Singh, A., Murdock, C. C., & Johnson, L. R. (2023). Mapping current and future thermal limits to suitability for malaria transmission by the invasive mosquito Anopheles stephensi. *Malaria Journal, 22*(1), 104. https://doi.org/10.1186/s12936-023-04568-z

Sá, E., Martins, H., Ferreira, J., Marta-Almeida, M., Rocha, A., Carvalho, A., & Borrego, C. (2016). Climate change and pollutant emissions impacts on air quality in 2050 over Portugal. *Atmospheric Environment, 131*, 209–224. https://doi.org/10.1016/j.atmosenv.2016.01.037

Saeed, F. (2024). Hajj in the heat: From sacred journey to survival test. *Climate Analytics*. https://climateanalytics.org/comment/hajj-in-the-heat-from-sacred-journey-to-survival-test

Safieddine, S., Clerbaux, C., Clarisse, L., Whitburn, S., & Eltahir, E. A. B. (2022). Present and future land surface and wet-bulb temperatures in the Arabian Peninsula. *Environmental Research Letters, 17*(4), 044029. https://doi.org/10.1088/1748-9326/ac5b16

Schmidt, S. (2023). Precision public health opportunity: Heat-associated mortality greater among pilgrims during Hajj. *Environmental Health Perspectives, 131*(6), 064001. https://doi.org/10.1289/EHP13055

Shin, H. I., Ku, B., Jung, H., Lee, S. D., Lee, S. Y., Ju, J. W., et al. (2024). *2023 World malaria report (status of world malaria in 2022)*. World Health Organization.

Smith, L. C., & Haddad, L. (2015). Reducing child undernutrition: Past drivers and priorities for the post-MDG era. *World Development, 68*, 180–204. https://doi.org/10.1016/j.worlddev.2014.11.014

Song, X., Guo, X., Hu, X., Zhang, Y., Wei, D., Hu, Y., et al. (2023). Human exposure risk assessment for infectious diseases due to temperature and air pollution: An overview of reviews. *Environmental Science and Pollution Research, 30*(38), 88272–88280. https://doi.org/10.1007/s11356-023-27670-7

Sun, S., Chang, Q., He, J., Wei, X., Sun, H., Xu, Y., et al. (2023). The association between air pollutants, meteorological factors and tuberculosis cases in Beijing, China: A seven-year time series study. *Environmental Research, 216*, 114581. https://doi.org/10.1016/j.envres.2022.114581

Teather, K., Hogan, N., Critchley, K., Gibson, M., Craig, S., & Hill, J. (2013). Examining the links between air quality, climate change and respiratory health in Qatar. *Avicenna, 2013*(1), 9. https://doi.org/10.5339/avi.2013.9

Teymouri, P., & Dehghanzadeh, R. (2022). Climate change and water-related diseases in developing countries of Western Asia: A systematic literature review. *Climate and Development, 14*(3), 222–238. https://doi.org/10.1080/17565529.2021.1928230

Thornton, P., Al-Juaidi, A., & Habib, M. A. (2021). Energy demand, air conditioning, and climate adaptation in the Middle East. *Energy Policy, 156*, 112413. https://doi.org/10.1016/j.enpol.2021.112413

Tyrovolas, S., El Bcheraoui, C., Alghnam, S. A., Alhabib, K. F., Almadi, M. A. H., Al-Raddadi, R. M., et al. (2020). The burden of disease in Saudi Arabia 1990–2017: Results from the global burden of disease study 2017. *The Lancet Planetary Health, 4*(5), e195–e208. https://doi.org/10.1016/S2542-5196(20)30084-5

Ullah, S., Aldossary, A., Ullah, W., et al. (2024). Augmented human thermal discomfort in urban centers of the Arabian Peninsula. *Scientific Reports, 14*, 3974. https://doi.org/10.1038/s41598-024-58495-3

UNICEF. (2024). *UNICEF yemen humanitarian situation report*. https://www.unicef.org/documents/unicef-yemen-humanitarian-situation-report-2024

Walsh, J. E., Ballinger, T. J., Euskirchen, E. S., Hanna, E., Mård, J., Overland, J. E., et al. (2020). Extreme weather and climate events in northern areas: A review. *Earth-Science Reviews, 209*, 103324. https://doi.org/10.1016/j.earscirev.2020.103324

Ward, M. P., Liu, Y., Xiao, S., & Zhang, Z. (2021). Challenges in the control of COVID-19 outbreaks caused by the delta variant during periods of low humidity: An observational study in Sydney, Australia. *Infectious Diseases of Poverty, 10*(6), 38–47. https://doi.org/10.1186/s40249-021-00843-2

Watts, N., Adger, W. N., & Agnolucci, P. (2015). Health and climate change: Policy responses to protect public health. *The Lancet, 386*, 1861–1914. https://doi.org/10.1016/S0140-6736(15)60854-6

World Health Organization. (2016). *Urban air pollution database*. World Health Organization.

World Health Organization. (2021). *Climate change and health: Key facts*. https://www.who.int/news-room/fact-sheets/detail/climate-change-and-health

World Health Organization. (2022). *Ambient (outdoor) air pollution*. World Health Organization. https://www.who.int/news-room/fact-sheets/detail/ambient-(outdoor)-air-quality-and-health

Yeatts, K. B., El-Sadig, M., Leith, D., Kalsbeek, W., Al-Maskari, F., Couper, D., et al. (2012). Indoor air pollutants and health in the United Arab Emirates. *Environmental Health Perspectives, 120*(5), 687–694. https://doi.org/10.1289/ehp.1002927

Yezli, S., Khan, A. H., Yassin, Y. M., Khan, A. A., Alotaibi, B. M., & Bouchama, A. (2023). Human tolerance to extreme heat: Evidence from a desert climate population. *Journal of Exposure Science and Environmental Epidemiology, 33*(4), 631–636. https://doi.org/10.1038/s41370-023-00572-0

Yezli, S., Yassin, Y., Ghallab, S., Abdullah, M., Abuyassin, B., Vishwakarma, R., & Bouchama, A. (2024). Diagnosing and managing heat exhaustion: Insights from a systematic review of cases in the desert climate of Mecca. *Reviews on Environmental Health, 39*(4), 729–736. https://doi.org/10.1515/reveh-2023-0084

Yuan, Y., Wang, K., Sun, H. Z., Zhan, Y., Yang, Z., Hu, K., & Zhang, Y. (2023). Excess mortality associated with high ozone exposure: A national cohort study in China. *Environmental Science and Ecotechnology, 15*, 100241. https://doi.org/10.1016/j.ese.2023.100241

Zhang, S. X., Yang, G. B., Zhang, R. J., Zheng, J. X., Yang, J., Lv, S., et al. (2024). Global, regional, and national burden of dengue, 1990–2021: Findings from the Global Burden of Disease Study 2021. *Decoding Infection and Transmission, 2*, 100021. https://doi.org/10.1016/j.dit.2024.100021

Zhang, Y., Ma, Y., Feng, F., Cheng, B., Shen, J., Wang, H., et al. (2021). Respiratory mortality associated with ozone in China: A systematic review and meta-analysis. *Environmental Pollution, 280*, 116957. https://doi.org/10.1016/j.envpol.2021.116957

Zhang, Y., Murray, F., & Turner, D. (2022). High resolution modeling of the impact of urbanization and green infrastructure on the water and energy balance. *Urban Climate, 41*, 101051. https://doi.org/10.1016/j.uclim.2021.101051

Zhu, Z., Deng, T., & Pan, X. (2024). Global climate change: The dangers of heatwaves for chronic obstructive pulmonary disease patients cannot be ignored. *Journal of Global Health, 14*, 03032. https://doi.org/10.7189/jogh.14.03032

Ziou, M., Tham, R., Wheeler, A. J., Zosky, G. R., Stephens, N., & Johnston, F. H. (2022). Outdoor particulate matter exposure and upper respiratory tract infections in children and adolescents: A systematic review and meta-analysis. *Environmental Research, 210*, 112969. https://doi.org/10.1016/j.envres.2022.112969

Ziter, C. D., Pedersen, E. J., Kucharik, C. J., & Turner, M. G. (2019). Scale-dependent interactions between tree canopy cover and impervious surfaces reduce daytime urban heat during summer. *Proceedings of the National Academy of Sciences, 116*(15), 7575–7580. https://doi.org/10.1073/pnas.1817561116

Chapter 9
Building Climate Resilience in the Hyper-Arid Arabian Peninsula: Policy Pathways, Adaptation Measures and the Water-Energy-Food Nexus

Mir Khursheed Alam, Umra Rashid, and Aarifah Jan

9.1 Introduction

Climate risk emerges from the interaction of hazards, exposure and systemic vulnerabilities inherent in specific systems, sectors, or regions. While the implementation of mitigation measures serves to reduce the likelihood of long-term hazard occurrence, climate adaptation remains essential to address vulnerability and exposure in the short to medium term, thereby contributing to an overall reduction in climate risk across various regions, sectors and systems (IPCC, 2022). In the hyper-arid Arabian Peninsula, climate change functions as a potent risk multiplier. It intensifies both environmental and socio-economic vulnerabilities and presents significant threats to water, energy and food (WEF) security, along with the broader objectives of sustainable development in the region (Cooper, 2020; Waha et al., 2017). The risks confronting the Arabian Peninsula are inherently multidimensional, as the region is exposed to a combination of climatic and non-climatic stressors. These include adverse impacts on agricultural productivity, water availability and security, human health and the overarching goals of sustainable development. Beyond these critical development challenges, the region experiences acute environmental threats resulting from escalating water scarcity, extreme temperatures and increasing climate variability. Consequently, the Arabian Peninsula faces heightened climate risks, reflected in recurrent hazards, amplified exposure and entrenched systemic

M. K. Alam (✉) · U. Rashid
School of Business, UPES, Dehradun, Uttarakhand, India
e-mail: mirkhursheed.alam@ddn.upes.ac.in

A. Jan
School of Humanities and Social Sciences, Indian Institute of Technology Mandi, Kamand, Himachal Pradesh, India

F. Mushtaq et al. (eds.), *Climate Change in the Arabian Peninsula*,
https://doi.org/10.1007/978-3-032-02481-7_9

vulnerabilities, all of which are projected to deepen the region's socio-economic and environmental fragility.

The objective of climate action encompassing efforts such as the reduction of greenhouse gas emissions (GHG) and the strengthening of climate resilience is advancing across the region. Nevertheless, these efforts continue to confront a range of environmental and socio-economic challenges. These include vital considerations related to social equity, the persistence of widening disparities, the pursuit of gender inclusion and issues concerning public health, human well-being and employment within and across countries of the Arabian Peninsula (Zakzak et al., 2024). The pursuit of greener transitions in response to the climate crisis is intrinsically linked to broader development priorities, including the realization of sustainable and climate-resilient development, the implementation of effective adaptation policies and the resolution of complex challenges within the WEF nexus. Importantly, the risks associated with climate change are not evenly distributed but are socially differentiated across sub-regions. These disparities are likely to manifest in unequal impacts on human well-being (Verner, 2012). The literature specific to the Arabian Peninsula increasingly emphasizes the necessity of transboundary governance in climate action (Gunel, 2016), particularly with regard to addressing systemic vulnerability and exposure. In this context, adaptation efforts in the region must reflect the deepening interdependence among water, energy and food systems an interconnection that is essential to sustaining regional stability (Dawoud & Al-Mulla, 2012). Therefore, strengthening climate resilience requires more than just technical interventions; it necessitates robust regulatory frameworks that can embed adaptation within long-term planning, while upholding principles of socio-economic equity and sustainability (Saudi Water Partnership Company, 2021).

The Arabian Peninsula, characterized by its hyper-arid climate and extreme temperature regimes, stands at the forefront of global climate vulnerability. It comprises countries located in predominantly arid to semi-arid agro-climatic zones. Under high-emission scenarios, mean annual temperatures across the region are projected to rise by 2–4 °C by mid-century (IPCC, 2022), exposing nations to increasing risks from heatwaves, prolonged droughts, flash floods and desertification. Observational and gridded data provide compelling evidence of statistically significant increases in temperature and declines in precipitation across the region, with notable emphasis on Saudi Arabia (Almazroui et al., 2012). This dual climatic threat is not confined to the Arabian Peninsula but extends across the broader North Africa and Arabian Peninsula (NAAP) region, compounding extreme heat events (Hamed et al., 2024). Future projections, based on high-resolution datasets, indicate that summer maximum and winter minimum temperatures are expected to increase by 0.11–0.67 and 0.09–0.70 °C per decade, respectively. Simultaneously, warm temperature extremes are projected to rise, while cold extremes are expected to decline under three Shared Socioeconomic Pathways (SSPs): SSP1-2.6, SSP2-4.5 and SSP5-8.5 (Vinodh Kumar et al., 2024). Over a longer temporal horizon, projections suggest that by 2100 the region may experience average warming of 2.6 °C under low-emission scenarios and up to 7.6 °C under high-emission scenarios (Vinodh Kumar et al., 2024). These anticipated shifts in temperature and rainfall patterns signal a likely increase in drought frequency and severity across various scenarios and timeframes (Saharwardi et al., 2024), further underscoring the escalating threats to socio-economic systems, particularly in terms of public health burdens (Akasha et al., 2023) and broader well-being (Hamed et al., 2024).These emerging challenges are further aggravated by rapid urban expansion, population growth and resource-intensive economic models, all of which exert immense pressure on already limited freshwater reserves and arable land (Francis & Fonseca, 2024). In a region dominated by arid and semi-arid conditions and characterized by highly constrained freshwater availability and extreme climatic variability, the implications of climate change for food and water security are profound. Recurrent droughts, intensifying heatwaves and frequent sandstorms increasingly jeopardize agricultural productivity (Aleisa & Al-Zubari, 2017; IPCC, 2022).

As climate-induced impacts continue to intensify, adaptation has emerged as a critical imperative for protecting human health, ensuring economic stability and maintaining ecological balance in this region of considerable geopolitical importance. Against this backdrop, the present chapter pursues two key objectives:

1. First, it seeks to examine the long-term and country-specific dynamics of vulnerability and adaptive readiness.
2. Second, it investigates the policy responses and adaptation measures undertaken across the Arabian Peninsula, with particular focus on advancing water, energy and food security amid ongoing climatic transformations.

9.2 Assessing Climate Vulnerability and Readiness (to Adapt) in the Arabian Peninsula

The Arabian Peninsula faces a diverse set of challenges in reducing climate vulnerability and advancing adaptation, leading to differential policy responses across countries. These differences are shaped not only by climatic factors but also by widespread socio-economic, demographic and institutional disparities. Consequently, countries in the region experience varying levels of vulnerability to climate change impacts. Some nations are more exposed and less equipped to respond effectively, while others exhibit greater readiness to undertake adaptation actions through public and private sector investments, government initiatives, community engagement and enabling environments for private sector participation.

To examine these dynamics, this section utilizes long-term data (1995–2022) on vulnerability and readiness scores derived from the Notre Dame Global Adaptation Initiative (ND-GAIN, 2024). The ND-GAIN framework offers a robust and comprehensive set of indicators based on peer-reviewed literature, the Intergovernmental Panel on Climate Change (IPCC) review process and feedback from policymakers, development practitioners and corporate stakeholders. Table 9.1 presents the list of indicators used in constructing the vulnerability index. The framework includes 36 indicators, capturing the dimensions of exposure ($n = 12$), sensitivity ($n = 12$) and adaptive capacity ($n = 12$), across six life-supporting sectors: food, water, health, ecosystem services, human habitat and infrastructure. These sectors are critical to regional sustainability and are analyzed through three cross-cutting dimensions: exposure to climate-exacerbated hazards, sensitivity to these hazards and capacity to adapt.

Table 9.1 Indicators considered for the construction of vulnerability

Dimension/sector	Exposure	Sensitivity	Capacity
Food	Projected change in cereal yields	Food import dependency	Agricultural capacity (fertilizer, irrigation, pesticide, tractor use)
	Projected population change	Rural population	Child malnutrition
Water	Projected change in annual groundwater runoff	Freshwater withdrawal rate	Access to reliable drinking water
	Projected change of annual groundwater recharge	Water dependency ratio	Dam capacity
Health	Projected change in vector-borne diseases	Slum population	Medical staff (physicians, nurses and midwives)
	Projected change in deaths from climate change-induced diseases	Dependency on external resources for health services	Access to improved sanitation facilities
Ecosystem service	Projected change in biome distribution	Dependency on natural capital	Protected biomes
	Projected change in marine biodiversity	Ecological footprint	Engagement in international environmental conventions
Human habitat	Projected change of warm periods	Urban concentration	Quality of trade and transport-related infrastructure
	Projected change in flood hazard	Age dependency ratio	Paved roads
Infrastructure	Projected change in hydropower generation capacity	Dependency on imported energy	Electricity access
	Projection of sea level rise impacts	Population living under 5 m above sea level	Disaster preparedness
Total indicators	12	12	12

Source Compiled from ND-GAIN (2025)

Under the ND-GAIN framework, the conceptualization of vulnerability aligns with the Fourth Assessment Report (AR4) of the IPCC (2007), wherein vulnerability is a function of sensitivity, adaptive capacity and exposure. It is defined as the "propensity or predisposition of human societies to be negatively impacted by climate hazards." The ND-GAIN vulnerability index captures this construct through indicators specific to six critical sectors. However, the Fifth and Sixth Assessment Reports (IPCC,) reconceptualize vulnerability as "the propensity or predisposition to be adversely affected by climate change," encompassing sensitivity and limited capacity to cope or adapt, while treating exposure as a co-determinant of climate risk. In this revised framing, climate risk is understood to emerge from the interaction of three core elements: hazard, exposure and vulnerability (IPCC,). Table 9.2 outlines the indicators used to assess readiness under the ND-GAIN framework. Readiness reflects a country's ability to effectively leverage investments toward adaptation efforts and encompasses three components (ND-GAIN, 2024):

1. Economic readiness
2. Social readiness
3. Governance readiness

The readiness score is based on a country's efforts to promote climate change adaptation through individual actions, government programs, bilateral and multilateral initiatives, NGO activities and private sector investments. Economic readiness is evaluated through a composite "Doing Business" indicator comprising 10 sub-indicators. Social and governance readiness are assessed through four indicators each, as presented in Table 9.2.

The selection of indicators for both vulnerability and readiness under ND-GAIN is methodologically sound. The indicators are dynamic (drawn from time-series data), globally consistent, transparent, conceptually clear, quality-controlled and representative of the constructs they measure. Vulnerability indicators particularly those under

Table 9.2 Indicators considered for the construction of readiness

Component	Indicators			
Economic readiness	Doing business (one indicator considering the following 10 sub-indicators)			
	Starting a business	Dealing with construction permits	Getting electricity	Registering prosperity
	Getting credit	Protecting investors	Paying taxes	Trading across borders
	Enforcing contracts		Resolving insolvency	
Governance readiness	Political stability and non-violence	Control of corruption	Rule of law	Regulatory quality
Social readiness	Social inequality	ICT infrastructure	Education	Innovation

Source Compiled from ND-GAIN (2025)

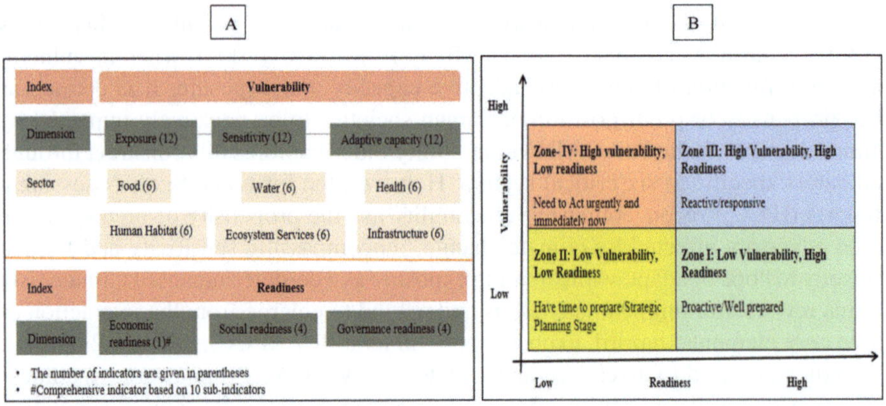

Fig. 9.1 Structure of vulnerability and readiness score, adapted from ND GAIN and adaptation policy framework through vulnerability-readiness matrix, adapted from Buchir and Detzel (2023). *Source* Buchir and Detzel (2023)

the exposure dimension incorporate projected changes in climate, hazard patterns, population trends and ecosystem impacts. Other dimensions focus on actionable indicators for adaptation. Likewise, readiness indicators reflect actionable domains for enhancing adaptation, indicating the potential effectiveness of national adaptation policies and frameworks. These indicators can respond to interventions by governments, private actors, or communities, thereby enabling progress monitoring. Higher vulnerability scores indicate greater vulnerability ("worse"), while higher readiness scores signal better adaptive capacity ("better"). Figure 9.1a provides a schematic representation of the construction of the vulnerability and readiness scores under ND-GAIN. In addition, Fig. 9.1b illustrates a vulnerability-readiness matrix adapted from Buchir and Detzel (2023), which jointly interprets the indices within a rational adaptation policy framework. The matrix delineates four zones that can guide policymakers in prioritizing areas for climate risk management, climate-resilient development and tailored adaptation responses.

Temporal Dynamics and Regional Disparities in Vulnerability and Readiness in the Arabian Peninsula

Figures 9.2 and 9.3 illustrate the trends in vulnerability and readiness scores, respectively, for countries located in the Arabian Peninsula over the period 1995–2022. These figures, in conjunction with the data summarized in Table 9.3, provide valuable insights into the long-term dynamics of climate vulnerability and adaptive readiness across the region. The vulnerability scores in Fig. 9.2 highlight a gradually declining trend across most countries in the Arabian Peninsula, with the notable exception of Yemen. The decrease in vulnerability measured across six life-supporting sectors (food, water, health, ecosystem services, human habitat and infrastructure) suggests incremental progress in reducing systemic exposure and sensitivity to climate hazards. Specifically, Qatar and Kuwait demonstrate the lowest average vulnerability over the study period, with scores of 0.36 and 0.38, respectively. These countries exhibit consistently low vulnerability (<0.4) throughout the 2000–2020 period, reflecting successful investments in infrastructure, governance and sectoral adaptation strategies. In contrast, Yemen stands out with the highest average vulnerability (0.52), attributed primarily to its prolonged conflict, socio-political instability and constrained institutional capacity. Saudi Arabia and the United Arab Emirates (UAE) show a notable decline in vulnerability, particularly after 2017 and 2018, respectively. However, a slight reversal is observed in Saudi Arabia post-2020, indicating emerging stressors or setbacks in adaptive progress. Oman and Bahrain also reflect modest improvements in vulnerability scores, although coastal exposure and water-related challenges remain significant. These observed trends suggest that while climate-related vulnerability is gradually declining across most nations, structural and sectoral disparities continue to shape country-specific trajectories.

Figure 9.3 depicts the readiness scores of Arabian Peninsula countries over the same period. The upward trend across most countries except Yemen indicates growing capacity to leverage investments, implement adaptation measures and enhance institutional and governance mechanisms. Readiness, in this context, encapsulates economic, social and governance readiness dimensions, each contributing to a country's ability to translate resources into effective climate adaptation. The UAE emerges as the most adaptive-ready nation, with the highest average readiness score of 0.51, followed closely by Qatar (0.47) and Bahrain (0.46). These countries have implemented progressive policies that promote climate resilience, sustainable infrastructure and technological innovation. Saudi Arabia and Oman register intermediate readiness levels (0.44 each), which are comparable and suggest moderately robust

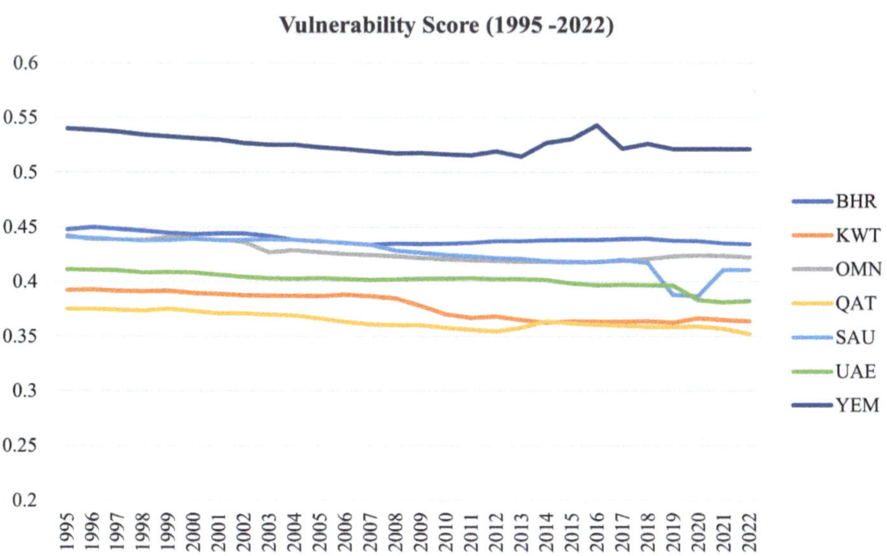

Fig. 9.2 Vulnerability score of the countries located in the Arabian Peninsula anlyzed by authors

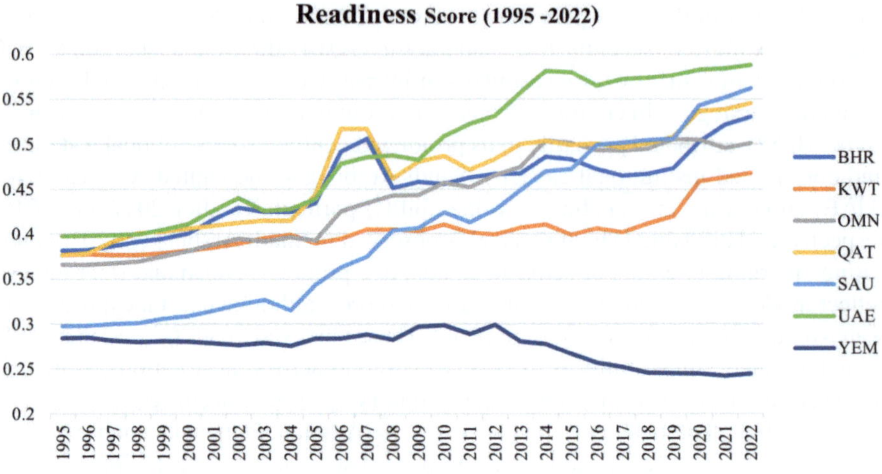

Fig. 9.3 Readiness score of the countries located in the Arabian Peninsula anlyzed by authors

adaptation frameworks. Kuwait, with a readiness score of 0.41, shows substantial improvement in recent years, signaling a shift toward climate-responsive planning and governance. Yemen, on the other hand, remains critically unprepared, with the lowest readiness score (0.27), reflecting significant institutional deficits, political fragility and minimal investment in climate adaptation.

The comparative insights from Table 9.3 support the visual interpretation of Figs. 9.2 and 9.3. The table outlines each country's average scores and provides brief contextual descriptions of vulnerability and readiness status. The UAE and Qatar exemplify effective climate governance, demonstrating both low vulnerability and high readiness indicators of long-term resilience. Saudi Arabia's decline in vulnerability and improved readiness post-2017 reflects strategic investments in water management, renewable energy and desertification control. Oman's alignment of socio-economic reforms with climate adaptation goals is reflected in its balanced scores. Conversely, Yemen's high vulnerability and low readiness highlight the urgent need for targeted, externally supported adaptation interventions. The divergence in trajectories among these countries emphasizes the heterogeneity of climate risk across the region. It also reinforces the value of the ND-GAIN framework in facilitating comparative assessment and guiding evidence-based policy responses. These findings underscore the importance of tailored adaptation planning, transboundary cooperation and inclusive governance to address the differentiated needs of countries within the Arabian Peninsula.

Table 9.3 Average vulnerability and readiness score of countries located in the Arabian Peninsula

Country	Average vulnerability score (1995–2022)	Average readiness score (1995–2022)	Vulnerability	Readiness
Bahrain (BHR)	0.44	0.46	Bahrain shows a gradual decline in vulnerability. However, vulnerabilities remain due to coastal exposure	It witnessed a steady increase in readiness, indicating moderate adaptation progress
Kuwait (KWT)	0.38	0.41	Kuwait has relatively low vulnerability (<0.4 throughout 2000–2020) and has reduced significantly in the last two decades. However, water scarcity remains a challenge	It has moderate readiness, improving significantly in recent years, highlighting better climate adaptation policies
Oman (OMN)	0.43	0.44	Oman has witnessed a slight decline in vulnerability throughout 2000–2020	Oman's improved readiness through the implementation of socio-economic and governance-related adaptation strategies in water resource management has contributed to this trend
Qatar (QAT)	0.36	0.47	Qatar consistently has the lowest average vulnerability (0.36). It has low vulnerability throughout 2000–2020	Qatar has strong readiness and is among the best performers, reflecting its significant investments in climate resilience and infrastructure. It has managed to shift upwards in readiness in the last 2 decades
Saudi Arabia (SAU)	0.42	0.44	Saudi Arabia has a declining trend in vulnerability, especially post-2017. However, witnessed a rise in vulnerability post-2020	It has witnessed an increasing readiness score, driven by investments in renewable energy and desertification control

(continued)

Table 9.3 (continued)

Country	Average vulnerability score (1995–2022)	Average readiness score (1995–2022)	Vulnerability	Readiness
United Arab Emirates (UAE)	0.40	0.51	Overall, the UAE stands in a relatively low vulnerability category and has managed to reduce it significantly post-2018	The UAE has the highest readiness and is the best performer, highlighting strong investments in sustainability initiatives and technological advancements
Yemen (YEM)	0.52	0.27	Yemen has the highest vulnerability. It gets exacerbated largely due to associated socio-economic challenges and political factors that further hinder adaptation efforts	Yemen has the lowest readiness, as it is plagued by political instability and economic challenges. It limits both private and public adaptation capacity

Source Author's analysis

9.3 Strategic Adaptation Responses Based on Vulnerability–Readiness Profiles in the Arabian Peninsula

Figure 9.4 illustrates the Vulnerability–Readiness Matrix, which categorizes countries in the Arabian Peninsula into four distinct strategic response zones. These quadrants are derived from joint interpretations of average vulnerability and readiness scores and serve as a practical framework for profiling adaptation strategies and prioritizing policy interventions. The first zone characterized by low vulnerability and high readiness includes Qatar and the United Arab Emirates (UAE). These countries fall into the category of proactive governance, reflecting robust institutional capacity, anticipatory planning and long-term investment in adaptive infrastructure. Their positioning in this quadrant suggests that they have not only succeeded in minimizing systemic vulnerability but also developed substantial institutional readiness to implement further adaptation measures. The presence of Qatar and UAE in this category highlights the effectiveness of early, well-funded adaptation strategies, including technological innovation, integrated planning and international partnerships. The second zone low vulnerability but low readiness is represented by Kuwait, which is categorized under strategic planning. Although Kuwait exhibits relatively low vulnerability, its limited readiness signals the need to enhance institutional frameworks, governance capacity and resource mobilization mechanisms. The country demonstrates foundational strengths but requires forward-looking policies

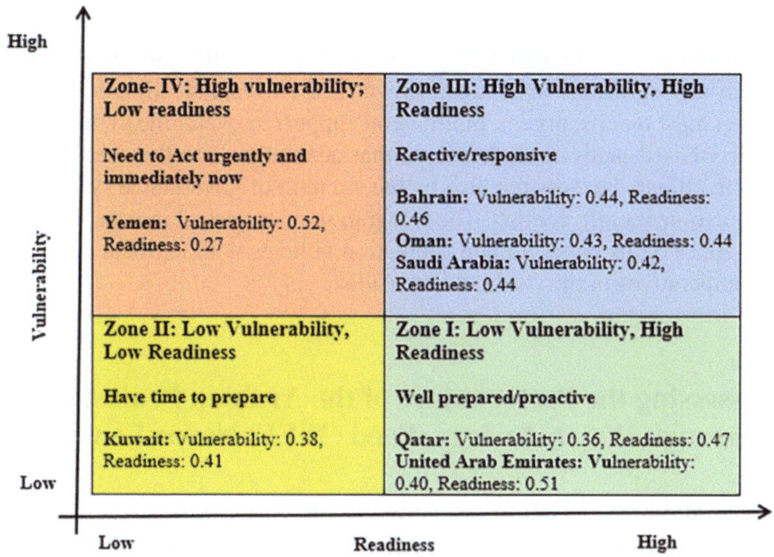

Fig. 9.4 Vulnerability–readiness matrix illustrating four strategic response zones in the Arabian Peninsula

and investments to translate its low vulnerability into long-term climate resilience. Strengthening climate governance structures and expanding adaptation financing could accelerate Kuwait's transition into the proactive zone.

The third zone encompasses Bahrain, Oman and Saudi Arabia, which exhibit high vulnerability and high readiness. These countries fall into the responsive governance category. Despite being significantly exposed to climate-related risks, they possess the institutional and economic readiness to respond effectively. Their readiness indicates that systems and resources are in place; however, policy implementation may still be reactive rather than anticipatory. This zone reflects a critical window for these countries to advance toward proactive governance by translating readiness into transformative adaptation policies, particularly in areas such as water conservation, health infrastructure and sustainable agriculture. The fourth and most critical zone includes Yemen, which displays high vulnerability and low readiness. This configuration positions Yemen in the policy under stress category, where the country faces compounded risks due to structural fragility, conflict and minimal institutional capacity for climate adaptation. The urgent and systemic nature of Yemen's climate challenges calls for immediate intervention. National-level strategies must be supported by international assistance, capacity-building programs and cross-border cooperation to strengthen adaptive systems. The policy priority here is to initiate foundational frameworks that can address both urgent humanitarian needs and long-term climate resilience. Collectively, Fig. 9.4 and the associated categorization offer a practical and policy-relevant lens to interpret empirical trends and identify strategic adaptation priorities. Countries across different zones require differentiated approaches: while Qatar and the UAE can focus on advancing innovation and global leadership in adaptation, countries like Oman and Saudi Arabia must work to close the gap between readiness and proactive governance. In contrast, Kuwait must invest in readiness infrastructure and Yemen must receive urgent, multilateral support to overcome systemic barriers. This matrix-based analysis underscores that adaptation is influenced not only by exposure to climate hazards but also by the strength of governance systems and the broader socio-economic context. The integration of empirical indices with strategic response zones helps align climate adaptation policies with national capacities and regional imperatives in the Arabian Peninsula.

9.4 Assessing the Performance of the Arabian Peninsula Through Water–Energy–Food (WEF) Nexus Lens

To evaluate the performance of countries in the Arabian Peninsula with respect to climate adaptation and sustainable development, this section adopts the Water–Energy–Food (WEF) nexus framework, aligned with relevant Sustainable Development Goals (SDGs). The interdependence between water, energy and food systems is especially critical in this hyper-arid and resource-constrained region. Accordingly, six SDGs have been selected to assess the performance and track enabling trends: three core WEF nexus SDGs; SDG 2 (Zero Hunger), SDG 6 (Clean Water and Sanitation) and SDG 7 (Affordable and Clean Energy) and three cross-cutting and enabling SDGs; SDG 12 (Responsible Consumption and Production), SDG 13 (Climate Action) and SDG 15 (Life on Land). Figure 9.5 presents a consolidated assessment of the performance and directional trends of WEF-specific and enabling SDGs across the seven countries of the Arabian Peninsula, highlighting both challenges and progress.

Performance on Core WEF Nexus SDGs (SDG 2, 6 and 7)

The analysis reveals that SDGs directly related to the WEF nexus remain significantly unachieved across most countries, with varying degrees of stagnation or slow improvement. Yemen, in particular, faces severe challenges across all WEF-related goals, with the lowest overall SDG score (48.2), underscoring systemic development deficits. While most countries are either stagnant or only slightly improving, some progress is notable. Kuwait and Oman, for instance, perform relatively better, especially in terms of maintaining or advancing SDG 6, related to water security. These improvements can be attributed to policy efforts in water governance and resource management. In terms of SDG 2 (Zero Hunger), all countries encounter significant challenges due to high dependency on food imports, limited arable land and climate-induced agricultural constraints. Food security remains a persistent concern, despite investments in technology-driven agricultural strategies. For SDG 6 (Clean Water and Sanitation), the region continues to experience critical issues stemming from water scarcity, over-reliance on desalination and limited renewable water sources.

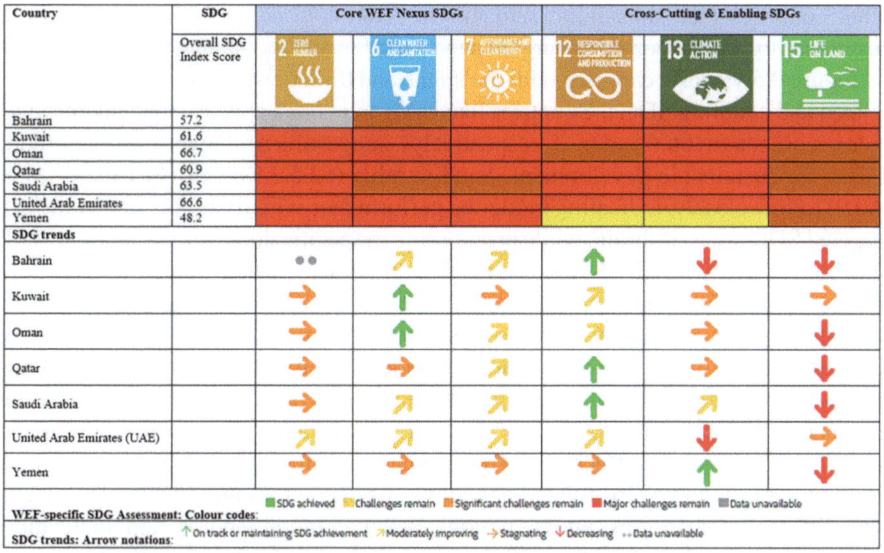

Fig. 9.5 Performance and trend of core water–energy–food (WEF) nexus SDGs and cross-cutting/enabling SDGs in the Arabian Peninsula, based on SDG index score and directional trend indicators. *Source* Author's analysis and adapted from world government summit, the Arab Region SDG index and dashboards 2023/2024

Nevertheless, Oman and the UAE show positive trends, reflecting improvements in water infrastructure and institutional planning. Regarding SDG 7 (Affordable and Clean Energy), modest advancements have been observed in Bahrain, Qatar and the UAE, largely due to renewable energy initiatives and improvements in energy access. However, significant challenges remain in decarbonizing the energy sector and improving affordability.

Trends in Cross-Cutting and Enabling SDGs (SDG 12, 13 and 15)

The performance on cross-cutting and enabling SDGs also reflects mixed progress and ongoing systemic challenges. Particularly concerning are SDG 13 (Climate Action) and SDG 15 (Life on Land), where most countries demonstrate either stagnant or negative trends. Bahrain, Saudi Arabia and Qatar exhibit improvements in SDG 12, indicating steps toward sustainable consumption and production, including circular economy initiatives and environmental regulations. A noteworthy exception is Yemen, which despite broader development deficits is maintaining its SDG 13 performance. This suggests continuity in climate policy engagement, possibly through international support or alignment with NDC frameworks. In contrast, countries like Bahrain and the UAE are witnessing a declining trend in climate action performance, raising concerns about the sustainability of current policies. SDG 15 (Life on Land) shows a widespread decline in performance across the region, driven by factors such as land degradation, biodiversity loss and expanding urbanization. Only UAE and Kuwait show relatively stable or slightly improving trends. Most other countries face substantial setbacks in ecosystem conservation and land-use sustainability, underscoring the urgent need for stronger land management strategies.

Figure 9.5 visually captures both the status and directional trends of WEF nexus-related and enabling SDGs. Countries like UAE and Oman appear in the higher SDG index score range (66.6 and 66.7, respectively), suggesting a relatively stronger foundation for sustainable development. Saudi Arabia (63.5), Kuwait (61.6) and Qatar (60.9) follow closely, indicating moderate performance with room for acceleration in targeted SDG areas. Bahrain, while progressing in specific dimensions, lags in overall WEF alignment with an SDG score of 57.2. As noted earlier, Yemen continues to face severe multidimensional challenges, reflected in its low score (48.2). The arrow indicators in Fig. 9.5 denote SDG trends whether they are improving, stagnant, or declining. These trend lines are crucial for identifying not only the current status but also the momentum (or regression) of climate-resilient development. The visual summary provides a diagnostic overview for policymakers to prioritize action and allocate resources more effectively. The analysis confirms that, despite some positive developments, major WEF-related challenges persist across the Arabian Peninsula, with differentiated progress among nations. Water and food security remain the most pressing issues, while enabling SDGs particularly SDG 13 and SDG 15 require renewed focus. The integrated WEF-SDG approach adopted in this section allows for a nuanced understanding of national performance and helps identify leverage points for climate-resilient development strategies.

9.5 Government Initiatives and Policies Addressing Climate Change

While dealing with climate change through adaptation interventions, there is strong empirical evidence that countries in the Arabian Peninsula individually and collectively adhere to the Paris Agreement. The region has progressively deepened its commitment to climate action and is aligning adaptation responses with global frameworks. Several governments have launched targeted policies and initiatives that span key sectors such as water management, agriculture, urban planning and disaster risk reduction. A key focus has been on the integration of water–energy–food (WEF) security strategies and governance frameworks. As one of the most water-stressed regions in the world, the Arabian Peninsula relies heavily on non-renewable groundwater extraction, energy-intensive desalination and limited renewable freshwater resources (UNESCWA, 2021). The combined pressures of climate change, population growth and urban expansion have intensified stress on these interlinked sectors. Rising temperatures, erratic precipitation and frequent droughts exacerbate vulnerabilities in water and food security, thereby distorting the WEF nexus (IPCC, 2022).

In response, regional governments have implemented a range of adaptation strategies and policy reforms. These include efforts to enhance water governance, promote resource and energy efficiency and support integrated sustainable development (World Bank, 2021a, 2021b). However, the effectiveness of these interventions is often constrained by fragmented governance systems, unresolved transboundary water disputes and socio-economic disparities (Daoudy et al., 2024). The unique biophysical and climatic constraints of the region further complicate adaptation planning and implementation. The Arabian Peninsula continues to experience water stress and high precipitation variability due to geographical and climatic factors. The imbalance between water demand and supply has become a central issue in the region's adaptation agenda. Over 85% of freshwater demand is met through desalination and depleting groundwater reserves (World Bank, 2021a, 2021b). At the same time, rapid economic development has significantly increased per capita water consumption, driven by growth in agriculture, urbanization and population (Abdella et al., 2024). Literature highlights that per capita water availability is declining, due in part to water-intensive cropping systems, industrialization and high consumption in the service sector (Miralles-Wilhelm et al., 2014). These systemic pressures necessitate integrated approaches to managing water, energy and food resources. Such approaches must also strengthen institutional frameworks, promote governance reforms and reinforce sustainable consumption and climate action strategies (D'Odorico et al., 2018; Miralles-Wilhelm et al., 2018).

Government-led initiatives are gradually evolving to address these complexities. Countries such as Saudi Arabia, the UAE, Oman and Qatar have introduced national-level adaptation policies focused on institutional coordination, regulatory development and community resilience. Notably, Saudi Arabia's National Water Strategy 2030 promotes integrated resource management, reduced reliance on non-renewable groundwater and improvements in water-use efficiency through desalination, wastewater recycling and irrigation reforms (Saudi MEWA, 2021). The strategy incorporates climate resilience through regulatory reforms, digital water monitoring systems and private sector participation, consistent with broader sustainable development objectives under Vision 2030 (Saudi Vision, 2023). Urban adaptation policies are also advancing in cities like Dubai and Doha. These include green building regulations, cool roof technologies and nature-based solutions (NbS) that address rising urban heat stress and the urban heat island effect (Salimi & Al-Ghamdi, 2020). While these initiatives reflect a growing recognition of sustainable urban design, their scalability remains limited by institutional and financial barriers.

In the context of water resource governance, recent literature recommends a conceptual policy framework to define sectoral objectives, service standards and performance monitoring systems (Turk & Zeineldin, 2024). Several supply and demand-side measures have been proposed, including the use of treated wastewater for irrigation, enhanced water-use efficiency and investment in water supply infrastructure (Amin Mir & Waqar Ashraf, 2023). Additionally, policy directions now emphasize renewable-energy-powered desalination and foreign direct investment in agricultural land as viable strategies to address future water demand (Saif et al., 2014). Given the varying levels of vulnerability and adaptive capacity across the Peninsula, the literature recommends that adaptation efforts prioritize the most vulnerable countries and sectors, especially Yemen, Bahrain, Saudi Arabia and Oman

(Aldlbahi, 2024). Strengthening community-led and nature-based adaptation is critical to support climate-resilient development and reduce both current and long-term climate vulnerabilities. Governance reforms and hazard-specific adaptation strategies along with efforts to minimize maladaptation are essential for building climate resilience. The way forward must involve integrated risk assessments, robust WEF nexus analysis and the incorporation of loss and damage evaluations into adaptation planning. Scientific research is needed to assess the determinants and effectiveness of transformative adaptation actions. Ultimately, the region's ability to respond to climate change will depend on the integration of adaptation strategies into national development plans, supported by inclusive governance, technological innovation and evidence-based policymaking.

Role of Nationally Determined Contributions (NDCs) in Adaptation Planning

The Paris Agreement, adopted in 2015, established a global framework for climate action through Nationally Determined Contributions (NDCs), which are country-specific climate action plans submitted to the United Nations Framework Convention on Climate Change (UNFCCC). NDCs serve as essential instruments through which nations outline their commitments to reduce greenhouse gas (GHG) emissions and enhance climate resilience. Within the Arabian Peninsula, NDCs form the central policy vehicles for integrating climate change adaptation into national development agendas. As of the second round of submissions under the Paris Agreement, all countries in the Arabian Peninsula have submitted updated NDCs that include adaptation components. These updated submissions demonstrate a growing recognition of the urgent need to address both sectoral and cross-sectoral vulnerabilities, especially in water, agriculture, energy and health. However, the depth, specificity and institutional arrangements for adaptation vary significantly among countries.

For example, Saudi Arabia's updated NDC 2021 places strong emphasis on ecosystem-based adaptation, sustainable land use and water resource management, aligning these objectives with the broader Vision 2030 framework. These commitments are further reinforced by initiatives such as the Saudi Green Initiative (SGI), which aims to expand vegetation cover, combat desertification and invest in renewable energy as part of a broader national strategy for climate resilience. The United Arab Emirates (UAE) has committed to enhancing climate resilience through investments in nature-based solutions, clean energy and sustainable urban development. Oman's NDC outlines adaptation priorities in coastal management, water resource conservation and food security. Meanwhile, Qatar and Bahrain have begun to expand adaptation actions, albeit with varying levels of implementation maturity and institutional capacity. Yemen's NDC, while limited in scope due to ongoing conflict, still identifies critical sectors for climate resilience support, particularly agriculture and water.

Despite these commitments, several gaps persist. Many NDCs lack detailed implementation strategies, robust monitoring and evaluation (M&E) frameworks and concrete financing mechanisms. Furthermore, the level of integration between adaptation and disaster risk reduction (DRR) policies remains uneven, which may hinder long-term resilience planning. In terms of regional alignment, there is a notable absence of a cohesive transboundary adaptation framework, even though countries share common climate vulnerabilities—such as water scarcity, desertification and rising temperatures. This fragmentation risks duplication of efforts and reduces opportunities for shared infrastructure, coordinated data systems and joint adaptation investments. Moreover, while the Water–Energy–Food (WEF) nexus is recognized in some NDCs (notably in the UAE, Saudi Arabia and Oman), few have operationalized this interdependence through cross-sectoral coordination mechanisms. The limited mainstreaming of the WEF nexus in adaptation planning represents a missed opportunity to enhance systemic resilience. To improve the effectiveness of NDCs as tools for adaptation governance in the Arabian Peninsula, countries must move beyond commitment statements toward implementation-focused planning. This includes developing clear sectoral targets, mobilizing domestic and international climate finance and enhancing institutional capacity for cross-sector coordination. Strengthening science-policy interfaces, investing in data-driven decision-making and fostering inclusive participation will also be essential for transforming NDCs into effective instruments for climate-resilient development.

9.6 Water Resource Management: Innovations and Challenges

As highlighted in previous sections, the Arabian Peninsula ranks among the most water-scarce regions globally due to its hyper-arid climate and minimal renewable freshwater resources. This scarcity has been further exacerbated by rapid population growth, urbanization and the impacts of climate change (FAO, 2021a, 2021b). The region receives <100 mm of average annual rainfall and is characterized by high evapotranspiration rates and a lack of significant surface water bodies. To meet growing water demands, it relies extensively on non-renewable groundwater, desalinated seawater (UNESCWA, 2021) and the import of virtual water (Aleisa & Al-Zubari, 2017).

In this context, effective water resource management in the Arabian Peninsula necessitates balancing technological innovation with biogeographic, socio-economic and environmental constraints. While notable improvements have been made, the region continues to exhibit a paradox rapid technological advancement coexisting with deep-rooted systemic vulnerabilities. Innovations such as desalination, cloud seeding and circular water economy practices provide partial relief, yet ongoing challenges particularly groundwater depletion, fragmented governance and heightened climate risks call for enhanced regional cooperation and adaptive policy frameworks. Achieving water security will require not only technological interventions such as desalination, wastewater reuse and smart monitoring systems but also the development and enforcement of targeted policy reforms grounded in sustainable climate adaptation strategies. Moving forward, priority should be given to demand-side water management, integration of renewable energy technologies and the establishment of robust transboundary data-sharing frameworks.

Desalination Technologies and Policy Innovations

Desalination has emerged as a cornerstone of water resource management in the Arabian Peninsula, particularly among the Gulf Cooperation Council (GCC) countries. The region accounts for over 50% of the world's desalination capacity, with Saudi Arabia, the United Arab Emirates (UAE) and Kuwait leading in the production of desalinated water (World Bank, 2021a, 2021b). Recent technological advancements in reverse osmosis (RO) and membrane distillation techniques have improved energy efficiency and reduced operational costs (Alkaisi et al., 2017). However, desalination remains an energy-intensive process and contributes to carbon emissions, making the integration of renewable energy into desalination systems a key area of ongoing research (Kesieme et al., 2013). To address water scarcity, GCC nations have pursued decentralized innovations. The UAE's cloud-seeding program, operational since the 1990s, has increased annual rainfall by 10–30% through the dispersion of hygroscopic particles (NCMS, 2019). Additionally, wastewater recycling has gained prominence, with Abu Dhabi now reusing 90% of treated sewage effluent for landscaping and agricultural purposes (Abdulrahman & Rizk, 2020).

These initiatives demonstrate the region's evolving capacity to explore alternative water sources and enhance supply resilience. The persistent water scarcity challenge also presents opportunities to integrate desalination systems with renewable energy technologies. By leveraging solar and wind energy, the GCC countries can not only mitigate the environmental impacts of desalinationparticularly high energy consumption and brine discharge but also promote long-term economic and environmental resilience in hyper-arid regions (IEA, 2023; RENA, 2022). Such integrated systems support carbon emissions reduction while contributing to broader sustainable development goals. Nonetheless, desalination continues to face challenges, including high capital investment requirements, technical complexities and limited funding for research and development. To overcome these barriers, enhanced investment in renewable-powered desalination technologies and regional research collaboration is essential. Strengthening these areas will be critical to ensuring sustainable water security and development across the Arabian Peninsula (Alkaisi et al., 2017).

Groundwater Management, Artificial Recharge and Digital Innovations

Groundwater remains a crucial water source in the Arabian Peninsula but is being extracted at unsustainable rates. To enhance groundwater sustainability, countries like Oman and Saudi Arabia have adopted Managed Aquifer Recharge (MAR) techniques, including infiltration basins and subsurface dams (Ajjur & Baalousha, 2021). Despite these efforts, over-extraction continues to pose a critical challenge, largely due to weak regulatory enforcement and insufficient institutional oversight (UNESCWA, 2019). Recent advancements in smart water management technologies are transforming groundwater governance across the region. Innovations such as Internet of Things (IoT)-based monitoring systems, remote sensing and artificial intelligence (AI)-driven predictive models are increasingly being deployed to optimize water usage and reduce losses (Kharraz et al., 2021). In particular, Saudi Arabia's National Water Strategy (NWS) incorporates real-time data analytics to manage water distribution networks and minimize leakage (Saudi Water Partnership Company, 2021).

Complementing these digital innovations, governments have also launched key national water strategies focused on Integrated Water Resources Management (IWRM), substantial investments in wastewater reuse and the promotion of smart irrigation techniques (Kharraz et al., 2021). These measures are central to adapting to growing resource constraints while ensuring efficient water allocation. Furthermore, technologies such as AI-based water monitoring, remote sensing for aquifer assessment and renewable energy-powered desalination systems are proving instrumental in addressing water challenges in hyper-arid environments. These advancements align with broader efforts to improve water governance, reduce environmental stress and build climate-resilient water infrastructure (Aleisa & Al-Zubari, 2017).

Challenges in Water Resource Management

In the context of the Arabian Peninsula, several persistent challenges continue to affect water resource management challenges that are further aggravated by the impacts of climate change, water-intensive agricultural practices, rapid urbanization and population growth (IPCC, 2022; UNESCWA, 2019). The UNESCWA report calls for urgent groundwater management reforms, including stricter regulations, improved monitoring systems and the adoption of sustainable water conservation practices to mitigate long-term water scarcity. Some of the key challenges are briefly outlined below:

Over-Dependence on Groundwater Reserves

The excessive reliance on groundwater extraction remains one of the most pressing challenges in the region (UNESCWA, 2019). The report highlights that unregulated and unsustainable withdrawal has resulted in rapid depletion of aquifers and declining groundwater tables. This over-extraction has also intensified salinization, compromising both water quality and long-term availability. The problem is compounded by the lack of adequate regulatory frameworks, limited implementation mechanisms and continued reliance on groundwater for agriculture and urban expansion. Desalination has been used to offset water scarcity, currently supplying approximately 50% of the domestic water needs of GCC states. However, the process raises additional concerns due to its high energy intensity and environmental impacts. Brine discharge, with salinity levels averaging 5–7%, disrupts marine ecosystems, while the dependence on fossil fuels for desalination exacerbates carbon emissions (Dawoud & Al-Mulla, 2012; Elsaid et al., 2020). Furthermore, managing shared water resources—particularly transboundary aquifers and basins remains complex, especially in areas such as Saudi Arabia, Oman and Yemen, where varying water demands and consumption patterns create institutional and diplomatic tensions (Sowers et al., 2017).

Policy and Governance Issues

Weak institutional frameworks, fragmented governance structures and the lack of effective transboundary cooperation hinder integrated water resource management (UNEP, 2020). These institutional limitations are further complicated by disputes over access to and management of shared water bodies, such as aquifers and rivers that traverse national boundaries. The absence of binding regional agreements often results in inefficient allocation and use of shared water resources. Additionally, the limited use of alternative water sources remains a critical issue across the region. Infrastructure for desalination, wastewater treatment and groundwater recharge demands high capital investment and often remains inaccessible to lower-income Gulf states (World Bank, 2021a, 2021b). Despite the potential of techniques such as rainwater harvesting, wastewater reuse and advanced desalination, these strategies are underutilized due to financial and technological constraints (UNESCWA, 2019). Desalination and water pumping are energy-intensive processes that significantly increase fossil fuel consumption and carbon emissions. The continued dependence on conventional energy sources in water production amplifies the region's environmental footprint and highlights the need for a transition to cleaner technologies and better-integrated resource planning.

9.7 Agricultural Adaptation and Food Security

While climatic factors pose significant threats to agricultural productivity, non-climatic factors such as a growing dependence on food imports further exacerbate food insecurity in the Arabian Peninsula. The region largely relies on desalination, groundwater extraction and treated wastewater for irrigation (Gleick, 2018). Recent studies have also highlighted a concerning trend of soil degradation and desertification. Intensive farming practices combined with climate-induced land degradation are depleting soil fertility, thereby reducing crop yields and agricultural output (Food and Agriculture Organization [FAO], 2019). More than 80% of food requirements in GCC countries are met through imports, rendering them highly vulnerable to global supply chain disruptions. In response, governments are implementing food security strategies, such as overseas land acquisitions and public–private partnerships aimed at boosting local production (Woertz, 2017).

To address these challenges, several public adaptation programs have emerged, including the deployment of advanced irrigation methods such as drip irrigation, hydroponics and aquaponics which are considered among the most water-efficient systems. These technologies are further strengthened by measures to reduce overall water consumption. Given the acute scarcity of freshwater resources, countries like the UAE have begun investing in innovative agricultural technologies. One notable initiative is the seawater greenhouse system, which utilizes seawater for evaporative cooling and humidity regulation, creating favorable conditions for crop cultivation in arid environments. This approach not only conserves freshwater but also facilitates food production in otherwise barren landscapes. The UAE has also taken the lead in agroforestry initiatives, including large-scale date palm plantations that contribute to both food security and soil stabilization. Complementary water-saving strategies, such as drip irrigation and the reuse of treated wastewater, are being implemented to maximize water use efficiency. Additionally, countries like Oman and Saudi Arabia are promoting climate-smart agriculture (CSA), which includes the use of drought-resistant crops and hydroponic farming systems to improve agricultural sustainability (Verner, 2012).

Further advancements include the introduction of climate-resilient crops, such as drought- and heat-tolerant varieties like quinoa and salt-tolerant barley. Controlled Environment Agriculture (CEA), including vertical farming and greenhouse cultivation, is gaining traction for its potential to enhance food production while minimizing land and water use. On the technological front, innovations such as solar-powered desalination and smart farming systems are playing a vital role in optimizing resource efficiency (Al-Karaghouli & Kazmerski, 2013). International organizations are also supporting regional adaptation efforts. The International Centre for Biosaline Agriculture (ICBA) and the International Centre for Agricultural Research in the Dry Areas (ICARDA) have been instrumental in advancing agricultural resilience. ICBA has successfully introduced salt-tolerant crops such as quinoa and *Salicornia*, which thrive in saline environments and support localized food production (ICBA, 2023). Meanwhile, ICARDA promotes water-efficient farming techniques and crop diversification strategies to aid smallholder farmers in arid regions (ICARDA, 2022). Table 9.4 presents a summary of key country-wise agricultural adaptation measures implemented across the Arabian Peninsula.

9.8 Conclusion

The Arabian Peninsula faces intensifying climate risks driven by rising temperatures, declining precipitation and water scarcity, all of which are compounded by socio-economic pressures such as population growth, rapid urbanization and resource-intensive development. This chapter examined the region's climate vulnerability and adaptation readiness through a multi-dimensional analysis incorporating vulnerability indices, national adaptation frameworks, sectoral interventions and SDG-linked performance metrics. Despite growing political commitment and sector-specific progress particularly in water resource management, renewable-powered desalination and climate-smart agriculture systemic vulnerabilities persist. The analysis revealed wide disparities in readiness and adaptive capacity across countries, with Yemen facing compounded development and climate challenges, while nations like the UAE and Saudi Arabia are advancing proactive adaptation strategies. The study highlights the importance of integrated policy frameworks that align national climate actions with sustainable development goals, particularly within the Water–Energy–Food (WEF) nexus. Institutional coordination, inclusive governance and transboundary cooperation remain critical gaps that must be addressed to achieve

Table 9.4 Agricultural adaptation measures in the Arabian Peninsula

Country	Agricultural adaptation measure	Description
Bahrain	Hydroponic greenhouses and aquaponics	Focus on urban farming and integrating fish-crop farming to optimize water use. It aims to promote greenhouse farming and climate-resilient crops to reduce import dependence
Kuwait	Biochar soil enhancement; saline agriculture and urban greening with drought-tolerant trees	Extensive climate research resilient; and heat and salinity tolerant crops (barley and vegetables); and tree planting (e.g., Prosopis) to reduce heat exposure and improve climate resilience
Oman	Traditional irrigation systems and treated water use	Aflaj restoration for receiving ancient underground water channel systems and recycling water for crops (dates and alfalfa) for sustainable agriculture
Qatar	Tech-based farming and renewable energy-driven water use	Use of solar-powered desalination; hydroponics; indoor farming; and smart irrigation to reduce water use and tailor local food sustainability
Saudi Arabia	Climate resilient agriculture	Drought resilient crops (Quinoa and millet suited to arid conditions); drip irrigation systems for water-intensive crops and reduce groundwater depletion. It also contributes to land restoration and drought resilience to promote efficient irrigation and sustainable land use
United Arab Emirates	Vertical farming; desalinated water for irrigation and salt-tolerant crop research	Focus on the adoption of climate-smart agri-innovations to support smallholder farmers and controlled environment agriculture (hydroponics) to reduce water use and increase productivity
Yemen	Rainwater harvesting and traditional practices	Investment in infrastructure (water tanks and irrigation spillways) to support drought-prone agricultural land and revival of farming terraces for soil and water conservation. It also focuses on the revival of Indigenous seeds

climate-resilient development. Moving forward, countries across the Arabian Peninsula must prioritize investment in nature-based solutions, digital innovation for water and agriculture and cross-border data sharing. Enhancing science-policy linkages, securing climate finance and embedding adaptation into long-term planning will be key to building resilience in this hyper-arid and geopolitically significant region.

References

Abdella, F. I. A., El-Sofany, W. I., & Mansour, D. (2024). Water scarcity in the Kingdom of Saudi Arabia. *Environmental Science and Pollution Research, 31*, 27554–27565. https://doi.org/10.1007/s11356-024-33024-3

Abdulrahman, M. A., & Rizk, S. (2020). *Wastewater reuse in the Arab region: Case studies from the UAE*. Springer. https://doi.org/10.1007/978-3-030-31684-6

Ajjur, S. B., & Baalousha, H. M. (2021). A review on implementing managed aquifer recharge in the Middle East and North Africa region: Methods, progress and challenges. *Water International, 46*(4), 578–604. https://doi.org/10.1080/02508060.2021.1889192

Akasha, H., Ghaffarpasand, O., & Pope, F. D. (2023). Climate change, air pollution and the associated burden of disease in the Arabian Peninsula and neighboring regions: A critical review of the literature. *Sustainability, 15*(4), 766. https://doi.org/10.3390/su15043766

Aldlbahi, S. M. M. (2024). Environmental adaptation tools as an approach to meet challenges in urban planning in the Kingdom of Saudi Arabia. *IOP Conference Series: Earth and Environmental Science, 1283*(1), 012003. https://doi.org/10.1088/1755-1315/1283/1/012003

Aleisa, E., & Al-Zubari, W. (2017). Wastewater reuse in the countries of the gulf cooperation council (GCC): The lost opportunity. *Environmental Monitoring Assessements, 189*, 553. https://doi.org/10.1007/s10661-017-6269-8

Alkaisi, A., Mossad, R., & Sharifian, B. (2017). A review of the water desalination systems integrated with renewable energy. *Energy Procedia, 110*, 268–274. https://doi.org/10.1016/j.egypro.2017.03.138

Al-Karaghouli, A., & Kazmerski, L. L. (2013). *Renewable energy applications in water desalination*. National Renewable Energy Laboratory, Golden. https://nmwaterconference.nmwrri.nmsu.edu/wp-content/uploads/2011/Relevant-Papers/renewable_energy_water_desal.pdf

Almazroui, M., Islam, M. N., Jones, P. D., Athar, H., & Rahman, M. A. (2012). Recent climate change in the Arabian Peninsula: Seasonal rainfall and temperature climatology of Saudi Arabia for 1979–2009. *Atmospheric Research, 111*, 29–45. https://doi.org/10.1016/j.atmosres.2012.02.013

Amin Mir, M., & Waqar Ashraf, M. (2023). The challenges and potential strategies of Saudi Arabia's water Resources: A review in an analytical way. *Environmental Nanotechnology, Monitoring and Management, 20*, 100855. https://doi.org/10.1016/j.enmm.2023.100855

Buchir, L. M. S. T., & Detzel, D. H. M. (2023). The role of the governance on the climate vulnerability index definition in Mozambique. *GeoJournal, 88*, 1687–1702. https://doi.org/10.1007/s10708-022-10711-7

Cooper, R. (2020). Climate change risks and opportunities in the middle east and north Africa. The institute of development studies and partner organisations. Report. https://hdl.handle.net/20.500.12413/15798

Daoudy, M., Al-Saidi, M., Al Manji, A., Ayoub, J., Bateh, F., Dajani, M., et al. (2024). *Troubled waters in conflict and a changing climate: Transboundary basins across the middle east and north Africa*. Carnegie Endowment for International Peace. https://carnegie-production-assets.s3.amazonaws.com/static/files/Daoudy_et_al_ME_Climate_Change_draft.pdf

Dawoud, M. A., & Al-Mulla, M. M. (2012). Environmental impacts of seawater desalination: Arabian Gulf case study. *International Journal of Environment and Sustainability, 1*(3), 22–37.

D'Odorico, P., Davis, K. F., Rosa, L., Carr, J. A., Chiarelli, D., Dell'Angelo, J., et al. (2018). The global food-energy-water nexus. *Reviews of Geophysics, 56*(3), 456–531. https://doi.org/10.1029/2017RG000591

Elsaid, K., Kamil, M., Sayed, E. T., Abdelkareem, M. A., Wilberforce, T., & Olabi, A. (2020). Environmental impact of desalination technologies: A review. *Science of the Total Environment, 748*, 141528. https://doi.org/10.1016/j.scitotenv.2020.141528

ESCWA. (2022). *Climate change adaptation in the Arab region: Policy frameworks and implementation*. United Nations Economic and Social Commission for Western Asia. Accessed 5 March 2025

FAO. (2019). *The state of food and agriculture*. Food and Agriculture Organization of the United Nations.

FAO. (2021a). *Water scarcity and agriculture in the Arab region*. Food and Agriculture Organization. Accessed 22 March 2025.

FAO. (2021b). *Groundwater management in arid regions: Case studies from Oman*. Food and Agriculture Organization. Accessed 7 April 2025.

Francis, D., & Fonseca, R. (2024). Recent and projected changes in climate patterns in the Middle East and North Africa (MENA) region. *Science of Reports, 14*, 10279. https://doi.org/10.1038/s41598-024-60976-w

Gleick, P. H. (2018). Water, drought, climate change, and conflict in the Syria. *Water Policy, 20*(5), 871–885. https://doi.org/10.1175/WCAS-D-13-00059.1

Günel, G. (2016). The infinity of water: Climate change adaptation in the Arabian Peninsula. *Public Culture, 28*(2), 291–315. https://doi.org/10.1215/08992363-3427463

Hamed, M. M., Al-Hasani, A. A. J., Nashwan, M. S., Saadi, Z., & Shahid, S. (2024). Assessing the growing threat of heat stress in North Africa and the Arabian Peninsula connected to climate change. *Journal of Cleaner Production, 447*, 141639. https://doi.org/10.1016/j.jclepro.2024.141639

ICARDA. (2022). *Sustainable farming in dry regions*. International Center for Agricultural Research in the Dry Areas.

ICBA. (2023). *Climate-resilient agriculture in drylands*. International Center for Biosaline Agriculture.

IEA. (2023). *Analysis and projections for renewables in 2023 through 2028*. Agency for International Energy. https://www.iea.org/reports/renewables-2023. Accessed 25 May 2025

IPCC. (2007). Climate change 2007: Impacts, adaptation, and vulnerability. In M. L. Parry, O. F. Canziani, J. P. Palutikof, P. J. van der Linden, & C. E. Hanson (Eds.), *Contribution of working groups II to the fourth assessment report of intergovernmental panel on climate change*. Cambridge University Press. Retrieved from https://www.ipcc.ch/site/assets/uploads/2018/03/ar4_wg2_full_report.pdf

IPCC. (2012). The special report for managing the risks of extreme events and disasters to advance climate change adaptation (SREX). In C. B. Field, et al. (Eds.), *Managing the risks of extreme events and disasters to advance climate change adaptation a special report of working groups I and II of the intergovernmental panel on climate change* (p. 582). Cambridge University Press.

IPCC. (2014). Climate change 2014: Impacts, adaptation, and vulnerability. Part A: Global and sectoral aspects. In C. B. Field, V. R. Barros, D. J. Dokken, K. J. Mach, M. D. Mastrandrea, T. E. Bilir, M. Chatterjee, K. L. Ebi, Y. O. Estrada, R. C. Genova, B. Girma, E. S. Kissel, A. N. Levy, S. MacCracken, P. R. Mastrandrea, & L. L. White (Eds.), *Contribution of working group II to the fifth assessment report of the intergovernmental panel on climate change* (p. 1132). Cambridge University Press.

IPCC. (2022). Climate change 2022: Impacts, adaptation and vulnerability. In H.-O. Pörtner, D. C. Roberts, M. Tignor, E. S. Poloczanska, K. Mintenbeck, A. Alegría, M. Craig, S. Langsdorf, S. Löschke, V. Möller, A. Okem, & B. Rama (Eds.), *Contribution of working group II to the sixth assessment report of the intergovernmental panel on climate change*. Cambridge University Press. https://doi.org/10.1017/9781009325844.

Kesieme, U. K., Milne, N., Aral, H., Cheng, C. Y., & Duke, M. (2013). Economic and environmental impacts of desalination technologies. *Desalination, 309*, 197–207.

Kharraz, J., El-Sadek, A., Ghaffour, N., & Mino, E. (2021). Water scarcity and technological innovations in the MENA region. *Water Science and Technology, 83*(1), 25–39.

Miralles-Wilhelm, F., Clarke, L., Hejazi, M., Kim, S., Gustafson, K., Muñoz-Castillo, R., & Graham, N. (2014). Physical impacts of climate change on water resources, water global practice water, poverty, and the economy. In *Discussion paper*. World Bank https://documents1.worldbank.org/curated/en/137141486625149687/pdf/112726-WP-PUBLIC-Physical-impacts-of-climate-change-on-water-resources.pdf

Miralles-Wilhelm, F., Hejazi, M. I., Kim, S., Yonkofski, C., Watson, D., Kyle, P. G., Liu, Y., Vernon, C. R., Delgado, A., Edmonds, J., & Clarke, L. (2018). Water for food and energy security.

National Center of Meteorology and Seismology (NCMS). (2019). *Cloud seeding program annual report*. UAE Government. Accessed 18 March 2025.

RENA. (2022). Technology brief on renewable energy for desalination. *International Agency for Renewable Energy: Renewable Energy for Desalination, 15*, 25.

Saharwardi, M. S., Dasari, H. P., Gandham, H., Ashok, K., & Hoteit, I. (2024). Spatiotemporal variability of hydro-meteorological droughts over the Arabian Peninsula and associated mechanisms. *Scientific Reports, 14*(1), 7. https://doi.org/10.1038/s41598-024-70869-7

Saif, O., Mezher, T., & Arafat, H. A. (2014). Water security in the GCC countries: Challenges and opportunities. *Journal of Environmental Studies Science, 4*, 329–346. https://doi.org/10.1007/s13412-014-0178-8

Salimi, M., & Al-Ghamdi, S. G. (2020). Climate change impacts on critical urban infrastructure and urban resiliency strategies for the Middle East. *Sustainable Cities and Society, 54*, 101948. https://doi.org/10.1016/j.scs.2019.101948

Saudi Water Partnership Company. (2021). *Saudi Arabia's national water strategy 2030*. Government Report.

Saudi Ministry of Environment, Water and Agriculture (MEWA). (2021). *National action plan on climate change in Saudi Arabia. Government Report.*

Saudi Vision. (2023). Retrieved from https://www.vision2030.gov.sa/en

Sowers, J., et al. (2017). *Climate change and water resources in the middle east*. Oxford University Press.

Turk, K. G. B., & Zeineldin, F. I. (2024). Water security in Saudi Arabia. In A. E. Ahmed, J. M. Al-Khayri, & A. A. Elbushra (Eds.), *Food and nutrition security in the Kingdom of Saudi Arabia* (Vol. 1). Springer. https://doi.org/10.1007/978-3-031-46716-5_4

UNESCWA. (2019). *Groundwater management in the Arab region*. United Nations Economic and Social Commission for Western Asia.

UNESCWA. (2021). *Water scarcity and sustainable development in the Arab region*. United Nations Economic and Social Commission for Western Asia.

UNFCCC. (2023b). *Saudi Arabia's third NDC submission*. United Nations Framework Convention on Climate Change.

United Nations Environment Programme (UNEP). (2020). *Water governance in the Arab region*. UNEP. Accessed 7 February 2025.

Verner, D. (2012). *Adaptation to changing climate in the Arab countries: A case for adaptation governance and leadership in building climate resilience*. World Bank. https://doi.org/10.15966/978-0-8213-9458-8

Vinodh Kumar, B., Ullah, S., Kumar, T. V. L., & Al-Ghamdi, S. G. (2024). Amplification of temperature extremes in the Arabian Peninsula under warmer worlds. *Scientific Reports, 14*(1), 1–13. https://doi.org/10.1038/s41598-024-67514-8

Waha, K., Krummenauer, L., Adams, S., et al. (2017). Climate change impacts in the Middle East and Northern Africa (MENA) region and their implications for vulnerable population groups. *Regional Environmental Change, 17*, 1623–1638. https://doi.org/10.1007/s10113-017-1144-2

Woertz, E. (2017). *Food security in the Gulf Cooperation Council*. Oxford University Press.

World Bank. (2021a). *Water and climate resilience in the Middle East and North Africa*. World Bank Publications. https://www.worldbank.org/en/country/gcc

World Bank. (2021b). *Water and climate resilience in the Middle East and North Africa*. World Bank Publications.

Zakzak, L., Luomi, M., Subramani, K., Fuller, G., Salem, F., & Lafortune G. (2024). *The arab region SDG index and dashboards 2023/2024: Towards just transitions in the Arab States*. Mohammed bin Rashid School of Government (MBRSG) and UN Sustainable Development Solutions Network (UN SDSN).

Chapter 10
Climate Adaptation Strategies and Policy Frameworks in the Arabian Peninsula

Md Arif Hasan, Fatema Hossain, and Syed Masiur Rahman

10.1 Introduction

Climate change is often regarded as the greatest threat to the existence of humans, as well as all other forms of life (Alvarez et al., 2021; OHCHR, 2021). Extreme weather events such as floods, cyclones, droughts, heatwaves and wildfires are becoming more frequent across the world due to rising global temperatures, which threaten human health, biodiversity, ecosystems and food and water security (OHCHR, 2021; Watts et al., 2018). The Arabian Peninsula (AP) is one of the most climate-vulnerable regions in the world owing to its semi- to hyper-arid climate, water scarcity, unpredictable rainfall patterns and climate-sensitive agriculture (Al-Sakkaf et al., 2024; Hamed et al., 2024; Vinodhkumar et al., 2024). Importantly, the geographical and political boundaries of the AP are characterised differently by researchers. According to Masoumi et al. (2024), the AP includes seven whole countries Yemen and all six Gulf Cooperation Council (GCC) members [Bahrain, Saudi Arabia, Kuwait, Qatar, Oman and the United Arab Emirates (UAE)]. Marcenaro (2024) used the same seven-country definition when analysing the impact of climate change on water stress in the AP. However, Akasha et al. (2023) argued that although Bahrain is not geographically on the AP, it should be included for geopolitical reasons; they also identified the southern parts of Iraq and Jordan as part of the AP. Almazroui et al. (2016)

M. A. Hasan (✉)
Climate Change Response Unit, Wellington City Council, 113 The Terrace, Wellington Central, Wellington 6011, New Zealand
e-mail: Arif.Hasan@wcc.govt.nz

F. Hossain
Intellectual Property Office of New Zealand, Ministry of Business, Innovation and Employment, 15 Stout Street, Wellington Central, Wellington 6011, New Zealand

S. M. Rahman
Department of Civil and Environmental Engineering, King Fahd University of Petroleum and Minerals, Dhahran 31261, Saudi Arabia

F. Mushtaq et al. (eds.), *Climate Change in the Arabian Peninsula*,
https://doi.org/10.1007/978-3-032-02481-7_10

excluded Iraq but included all of Jordan along with the other seven countries in their entirety. This study adopts the definition comprising Yemen, Bahrain, Saudi Arabia, Kuwait, Qatar, Oman and the UAE while excluding Jordan and Iraq. First, these seven countries were common to all definitions reviewed above. Second, because this research assesses climate-related government policy initiatives, including only parts of a country (e.g., southern Iraq or Jordan) would require sub-national data and policy documentation that are often unavailable. Finally, Akasha et al., (2023) did not delineate precise political boundaries for the southern regions of Iraq or Jordan an essential prerequisite given the aims and scope of this investigation.

Among the countries within the AP, Saudi Arabia's total greenhouse gas emissions are far higher than other countries. Saudi Arabia's total emissions increased from 159 million tonnes of carbon dioxide equivalent ($MtCO_2e$) in 1974 to 811 $MtCO_2e$ in 2022, a fivefold increase. Total greenhouse gas emissions from Bahrain, Yemen, Kuwait, Qatar, Oman and the United Arab Emirates (UAE) in 2022 were 70, 38, 168, 195, 137 and 295 $MtCO_2e$, respectively. Emissions have increased for all the countries within the AP between 1974 and 2022, but Oman's emissions increased the highest, around 9.3 times increase. Kuwait's emissions saw the smallest increase during this period, increasing by 2.8 times. Emissions increased by Bahrain, Yemen, Qatar and the UAE were 3.1, 5.4, 6 and 5.7 times, respectively (Fig. 10.1).

On a per capita basis, emissions from AP countries, except Yemen, are quite high compared to other countries. This is due to the high share of fossil fuels in the energy mix. The per capita greenhouse gas emissions of Bahrain, Saudi Arabia, Yemen, Kuwait, Qatar, Oman and the UAE in 2022 were 46 tonnes of CO_2e (tCO_2e), 25, 1, 37, 73, 29 and 29 tCO_2e, respectively (The World Bank, 2024). So, on the one hand, the AP countries are high emitters and significant contributors to climate

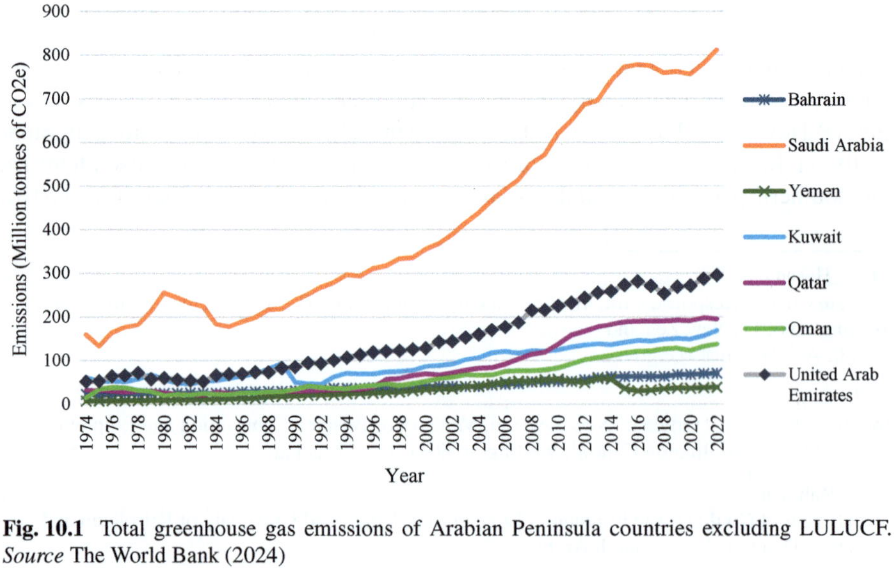

Fig. 10.1 Total greenhouse gas emissions of Arabian Peninsula countries excluding LULUCF. *Source* The World Bank (2024)

change, while on the other hand, they are highly vulnerable to climate change impacts. This dichotomy poses a unique policy challenge to the region and requires urgent and ambitious climate change mitigation and adaptation initiatives at the national, regional and international levels. Some regional and international collaborations have already been established. For example, the Middle East Green Initiative (MGI) is a Saudi Arabia-led regional collaboration that aims to plant 50 billion trees and thereby restore over 74 million hectares across the region (Saudi & Middle East Green Initiatives, 2024b). Out of the 50 billion trees, 10 billion (20%) trees will be planted within the borders of Saudi Arabia, while the remaining trees will be planted across the region. Besides carbon dioxide absorption, this project has several climate adaptation benefits, such as soil stabilisation and erosion control, protection from flood and dust storms and restoration of ecological functions (Saudi & Middle East Green Initiatives, 2024b). The UAE government supported to establish another Regional Collaboration Centre (RCC) for the Middle East and North Africa (MENA) and South Asia (SA) through signing a Memorandum of Understanding (MoU) between the World Green Economy Organisation (WEGO) and the United Nation (UN) Climate Change (UNFCCC, 2024). All countries of the AP are part of this international collaboration that aims to fill any knowledge gaps in climate change adaptation, develop strategies to secure climate finance, promote regional engagement in global climate actions, build youth capability to contribute effectively to the regional and global climate crisis and introduce novel techniques and resources to tackle climate change (World Green Economy Organization, 2024).

This research explores climate change mitigation and adaptation initiatives taken by the AP countries at the national, regional and international levels and identifies the progress made so far through analysing four indicators viz emissions intensity, energy intensity, per capita energy consumption and percentage of renewables in the electricity mix. The second section reviews the literature, while the third section describes the overall methodology of this research, including the selection of the indicators for evaluating existing climate initiatives. The results of the evaluation of climate initiatives are presented in section four and the analysis of climate change mitigation policies is highlighted in section five. Section six explores the climate change adaptation measures taken by the AP countries, while the last section presents any policy gaps with concluding remarks.

10.2 Review of Climate Impacts and Policy Responses in the Arabian Peninsula

There have been some climate change-related studies on the AP, but the focus of these studies was mainly on the impacts of climate change on this region. Rao et al. (2024) discussed the effects of climate change-induced global warming on human and natural systems in the Arabian Peninsula with a focus on the UAE and concluded that the country could see an increase in annual precipitation by up to 30%. As a result, the country could observe an increase in heavy rainfall events like flooding, which could significantly impact the country's economy and human life. Akasha et al. (2023) investigated the interrelating effects of climate change and air pollution on the inhabitants of the Arabian Peninsula and its adjacent regions. The study revealed that climate change impacts air pollution concentrations, which, together with extreme dust events, cause respiratory and cardiovascular health issues. Marcenaro (2024) investigated the interacting effects of extreme climate events (e.g. flash floods) on water stress in the Arabian Peninsula and highlighted that such flooding events could contaminate groundwaters, damage water and other infrastructures (pipelines, pumping stations, roads, bridges, etc.) and exacerbate agricultural productivity and human health in the region. The impacts of climate change on freshwater fishes in the Arabian Peninsula were investigated by Masoumi et al. (2024). They identified seven species that are highly vulnerable to climate change as they are likely to lose their climate-friendly habitats.

A number of researchers have investigated the changing patterns of temperature and precipitation in the Arabian Peninsula. Almazroui et al. (2016) assessed the temperature and precipitation changes in the northern and southern Arabian Peninsula based on the Assessment Report 4 (AR4) Global Climate Models (GCM) data and revealed that the northern Arabian Peninsula (NAP) would observe a higher increase in temperature compared to the southern Arabian Peninsula (SAP) during summer. The results also revealed that there is a robust decrease in precipitation in the NAP during winter, while the SAP region shows a robust increase in precipitation in summer but remains dry in winter. Similar results were found for both the NAP and SAP regions when Almazroui et al. (2017) used the AR5 GCM data. Kotwicki and Al Sulaimani (2009) predicted that the aridification process in the AP is proceeding. Hamed et al. (2024) assessed the impact of global warming on heat stress in the Arabian Peninsula and North Africa region and revealed that an increase in global temperature by 0.5° would result in a significant increase in the number of days with intense heat stress in the Arabian Peninsula. The study proposed urgent climate action strategies covering both mitigation and adaptation so that human life and well-being are protected from extreme weather events.

The importance of linking climate policies with a country's key non-climate strategies is highlighted by a few researchers who are looking to effectively tackle climate change in a sustainable way. Almulhim and Cobbinah (2024) highlighted the importance of integrating climate management with urban policy for Saudi Arabian cities to ensure climate resiliency in the country. Matar et al. (2024) examined the role of various energy policies in achieving Saudi Arabia's pledged emissions reduction target in its nationally determined contribution (NDC). Results revealed that the announced policies of Saudi Arabia, such as fuel price reforms, the use of renewable technology and natural gas in the electricity mix and the adoption of carbon capture and storage technologies in the fertiliser and petrochemical sectors would contribute to reducing 130 million tonnes of carbon dioxide equivalent (mtCO_2e) by 2030, which is around half of the country's total emissions reduction target pledged in the NDC. Malik et al. (2019) explored the drivers, barriers and emissions reduction potentials of various renewable energy policies in GCC countries and estimated that the deployment of renewable energy policies would reduce a total of 533 $mtCO_2e$ across all the GCC countries by 2030.

Rahman et al. (2022a) analysed the impacts of climate change on Saudi Arabia's food security and highlighted the importance of having appropriate food security and agricultural policies that will diversify the country's food system while increasing productivity and helping in combating climate change. Sharma et al. (2024) and Banhidarah et al. (2020) studied farming practices in the UAE and identified that innovative farming practices supported by appropriate agricultural and water-energy policies could help mitigate climate change in the country. Wiebelt et al. (2013) studied Yemen's food and income insecurity because of climate change and suggested appropriate policy initiatives to incentivise the production of water-efficient crops and promote investments in agricultural alternatives (e.g. coffee) and water-saving technologies through private investors. A study funded by the Qatar National Research Fund examined the relationship between climate change risks and the energy-water-food (EWF) nexus and highlighted the importance of having quantifiable EWF metrics to enhance resilience and balance resource security (Serdar et al., 2024). Lahlou et al. (2020, 2023) also highlighted the importance of EWF in Qatar's harsh climate.

Industrial policies are also important to curtail emissions from major industries in the AP. Khan et al. (2023) studied the emissions reduction opportunities in the cement industries of Saudi Arabia and emphasised the adoption of effective industrial policies related to carbon capture and storage (CCS), demand management, low-carbon alternatives, material efficiency standards, etc. Appropriate industrial policies, such as incentivisation of green hydrogen projects in Oman's northern-western region, are proposed to reduce emissions from various industries within the region (Sial et al., 2024). Al-Salem (2015) examined Kuwait's emissions from petroleum refining industries and emphasised on implementing new carbon capture strategies to reduce emissions from this sector. In the UAE, emissions from industries including the cement, steel, iron and aluminium sectors contribute to around one-third of the country's total energy emissions and policy options such as energy pricing, feed-in-tariff, clean and renewable energy policies, energy conservation and efficiency policies are anticipated to significantly reduce emissions from this sector (Khondaker et al., 2016). Babonneau et al. (2022) studied Qatar's transition to net-zero emissions and suggested introducing CCS facilities in every industrial sector to reduce emissions from this sector.

The importance of integrating climate policies with a country's transport policies is highlighted by several researchers, as transport is one of the biggest sources of emissions for most countries across the world (Hasan et al., 2019, 2022; Rahman et al., 2017, 2022b). Mendez et al. (2023) studied the advantages and disadvantages of various transport modes for Qatar and the GCC and suggested adopting hydrogen fuel cell vehicles for the region to reduce emissions. They also highlighted a few policy initiatives in this regard, including setting vehicle emissions standards, incentivising fuel cell vehicles and clean fuels, diversifying fuel types for public transport and gradually reducing the subsidy on fossil fuels. Waked and Afif (2012) studied transport emissions in the Middle East region, including Yemen and found that around

70% of the vehicles in the region (excluding GCC countries) are over 15 years old. They suggested regulating vehicle emissions standards to control emissions per vehicle in the region. Alkheder (2021) advocated for promoting public transport strategies in Kuwait to reduce emissions as well as congestion and suggested introducing dedicated bus lanes and sustainable road-pricing schemes. AlSabbagh et al. (2017) assessed various road transport emissions reduction policies for Bahrain and revealed that setting certain vehicle fuel economy standards is the most preferred emissions reduction option for the country.

The waste sector is another important source of emissions and an integrated waste-climate policy could reduce waste generation and help tackle climate change. Rahman et al. (2021) analysed Saudi Arabia's solid waste emissions and proposed policy options related to waste-to-energy, carbon capture and utilisation (CCU), waste reduction, recycling and composting. Attia (2020) studied the sustainable architecture of traditional multi-storey houses in Yemen and highlighted the use of solid waste as a fuel in the brick manufacturing industry. Abbas (2022) performed a cost–benefit analysis to evaluate the feasibility of anaerobic digestion for managing sewage sludge and biodegradable waste in Bahrain. Results revealed that anaerobic digestion is viable for Bahrain and this technology would reduce 535 kilotonnes of carbon dioxide equivalent ($ktCO_2e$) each year and generate an annual revenue of USD 4.3 million and USD 0.6 million from electricity and biofuel sales, respectively. Umar (2021) studied the electricity generation and emissions reduction potentials of the waste-to-energy initiative in Oman and revealed that this initiative could generate 2.9 gigawatts hour (GWh) of electricity each day and reduce annual emissions by 2.5 million tonnes of carbon dioxide equivalent ($MtCO_2e$). A recent study on food waste in Kuwait found that one ton of anaerobically digested food waste could generate 20 kg of liquid fuel and 207 kilowatts hour (KWh) of electricity (Aleisa & Alsaleh, 2024).

10.3 Methodological Approach: Policy Review and Indicator Framework

This study used secondary sources to explore the climate change mitigation and adaptation policy responses of respective government agencies in the AP. Governmental policy documents and reports of each AP country were investigated along with scientific research papers to understand various climate initiatives at the local, national, regional and international levels. In addition, website materials of various national and international climate change and environmental organisations and authorities were analysed, including the United Nations Framework Convention on Climate Change (UNFCCC), World Green Economy Organization (WGEO) and respective Ministries of Environment and Climate Change of AP countries. Examples of some of the public policy documents, international reports and regional/international agreements include national greenhouse gas mitigation and adaptation strategies, nationally determined contributions (NDCs), national mitigation and adaptation plans, national renewable energy policies, national development plans, visionary and long-term plans, sectoral development plans etc. As governments often revise their climate change-related policy initiatives based on new international agreements, updated data, political priorities and socio-economic considerations, the most up-to-date climate commitments and actions of the governments of the AP countries were investigated where available. Another rationale for adopting the most recent policy information was to represent innovation in the renewable energy sector, advances in carbon capture technologies and the availability of international funding for green infrastructure projects.

Besides identifying various climate change mitigation and adaptation initiatives of each of the AP countries, this study evaluated the effectiveness of various initiatives so far using four indicators adapted from Hasan et al. (2020) and Kameyama and Kawamoto (2018). These four indicators are:

1. Emissions intensity
2. Energy intensity,
3. Energy consumption per capita and
4. Share of renewables in the electricity mix.

A few factors are considered when selecting the indicators. Firstly, the availability of data to determine the indicators was considered. Secondly, emphasis was placed on the simplicity and comprehensibility of the indicators to facilitate the process of identifying policy gaps by individual countries. Finally, these indicators are not merely sectoral emissions but a set of options that better explain a country's policy portfolio, allowing for recommendations targeted at specific policy objectives. The details of the data used to develop the indicators are presented in Table 10.1.

10.4 Evaluation of Climate Initiatives

The results of the assessment of four indicators i.e. emissions intensity, energy intensity, energy consumption per capita and the percentage of renewables in the electricity mix are discussed below.

Emissions Intensity

The term emissions intensity refers to the total amount of emissions generated per unit of energy consumption. The unit of emissions intensity for this study is tonnes of carbon dioxide equivalent per terajoule (tCO_2e/TJ) (Hasan et al., 2020; Kameyama & Kawamoto, 2018). Oman's emissions intensity has dropped drastically from 408 tCO_2e/TJ in 2000 to 129 tCO_2e/TJ in 2022, a 68% decrease (Fig. 10.2). This indicates that despite an increased consumption of energy, Oman's emissions remained relatively stable. This could be due to an increased share of renewables in the energy mix, improved energy efficiency, adoption of new and upgraded low-carbon technologies,

Table 10.1 The details of the data used in this study

Variables	Time period	Units	Data sources
Total greenhouse gas emissions	1974–2022	Million tonnes of carbon dioxide equivalent ($MtCO_2e$)	The World Bank (2024)
Population	2000–2022	Individual	The World Bank (2024)
Energy consumption	2000–2022	Terajoule (TJ)	International Energy Agency (2025)
Renewables share in the electricity mix	2022	Percentage (%)	International Energy Agency (2025)
Gross domestic product	2000–2022	Current billion USD (US$)	The World Bank (2024)
Emissions intensity	2000–2022	Tonnes of carbon dioxide equivalent per terajoule (tCO_2e/TJ)	International Energy Agency (2025); The World Bank (2024)
Energy intensity	2000–2022	Terajoule per million USD (TJ/million US$)	International Energy Agency (2025); The World Bank (2024)

implementation of CCS technologies and the promotion of circular economy practices. The emissions intensities of Saudi Arabia and the UAE were the lowest among the AP countries and both of their emissions intensities in 2022 were 120 tCO_2e/TJ. However, the emissions intensities of these two countries remained almost unchanged over the past two decades. Saudi Arabia's emissions intensity increased by only 0.6% between 2000 and 2022, while the UAE's emissions intensity decreased by only 10% in this period. Qatar, Kuwait and Bahrain's emissions intensities decreased gradually in the past decades and their emissions intensities in 2022 were 195, 164 and 228 tCO_2e/TJ, respectively. Yemen's emissions intensity among the AP countries was the highest in 2022, at 371 tCO_2e/TJ. Between 2000 and 2022, the emissions intensity of Yemen increased by 55%, the highest among the AP countries. This indicates that the country needs to focus on renewable energy generation, energy efficiency, circular economy practices, low-carbon appliances, etc.

Energy Intensity

Energy intensity measures the energy efficiency of an economy and it shows how much energy is used for each unit of economic output (Hasan et al., 2020; Kameyama & Kawamoto, 2018). This study used the current gross domestic product (GDP) in US dollars as an economic output. Therefore, the unit of energy intensity in this study is terajoule per million US dollars (TJ/Million US$). From Fig. 10.3, it is evident that energy intensity decreased for all countries within the AP except Oman. Oman's energy intensity was the lowest among the AP countries in 2000 (5.7 TJ/Million US$), but it increased to 9.5 TJ/Million US$ in 2022, which was the highest among the AP countries. This indicates that Oman's GDP growth in

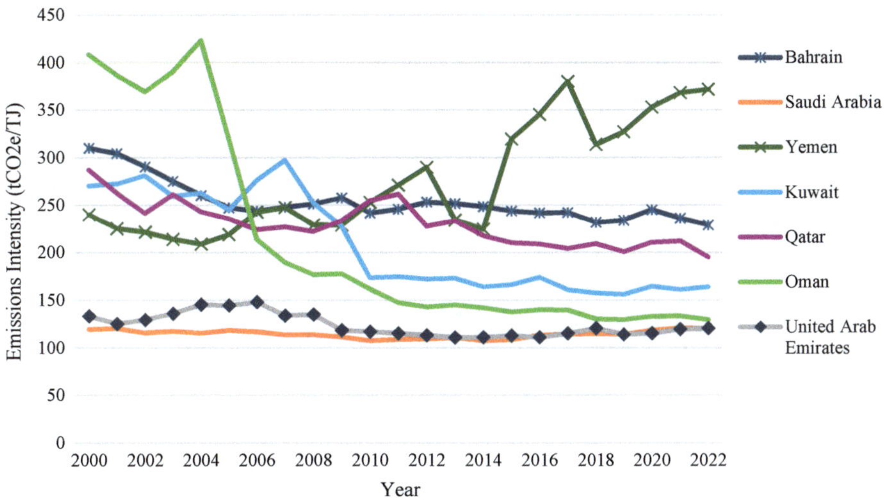

Fig. 10.2 Emissions intensity by country in the Arabian Peninsula. *Source* International Energy Agency (2025); The World Bank (2024)

the past two decades was highly dependent on energy-intensive sectors (e.g. fossil fuels). Saudi Arabia's energy intensity was the highest at 15.7 TJ/Million US$ in 2000, but it decreased to 6.1 TJ/Million US$ in 2022, over a 61% decrease. Both Qatar and Yemen observed a 69% decrease in emissions intensity between 2000 and 2022. Although both countries observed a significant decrease in energy intensity, the reasons could be different. For Qatar, their economic diversification (e.g. hosting the FIFA World Cup, tourism, real estate) could have contributed significantly to reducing their energy intensity, while Yemen's significant decrease in energy consumption in the past two decades due to political unrest and war could result in a decrease in energy intensity. Energy intensities of Qatar and Yemen in 2022 were 4.2 and 4.4 TJ/Million US$, respectively. Between 2000 and 2022, emissions intensities also decreased for Bahrain, Kuwait and the UAE by 54, 34 and 47%, respectively. In 2022, the emissions intensities of Bahrain, Kuwait and the UAE were 6.6, 5.6 and 4.9 TJ/Million US$, respectively.

Energy Consumption Per Capita

The total population of a country is often considered an essential factor for greenhouse emissions and energy consumption. So, this study uses per capita energy consumption as an indicator to assess each of the AP countries' progress in mitigating climate change. Figure 10.4 shows that per capita energy consumption was the highest in Qatar throughout the past two decades, while Yemen's per capita energy consumption was the lowest in the same period. Qatar's per capita emissions saw some major fluctuations between 2006 and 2012, but the country's per capita energy consumption remained almost unchanged between 2000 and 2022. In 2000, the per capita energy consumption was 378 gigajoules (GJ), while it was 375 GJ in 2022.

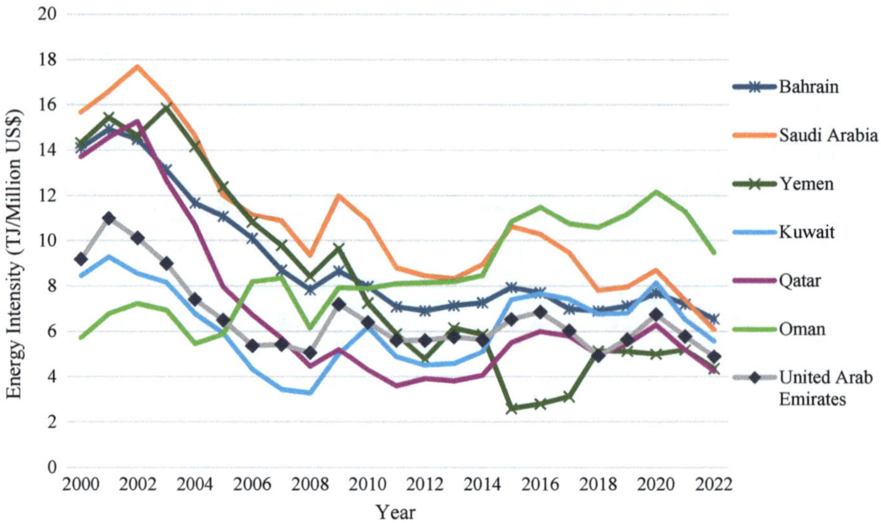

Fig. 10.3 Energy intensity by country in the Arabian Peninsula. *Source* International Energy Agency (2025); The World Bank (2024)

Similarly, Bahrain's per capita energy consumption remained unchanged between 2000 and 2022. The country's per capita energy consumption was 200.4 GJ both in 2000 and 2022. The UAE's per capita energy consumption decreased from 274 to 244 GJ in 2022, a decrease of 11%. Oman observed a fourfold increase in per capita energy consumption between 2000 and 2022, which is the highest among the AP countries: from 56 GJ in 2000 to 224 GJ in 2022. Saudi Arabia's per capita energy consumption increased by 14%, from 184 GJ in 2000 to 210 GJ in 2022. Kuwait's per capita energy consumption in 2000 was 163 GJ, but it increased by 37% to 223 GJ in 2022.

Share of Renewables in the Electricity Mix

As most AP countries are highly dependent on fossil fuels for their GDP and economic growth, the share of renewable energy in the electricity mix is an important factor for evaluating these countries' energy policies and measuring their climate change mitigation progress. Yemen's share of renewables in the electricity mix was the highest among all AP countries in 2022, 19.7% (Fig. 10.5). The UAE's share of renewables in the electricity mix was the second highest (5%), followed by Oman (2.5%), Qatar (1.1%), Saudi Arabia (0.6%), Kuwait (0.2%) and Bahrain (0.1%). Importantly, Saudi Arabia set a 50% renewable energy target by 2030, while Bahrain set a 5 and 10% renewable energy target by 2025 and 2035, respectively (Al-Sarihi, 2023). Kuwait and Qatar set a 15 and 20% renewable energy target, respectively, by 2030 (Al-Sarihi, 2023). Oman's (2025) renewable energy target is 10%, while the country's 2030 target is 30% (Al-Sarihi, 2023). The UAE's renewable energy target for 2050 is 44% (Al-Sarihi, 2023). This showcases that despite various renewable initiatives

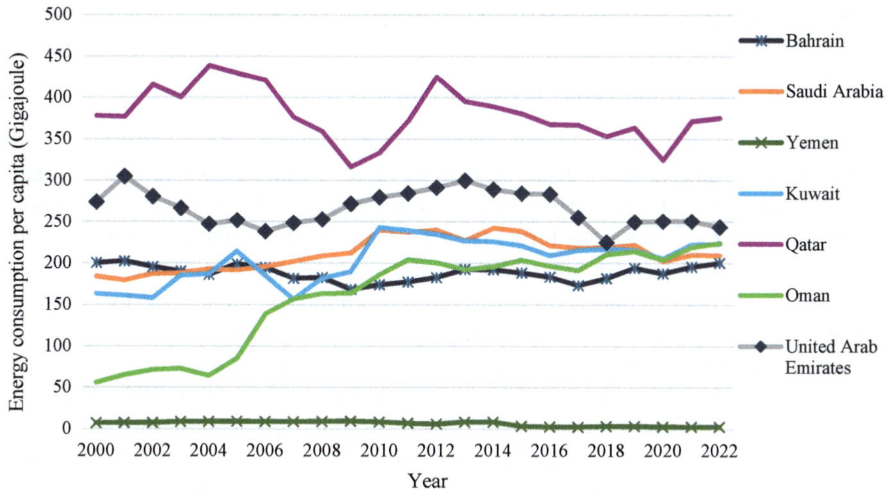

Fig. 10.4 Energy consumption per capita by country in the Arabian Peninsula. *Source* International Energy Agency (2025); The World Bank (2024)

by the GCC countries, the progress so far is significantly inadequate to achieve the targets they set by these countries, particularly for 2025 or 2030. Therefore, renewable energy initiatives must be rapidly and extensively implemented by each country to achieve their pledged targets without further delay.

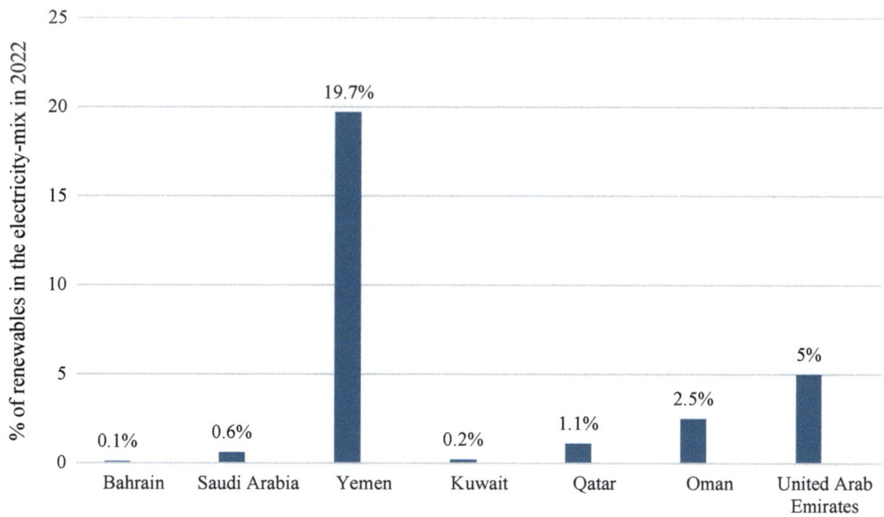

Fig. 10.5 Percentage of renewables in the electricity mix in 2022 by country in the Arabian Peninsula. *Source* International Energy Agency (2025)

10.5 Climate Change Mitigation Policy Initiatives

Net Zero and Interim Emissions Reduction Targets

One of the very first steps towards combating climate change is setting up a net zero emissions target. Without an ambitious target supported by relevant climate actions, it would not be possible to achieve climate goals. Realising the importance of having a net zero emissions target to drive climate works, most countries within the Arabian Peninsula set a net-zero emissions reduction target. Table 10.2 shows the emissions reduction targets for different AP countries. Oman and the UAE set a 2050 net zero target, while Saudi Arabia, Kuwait and Bahrain set their net zero emissions target by 2060 (Al-Sarihi, 2023). Qatar does not have a new zero target as of now, but they have set a 25% emissions reduction target by 2030 (Ministry of Municipality & Environment, 2021). Yemen's Intended Nationally Determined Contributions (INDC) proposed a 14% emissions reduction target (1% unconditional target and another 13% conditional target) by 2030 from the 2020 emissions level (UNFCCC, 2018).

Among the countries that have net-zero targets, Saudi Arabia, Oman, Bahrain, Kuwait and the UAE have interim emissions reduction targets. Saudi Arabia has committed to reducing, avoiding and removing greenhouse gas emissions by 278 million tonnes of carbon dioxide equivalent annually by 2030 (Saudi & Middle East Green Initiatives, 2024a). The interim targets for Oman and the UAE are to reduce emissions by 2 and 23.5%, respectively, by 2030 relative to their business-as-usual scenario (UNFCCC). Bahrain's interim emissions reduction target is 30% by 2035 from the 2016 levels (Mahmoud, 2024), while Kuwait's target is to reduce emissions by 7.4% between 2015 and 2035 (EPA, 2021).

Renewable Energy Policies

Renewable energy is the primary driver towards achieving the emissions reduction targets of the AP countries. Table 10.3 illustrates the renewable energy/electricity targets of the AP countries. Saudi Arabia's National Renewable Energy Program under the Saudi Vision 2030 aims to diversify the country's electricity mix by increasing renewables in the mix (Ministry of Energy, 2024a). By 2030, Saudi

Table 10.2 Emissions reduction pledged by AP countries

AP country	2030	2035	2050	2060
Bahrain	–	30%	–	100% (net emissions)
Kuwait	–	7.4%	–	100% (net emissions)
Oman	2%	–	100% (net emissions)	–
Qatar	25%	–	–	–
Saudi Arabia	278 $MtCO_2e$	–	–	100% (net emissions)
UAE	23.5%	–	100% (net emissions)	–
Yemen	1%	–	–	–

Arabia seeks to generate 50% of their electricity from various renewable sources. The production of 700 MW of electricity from the Sakaka solar photovoltaic (PV) project and the Dumat Al-Jandal wind project are examples of some of the many initiatives towards achieving the renewable electricity target of the country (Ministry of Energy, 2024b). In addition, a total of eleven additional solar PV projects were launched, which would generate a total of 4170 MW of electricity in the mix. The per-unit electricity cost from the Dumat Al-Jandal wind project and the Shuaiba solar PV project were the lowest for any wind and solar energy projects worldwide, respectively (Ministry of Energy, 2024b). Saudi Arabia is also a member of the International Solar Energy Alliance (National Platform GOV.SA, 2025). UAE has also developed targets and strategies to increase the share of renewable energy in their energy mix (Salim & Alsyouf, 2020). The UAE Energy Strategy 2050 was launched in 2017 and it aims to achieve a 44% share of renewables and a 6% share of nuclear in the electricity mix by 2050. In addition to the national level target, there are renewable energy plans and targets at the emirate level in the UAE. Dubai set a clean energy target of 25% by 2030 and 75% by 2050 (Salim & Alsyouf, 2020). The Government of Kuwait aims to increase renewable energy production from 15 to 30% by 2030 and 50% by 2050. The government announces the second and third phases of Kuwait's Shagaya renewable project to achieve these 2030 and 2050 targets (Kuwait Government Online, 2024).

Bahrain's National Renewable Energy Action Plan (NREAP) outlines a 5% renewable energy target by 2025 and a 10% target by 2035. These targets equate to an installed capacity of 255 megawatts (MW) and 710 MW by 2025 and 2035, respectively (Climate Change Laws, 2017b). Yemen is the only AP country that is a member of the Climate Vulnerable Forum (CVF). CVF aims to achieve a 100% renewable energy target between 2030 and 2050 (Climate Vulnerable Forum, 2016). The National Strategy for Renewable Energy and Energy Efficiency of Yemen set a 15% renewable electricity target by 2025, which equates to an overall installed capacity of 400 MW from wind firms, 160 MW from geothermal energy and 6 MW from landfill gas (Climate Change Laws, 2009). The Ministry of Energy and Minerals of Oman introduced the "Renewable Energy Policy for Self-Generation and Direct Sale" policy in 2025, where they set a target of generating the country's 90 to 100%

Table 10.3 Renewable energy/electricity targets by AP countries

AP country	2025	2030	2035	2050
Bahrain	5%	–	10%	–
Kuwait	–	30%	–	50%
Oman	–	30%	–	100%
Qatar	–	18% (electricity)	–	–
Saudi Arabia	–	50% (electricity)	–	–
UAE	–	–	–	44%
Yemen	15%	–	–	100%

electricity from renewables by 2050. This policy provides a regulatory framework for citizens, community groups, private sectors and government and non-government organisations to engage with projects related to electricity generation and direct sales (Ministry of Energy & Minerals, 2025). The "Residential PV Initiative" is another community-based action that the Oman government introduced to promote the use of solar PV in the country (International Energy Agency, 2025). Oman's energy transformation strategy is under development, but the strategy aims to increase the share of renewable energy in the energy mix to 30, 70 and 100% by 2030, 2040 and 2050, respectively (Oman Observer, 2024a). The Oman Electricity Transmission Company has recently introduced the "Green Financing Framework" to secure funding for renewable energy generation, green buildings, wastewater management and other key climate actions (Oman Petroleum and Energy Show, 2025).

Qatar General Electricity and Water Corporation launched the country's first renewable energy in 2024, where the country set a target of increasing the share of renewable energy in the electricity mix to 18% by 2030. This equates to about 4 gigawatts (GW) of utility-scale renewable energy capacity by 2030 (Qatar General Electricity & Water Corporation, 2024). The country has also adopted a National Renewable Energy Policy with the aim of carrying out the renewable energy transition in a sustainable manner so that the costs of electricity remain affordable to all (Qatar General Electricity & Water Corporation, 2020). Solar PV technologies are the government's focus in achieving the country's renewable energy generation targets. The government aims to generate 200 MW of renewable energy by supporting individuals, community groups and private sectors to install solar PV systems at their premises for self-use and export the excess power to the national grid (Qatar General Electricity & Water Corporation, 2024). Importantly, Qatar's first solar PV project, the 800-MW Al Kharsaah power plant, started operating in 2022 and the country aims to install two more plants with a combined capacity of 880 MW by 2025 (Oxford Business Group, 2025).

Energy Efficiency Policies

Energy efficiency is important for climate change mitigation and for a country's national energy security. AP countries have implemented several energy efficiency policies across various sectors. Saudi Arabia established the Saudi Energy Efficiency Centre (SEEC) in 2010 and launched the country's national energy efficiency program called the Saudi Energy Efficiency Program (SEEP) in 2012 (Ministry of Energy, 2025). The SEEP program mainly targeted three sectors i.e. construction, industry and transport. The program developed over 80 initiatives in collaboration with more than 30 government and semi-government authorities (Ministry of Energy, 2025). SEEC develops energy efficiency policies and regulations, provides licenses to energy efficiency service providers, inspects energy efficiency products at outlets, warehouses and local manufacturing sites to ensure compliance and carries out awareness campaigns. Some of the key energy efficiency policies and initiatives taken by the Saudi government include:

1. Introducing Corporate Average Fuel Economy Standards for light vehicles

2. Updating energy efficiency standards for air-conditioners
3. Introducing Saudi Building Code
4. Implementing energy efficiency labels of refrigerators, freezers and washing machines,
5. Introducing minimum energy performance standards for electric motors
6. Reducing energy loss and improving efficiency during power generation, transmissions and distributions through the National Energy Efficiency Program (International Energy Agency, 2025; UNDP, 2025).

UAE established the Energy Conservation and Energy Efficiency (ECEE) department in 2014 to establish a sectoral energy consumption database and enhance energy efficiency (United Arab Emirates' Government portal, 2025). The 'UAE Energy Strategy 2050' was launched in 2017 and it aims to increase energy consumption efficiency by 42–45% compared to 2019 levels (Ministry of Energy & Infrastructure, 2025). UAE's energy efficiency policies and strategies are also evident at the emirate level. Dubai Integrated Energy Strategy 2030, Dubai Clean Energy Strategy, Ras Al Khaimah Energy Efficiency and Renewables Strategy 2040, Abu Dhabi Demand Side Management and Energy Rationalization Strategy 2030 are examples of some of the local-level initiatives in the UAE (Department of Energy, 2019; United Arab Emirates Government portal, 2025). Some of the energy efficiency initiatives at the federal and local levels in the UAE include:

1. The introduction of Emirates Energy Star (EES)
2. The implementation of the Pearl Building Rating System (PBRS)
3. The introduction of efficiency-labelling scheme
4. Development of green building and sustainable building standards and
5. Organisation of national conservation campaign (United Arab Emirates Government portal, 2025).

Qatar National Renewable Energy Strategy highlighted the country's energy efficiency policy initiatives, including:

1. Organising a national conservation campaign (Tarsheed)
2. Introducing energy efficiency labelling for air conditioners
3. Phasing out of inefficient lamps
4. Introducing Kahramaa regulations for conserving electricity and water and
5. Enabling energy demand side management (Qatar General Electricity & Water Corporation, 2024).

In 2024, the Oman government issued a ministerial decision to establish the Oman Net Zero Centre, which will develop and implement energy efficiency strategies to reach carbon neutrality by the year 2050 (Ministry of Energy & Minerals, 2024). The government is also developing the country's national energy strategy with the aim of achieving the objectives outlined in Oman Vision 2040 as well as the net-zero carbon neutrality target by 2050. Energy efficiency, hydrogen, renewable energy and carbon capture technologies are identified as key pillars of their national energy strategy (Oman Observer, 2024a). The second NDC of Oman also highlighted the importance of energy efficiency improvements of existing facilities in the country to achieve its 2% emissions reduction target by 2030 (UNFCCC, 2021). Homeowner's Guide to Energy Efficiency is one of the initiatives taken by Oman to promote energy efficiency in the country (Authority for Public Services Regulation, 2024).

Bahrain's energy efficiency initiatives are guided by its National Energy Efficiency Action Plan (NEEAP). The plan was launched in 2017 and it outlines 22 initiatives across major sectors with an aim to reduce energy consumption by 6% by 2035 (Climate Change Laws, 2017a). In 2023, Kuwait adopted new energy efficiency requirements for electrical machines, washing machines, dryers, refrigerators and freezers (Bureau Veritas, 2023). Under the Energy Efficiency Technologies (EET) program, the Kuwait Institute for Scientific Research (KISR) developed the country's first energy conservation code of practice, introduced energy auditing of buildings and improved energy demand side management (KISR, 2025). In 2010, Yemen developed its National Strategy for Renewable Energy and Energy Efficiency (NSREEE), where the country set a target of a 15% increase in energy efficiency by 2025 in the power sector from 2009 levels (Food & Agriculture Organization, 2009). The energy efficiency labelling scheme and the minimum energy performance standard (MEPS) for air conditioners are some of the notable initiatives taken by the government to achieve the targets (TUV SUD, 2025).

Transport Policies

Transport is one of the major sources of emissions in the AP region. Therefore, the AP countries have been developing and implementing various policies and initiatives to reduce emissions from this sector. Saudi Arabia has made major investments in Electric Vehicles (EVs) through the Public Investment Fund (PIF) program and set a target of replacing 30% of all cars with electric cars on its capital Riyadh's roads by 2030. The PIF program has been helping Saudi Arabia to start manufacturing its own home-grown EV brand, 'Ceer', from 2025 (EVreporter, 2023; Phillips, 2024). Lucid established Saudi Arabia's inaugural EV manufacturing facility in 2023 (Phillips, 2024). The government has also set a target to manufacture 5000 EVs each year by 2030 (Phillips, 2024). In October 2023, Saudi Arabia launched the EV Infrastructure Company (EVIC) to build charging infrastructure across the country. By 2030, EVIC aims to install over 5000 fast chargers in cities across the country (SAIS Observer, 2024). Additionally, Saudi Arabia has enforced several incentives to advance the usage of EVs, which include subsidies on purchase prices, tax exemptions, free parking and toll waivers for EV owners (Swiss Saudi Tech, 2024). In 2023, the use of electric buses started in Jeddah, Dammam, Medina and Qatif. In October 2024, the Kingdom also trialled hydrogen-powered vehicles, such as taxis and trains (Phillips, 2024). This country has also been actively promoting walking, bike sharing, usage of cycles and e-scooters by building new bicycle and pedestrian paths (Phillips, 2024). The NEOM project is an example of an active transport friendly project where the residents could meet their daily needs by walking and/or cycling (Ministry of Media, 2025). As part of the Saudi Vision 2030, the country has been making significant investments in public transport. Riyadh Metro, Jeddah Metro, Mecca Metro, Haramain High-Speed Railway and the GCC Railways Project are some of the key public transport initiatives the country has taken to reduce car dependency and emissions in the country (Metro Rail Today, 2024; Railway Supply, 2024; Saudi Arabia Transport & Mobility, 2025). The Saudi Corporate Average Fuel Economy (CAFE) standard is another emissions reduction initiative taken by the government that aims for a 4% annual improvement in vehicle fuel efficiency to reduce emissions from light vehicles (International Energy Agency, 2024).

Oman's Ministry of Transport, Communications and Information Technology has set a three-phased emissions reduction target: 3% by 2030, 34% by 2040 and 100% by 2050 (Oman News Agency, 2023). In the first phase, the ministry aims to reach 7000 new light EVs, constituting 35% of new light vehicle registrations (Times of Oman, 2023). During phase 2, the government seeks to achieve 22,000 new light EVs, which is 65% of the new light vehicles in the country. Some other initiatives proposed during phase 2 are introducing age limits for public transport vehicles, establishing a regional hub for refuelling ships with green and sustainable fuels and upgrading public transport stations with solar panels. In the last phase, the government plans to decarbonise trucks and heavy machinery by adopting hydrogen and electric technologies (Times of Oman, 2023). UAE-Oman Rail Link project and Muscat Metro Project are some of Oman's key public transport projects (Atack, 2024; Zawya, 2025). In addition, the electrification of public transport and the introduction of biofuel in public transport are some of the recent initiatives taken by the Omani government to reduce transport emissions (Balushi, 2023; Times of Oman, 2024).

The Transportation Master Plan for Qatar-2050 prioritises sustainability and includes long-term initiatives aimed at reducing emissions from this sector (Ministry of Transport, 2024). The government of Qatar also introduced the Public Bus Infrastructure Programme, which aims to provide an integrated and sustainable public transit network and support the 100% electrification of public buses by 2030 (Alingal, 2024). Doha Metro is a state-of-the-art railway network that offers an alternative to traditional private vehicle uses and contributes to the country's transport emissions reduction (Haddad & Cordahi, 2022). The government aims for EVs to make up 10% of all new vehicle sales by 2030 (Al-Shaiba et al., 2023). The launch of VIM, Qatar's first EV brand, marks a major milestone in encouraging the adoption of EVs in the country (Qarjouli, 2023). The General Electricity and Water Corporation of Qatar (Kahramaa) intends to install 600 and 1000 EV charging stations by 2025 and 2030, respectively to support the future uptake of EVs in the country (Qarjouli, 2023). Hamad International Airport's 30% carbon efficiency target by 2030 and the production of electric tractors and other EVs as part of the electrification of Hamad port are some other initiatives taken by the government to curb emissions from the transport sector (Ministry of Transport, 2016; Qatar Free Zones Authority, 2020).

The UAE's updated 2023 NDC set a 1% transport emissions reduction target between 2019 and 2030 (Climate Action Tracker, 2023). Transition to EVs, increased use of public transport and improved walking and cycling facilities are the centre of UAE's transport emissions reduction efforts. The UAE's National Electric Vehicles Policy aims to increase the proportion of EVs to 50% of all vehicles on the country's roadways by 2050 (Ministry of Energy & Infrastructure, 2023). Dubai has launched a Green Mobility Strategy 2030 and a Clean Energy Strategy 2050, which include an EV Green Charger program aimed at expanding the EV charging networks and promoting the adoption of fully electric and hybrid vehicles throughout Dubai (PwC, 2024). The Green Charger initiative has increased EV ownership in Dubai from 14 to 11,000 between 2015 and mid-2023. The initiative aims to increase the number of EVs to 42,000 by 2030 (PwC, 2024). Realising the importance of EV charging infrastructure to transition to EV, the Abu Dhabi National Energy Company and the Abu Dhabi National Oil Company aim to install 70,000 EV charging stations by 2030 (PwC, 2024). Besides EVs, the UAE government has been investing in large public transport projects. Expansion of Dubai Metro, Abu Dhabi Metro, Abu Dhabi Bus Rapid Transit (BRT) and Light Rail Transit and Etihad Rail's Abu Dhabi to Dubai high-speed railway and GCC rail network are some of the future key public transport initiatives of the country (Bell, 2025; Deutsche Bahn, 2025; Gulf Business, 2024). The UAE government is also actively developing walking and cycling projects such as the Loop project (Dubai's 20-min city initiative) and the Dubai Cycle City 2040 project (URB, 2025a, 2025b).

Kuwait's transport emissions reduction policies mainly focus on increasing public transport use and supporting EV adoption (Generis Global Legal Services, 2024). The fourth Kuwait Master Plan 2040 (4 KMP) promotes mixed-use and transit-oriented development, emphasising the integration of BRT and Mass Rapid Transport (MRT) within the Kuwait Metropolitan Area (Ministry of Information, 2020). In alignment with this vision, the Kuwait National Development Plan 2020–2025 seeks to establish a comprehensive public transport infrastructure by introducing modern transit modes and enhancing smart transportation systems (Ministry of Information, 2020). The government of Kuwait is also advancing its EV sector through a number of initiatives as it aligns with Kuwait's Vision 2035 (Astute Analytica, 2024). In 2021, the Kuwait Ports Authority accepted a proposal to construct the first EV city in the Middle East (Astute Analytica, 2024). The EV city will offer all port and logistical services to international companies, including major EV manufacturers. As charging infrastructure is crucial for EV uptake, the government of Kuwait has been working to ensure that every fuel station has at least one fast-charging station (Astute Analytica, 2024).

Bahrain has been improving its public transport system and EV facilities through various policy initiatives. For example, Bahrain is developing the country's first LRT system, Bahrain Metro, as part of its Economic Vision 2030 (Arab Urban Development Institute, 2023). The country is also preparing a comprehensive EV strategy that includes regulations, policies and necessary infrastructure to increase EV adoption (Tamkeen, 2024). In 2023, the country engaged with ABB and Kanoo Power Solutions to further expand its EV charging network (Arab News, 2023). Unlike other AP countries, Yemen's transport emissions reduction initiatives are at a planning and conceptual stage due to the country's ongoing political conflict, which has created financial, institutional and technical challenges. However, a proposal for a BRT system has recently been developed in Yemen to reduce transport emissions, which aims to start implementing at least two BRT systems in two different cities by 2035 (UNEP, 2024a). Similarly, another proposal on the adoption of hybrid electric vehicle (HEV) technology has been developed with an aim to reach 30% of HEVs on Yemen's roads by 2030 (UNEP, 2024b). However, the success of these initiatives largely depends on the resolution of Yemen's ongoing political conflicts and the support from the international community.

Waste Policies

Saudi Arabia's latest Waste Management Law, published in 2021, regulates all waste-related activities such as transportation, storage, treatment and safe disposal of any type of waste. However, it excludes radioactive, nuclear and military waste. The Saudi government has established the National Centre for Waste Management to regulate the waste management sector. This centre is responsible for implementing the Strategic Waste Management Plan (SWMP), issuing licences and holding the Waste Management Service providers accountable (National Platform GOV.SA, 2025; Ministry of Media, 2021). By 2035, the government of Saudi Arabia aims to divert 82% of waste from landfills, with targets of 42% recycling, 35% composting and 19% incineration (National Platform GOV.SA, 2025). The National Centre for Waste Management (2023) notes that the country's Strategic Master Plan further aims to divert 90% of waste from landfills by 2040 through developing an integrated technical and economic plan. The first phase of the National Development Strategies of Qatar set three key targets for waste management. These are establishing a solid waste management plan, increasing the recycled solid waste to 38% and reducing waste generation per capita to 1.6 kg/day. The second phase of the strategy focuses on the application of waste recycling techniques that support infrastructure projects (Symms & Kler, 2024). The national climate change plan of Qatar also integrates waste management as a critical component for achieving the country's climate goals (The Peninsula, 2024).

The Integrated Waste Management Law of the UAE was introduced with the aim of regulating the country's waste management operation and unifying the ways of its disposal (UAE Legislation, 2018). UAE also launched the National Agenda for Integrated Waste Management to protect the environment from the harmful effects of waste-related activities. This agenda is working towards the achievement of some key targets by 2031. These key targets include treating 80% of municipal solid waste, 99% of hazardous waste and 80% of non-hazardous waste. Another key target is to ensure that sanitary standards are met by 90% of the UAE's landfills (Waste & Recycling, 2025). UAE's Net Zero 2050 Strategy has more than 25 programmes covering six key areas, including waste (Ministry of Climate Change & Environment, 2024c).

Kuwait's National Strategy for Integrated Waste Management outlines the country's strategy for implementing waste-related projects through 2040. This strategy includes five specific goals and 28 objectives. Some of the key objectives include promoting a circular economy, improving waste sorting, reducing waste generation, ensuring a hazard-free work setting, revising legislative frameworks, improving waste management capabilities, raising awareness and improving waste data quality for effective monitoring and informed decision-making (Arab Times, 2024). Bahrain developed the country's National Waste Management Strategy in 2018 and put emphasis on landfill management, waste-to-energy conversation and recycling of a variety of wastes, including green waste, household waste, construction waste and commercial waste (Bahrain News Agency, 2019). Bahrain's government is also looking into the privatisation of the waste management sector to enhance environmental sustainability (Jamal & El-Fattah, 2023). The quantity of municipal solid waste in Yemen is 0.6 kg/day per capita in urban areas, while it is 0.35 kg/day in rural areas. Yemen, like most developing countries, disposes of solid waste into landfills. Thus, problems such as the release of greenhouse gases, health concerns, odour issues and pollution are increasing (Al-Dailami et al., 2022). The national strategy for Solid Waste Management 2009–2013 focuses on reforming environmental conditions and health by improving the solid waste services that cover all Yemeni cities. Regrettably, Yemen's political crisis paralysed the implementation of this strategy (GIZ, 2014).

The National Solid Waste Management Strategy of Oman was created in 2015 to reduce waste generation, promote composting and recycling and improve the country's waste management infrastructure. The strategy's targets were to increase the recycling rate to 20% and reduce landfill waste by 35% by 2040 (Kanglebao, 2024). The government has recently proposed to formulate a new legislative framework to improve waste recycling (Oman Observer, 2024b). The Omani government is also seeking to introduce a law to privatise the Waste Sector (Oman Observer, 2024c). Oman has also set several goals that include diverting municipal solid waste into initiatives like recovery-recycling-reuse in a planned manner by 2030 and reducing the generation of waste to <1 kg/day per capita by 2040 (Invest Oman, 2025). The Sultanate of Oman has also taken initiatives to grow local recycling projects by imposing restrictions on the exportation of 11 different waste products, including motor oil and used cooking oil (Oman Observer, 2021).

Emissions/carbon Trading Schemes

Most countries in the AP region have recently initiated emissions/carbon trading schemes as an important tool to curb emissions from various sectors. In 2023, the Clean Development Mechanism Designated National Authority of Saudi Arabia launched the Greenhouse Gas Crediting and Offsetting Mechanism (GCOM) to reduce emissions and achieve the country's 2060 net-zero target. GCOM aims to offer economic incentives for developing activities related to emissions reduction and removal and it will be accessible to all entities within the country (Ministry of Energy, 2023). In 2022, the Saudi Tadawul Group Holding and the Public Investment Fund established the Regional Voluntary Carbon Market Company (RVCMC). In October 2022, RVCMC sold 1.4 million tonnes of carbon credits to companies like Aramco, Olayan Financing Company and Ma'aden (Middle East Institute, 2023). In 2023, the sovereign wealth fund of Bahrain (Mumtalakat) created 'Safa', a voluntary carbon offsetting platform (Middle East Council on Global Affairs, 2024). Oman's National Green Alliance (NGA) was also formed in 2023 with the aim of supporting afforestation and harnessing economic opportunities for carbon markets. NGA provides a policy framework for voluntary carbon markets and helps the government of Oman achieve climate goals (Middle East Council on Global Affairs. (2024).

In the UAE, carbon markets are evident at the national and emirate levels. The Abu Dhabi Global Market joined the AirCarbon Exchange to establish a carbon trading exchange in 2021 (Middle East Institute, 2023), while the UAE Carbon Alliance (UCA) was formed in 2023 to promote a carbon market in the country. The UCA is a coalition of companies that pledged to buy African carbon credits totalling $450 million by 2030 (Middle East Council on Global Affairs, 2024). Kuwait also has a carbon offsetting platform created by the Kuwait Finance House (Middle East Institute, 2023). Importantly, the longest-standing carbon market in the AP region was Qatar's Global Climate Council (Middle East Institute, 2023). Since 2016, the country has been hosting the Global Climate Council. The Global Climate Council received complete approval in 2021 under the Carbon Offsetting and Reduction Scheme for International Aviation and created its own crediting standards (Middle East Institute, 2023).

Adaptation Policies

From the analysis of the four indicators (i.e. emissions intensity, energy intensity, renewable energy share in the electricity-mix and per capita energy consumption), it is likely that despite various policies and pledges, the current efforts are insufficient to limit global warming to 1.5 or 2 °C. This emphasises the significance of adaptation policies for dealing with climate change impacts in the AP. Saudi Arabia's climate change adaptation efforts aim to enhance the country's resilience while protecting the biodiversity of land, coastlines and seas. Projects that have mitigation co-benefits and could help diversify the economy are prioritised. Under the Saudi Green Initiative (SGI), Saudi Arabia has already planted over 100 million trees and thereby rehabilitated 118 thousand hectares of land since 2021 (Saudi & Middle East Green Initiatives, 2024c). By 2030, Saudi Arabia aims to plant over 600 million trees. The country is currently preparing a strategic master plan based on scientific evidence to guide the different phases of the plantation project (Saudi & Middle East Green Initiatives, 2024c). Given the country's hyper-arid climate, the government of Saudi Arabia designed a water and wastewater management programme that aims to reduce, reuse and recycle water and wastewater from residential, commercial and industrial sectors (Clean Development Mechanism, 2025). Initiatives such as harnessing new freshwater sources and constructing additional dams for drinking water collection are also intended to develop as part of the climate change adaptation measures. With an aim to protect the marine ecosystem, the kingdom started implementing some of the actions listed in its coastal management strategies. For example, the country is planting mangrove forests along the coasts and restoring coral reefs throughout the north-west parts of the Arabian Gulf (Clean Development Mechanism, 2025). The government also intends to develop and implement integrated coastal zone management (ICZM) plans, which will protect key infrastructure (e.g. roads, seaports, residential buildings, desalination plants, industrial complexes, etc.) located along the coastline from sea level rise, coastal inundation and erosion. They also intend to develop and operationalise early warning systems (EWS), which are likely to reduce vulnerability and loss from floods, rainstorms and dust storms (Clean Development Mechanism, 2025).

The National Climate Adaptation Plan of the UAE aims to identify the impacts of climate change on the country's food system, natural ecosystems, businesses, health, water management, tourism, energy, trade, logistics and infrastructure (Ministry of Climate Change & Environment, 2024a). However, the assessment of vulnerability and risks through the National Climate Change Risk Assessment (NCCRA) are at an initial level. Parallel to NCCRA, the UAE government aims to develop a list of adaptation actions through establishing the National Adaptation Action Program (NAAP). The adaptation actions identified in the NAAP will be integrated with existing as well as future development planning policies at national, emirate and sector levels (Ministry of Climate Change & Environment, 2024b). The UAE government aims to monitor climate variables and review adaptation projects as well as the NAAP every two years through participatory processes. This biennial review is likely to help understand the progress made in managing climate risks and support in identifying any capacity gaps as well as the requirement for new innovative projects. In the short run, the UAE government aims to prioritise the projects that have the potential to generate co-benefits between mitigation and adaptation (Ministry of Climate Change & Environment, 2024b).

Climate change adaptation measures for Qatar have been highlighted in various national strategic documents. The Qatar National Vision 2030 aims to protect the environment by supporting efforts that help address climate change (Government Communications Office, 2024). The 2024–2030 strategy of the Qatar Ministry of Environment and Climate Change stated to protect 30% of the country's coastal land, restore 30% of climate-affected natural habitats and save 17 endemic and endangered species to protect biodiversity (Government Communications Office, 2024). The Climate Change Strategy (CCS) of the Qatar National Master Plan (QNMP) identified three major climate change risks for the country: flooding due to sea level rise, increasing temperatures and biodiversity losses (Qatar National Master Plan, 2025). The CCS recommended various initiatives to address each of the three risks. For example, the CCS proposed identifying vulnerable coastal zones, updating planning and regulatory frameworks and implementing emergency management actions to address risks associated with floods and sea level rises. The CCS puts emphasis on increasing green spaces and developing an emergency response plan for heat waves to tackle risks related to rising temperatures. To protect against biodiversity losses, the CCS recommended creating buffer zones from the built environment and introducing some kinds of environmental offsets to replace any ecological loss due to development works (Qatar National Master Plan, 2025). Qatar Ministry of Environment and Climate Change has recently engaged with the Global Green Growth Institute for the development of their National Adaptation Plan (NAP), which will integrate mid-term and long-term adaptation strategies into the country's national development plan (Ministry of Environment & Climate Change, 2024).

Kuwait is the only country in the MENA region that has submitted NAP to the United Nations Framework Convention on Climate Change (UNFCCC) as a standalone document (Carnegie Endowment for International Peace, 2024). Although drought and extreme heat are some of the widely recognised climate risks for the country, climate models show that there is an increasing risk of rising sea levels and flash floods (Mehryar & AlSahli, 2025). The NAP of Kuwait identified four main sectors through Climate Vulnerability Assessment and Climate Vulnerability

Index (UNFCCC, 2019). These are marine life and fisheries, coastal zones, water resources and human health. Some of the gaps identified through the NAP include the lack of integration between adaptation measures and the country's development plans and projects, poor management of water resources due to lack of information and technological support and inadequate legal restrictions and policies to protect coastal zones (UNFCCC, 2019). The NAP proposed short-term, medium-term and long-term initiatives at sectoral and national levels that need to be implemented in less than a year, in three years and in 3–5 years, respectively. A total of 56 initiatives were proposed in the NAP, where 21 initiatives were for the short term, 18 initiatives were for the medium term and the rest were for the long term (UNFCCC, 2019). Some of the short-term initiatives for protecting water resources include utilising water conservation technologies, conducting a risk assessment for water depletion and raising awareness, while some medium-term initiatives include an increase in water block tariff to reduce unnecessary consumption and adaptation of efficient irrigation methods and long-term initiatives include investing in water efficient device instalment in homes, designing effective policies to protect water resources etc. (UNFCCC, 2019).

The Environment Authority of Oman has started engaging with the Green Climate Fund (GCF) to develop a comprehensive NAP for the country (Muscat Daily, 2024). The NAP aims to identify Oman's adaptation needs and develop strategies to address climate effects across different sectors. In 2018, the Ministry of Environment and Climate Affairs endorsed the National Strategy for Adaptation and Mitigation to Climate Change for the Sultanate of Oman. This strategic document provides a roadmap for guiding Oman's climate efforts between 2020 and 2040 and assesses the vulnerability and adaptation measures for five key sectors of the country i.e. water resources, agriculture, public health, marine biodiversity and fisheries and urban areas, tourism and infrastructure (Environment Authority, 2018). Some of the strategic actions for addressing climate risks related to water resources include updating flood maps and flood warning systems, installing stormwater drainage infrastructure, assessing funding opportunities from GCF and improving the management of treated wastewater (Environment Authority, 2018). Oman's agricultural production along the coastline was identified as a key climate risk to this sector. Strategic actions proposed to tackle the risk include developing new varieties of crops, conserving indigenous plants and livestock, adopting climate-smart agricultural techniques, protecting agribusiness through developing various climate and non-climate policies and establishing international collaboration (Environment Authority, 2018). Importantly, the Oman Blue Carbon Project aims to plant 100 million mangrove trees and earn carbon credits (Oman Observer, 2024b). Oman's public health system is vulnerable to climate impacts due to heat stress, malaria outbreaks and sea level rise. Improving health surveillance, developing public health data and information management systems, building strong partnerships with concerned authorities at local, regional, national and international levels and increasing research capabilities in this area are some of the proposed strategic actions outlined in this strategic document to tackle climate impacts on public health (Environment Authority, 2018). Emphasis was given to impact analysis and governance improvements to protect marine biodiversity and fisheries. Some of the strategic actions proposed include assessing climate impacts on coral reefs, marine biodiversity, fisheries and endangered species, analysing the costs and benefits of various marine adaptation options and diversifying livelihoods from aquaculture and fishing (Environment Authority, 2018). Flash flooding, sea level rise and tropical cyclones are the climate events that were identified while assessing the strategic actions required to protect Oman's urban areas, tourism and infrastructure. Some of the strategic actions include evaluating existing urban areas and infrastructure exposed to climate risks, assessing climate impacts on tourism sectors and locations, developing building codes and design guidelines to promote climate resiliency and integrating climate risks into master plans and subdivision plans (Environment Authority, 2018).

Bahrain has a high climate risk of extreme heat and a medium climate risk of coastal floods, while the climate risks of urban floods, river floods and landslides are very low (Think Hazard, 2024). The third national communication of Bahrain published by the Supreme Council for Environment highlighted the country's key climate risks to mangrove habitats, coastal zones and water resources (UNFCCC, 2020a). The mangroves in Tubli Bay are at risk of climate change, although they provide coastal protection, ecotourism and habitats for bird and fisher species (UNFCCC, 2020a). The government aims to strengthen the current observation systems by mapping species, collecting relevant data and information and monitoring ecological services as an adaptation strategy. Decreasing water supply and increasing use of water due to saltwater intrusion into groundwater and low precipitation are the key climate risks to water resources in Bahrain (UNFCCC, 2020a). There are a few strategic adaptation actions proposed in the third national communication and some of these include removing subsidies on water tariffs, introducing new legislation to strengthen building codes and promote water-saving technologies, emphasising social awareness through social campaigns and strengthening leak detection and control programmes (UNFCCC, 2020a). Given the small size and location of Bahrain, coastal floods due to rising sea levels are often regarded as an existential risk to the country. The reclaimed land, infrastructure, industrial zones and wetlands are subject to coastal flooding (UNFCCC, 2020a). Some of the key actions outlined in the third national communication include modelling coastal risks, developing emergency preparedness strategies, integrating climate risks into building codes, introducing zoning offsets from the coastline through adjusting land use regulations and assessing the suitability of using hard and/or soft structural elements of coastal protection. Hard elements include breakwaters and/or dikes, while soft elements include mangrove plantations and/or beach nourishment (UNFCCC, 2020a).

Yemen is one of the least prepared countries to adapt to climate change. The country is extremely vulnerable to climatic events such as flooding, drought, disease outbreaks, pests, severe storms (cyclones) and sea level rise (ReliefWeb, 2023). These climatic events pose significant risks to the country's water, coastal areas, agriculture, food security and livelihoods. However, there are no dedicated national institutions to develop and finance climate strategies and action plans due to political instability and fragmented governance where the Yemeni government controls Aden and the Houthi controls Sanaa (Nasser, 2024). The Environment Protection Authority of Yemen is usually tasked with administering climate initiatives, but the institution has become weak due to the political unrest and chronic underfunding. Importantly, the country is not eligible for international funding, such as the GCF, because the country couldn't demonstrate the presence of adequate institutional capacity and a robust climate strategy (Nasser, 2024). So, it's important to build institutional capacity through some reforms and integrate climate adaptation into broader development and peacebuilding strategies (Nasser, 2024). This will strengthen Yemen's position in international aid efforts and funding negotiations. Once EPA secures steady funding, the country will be able to improve its early warning systems and disaster response measures (Nasser, 2024).

10.6 Policy Gaps and Conclusion

As the countries within the AP are among the top emitters on a per-capita basis and highly vulnerable to climate change impacts, climate change mitigation and adaptation policies of these countries are crucial to reducing greenhouse emissions and tackling climate change impacts. Although most AP countries have taken various policy initiatives in the areas of climate change mitigation and adaptation, nonetheless, further action is needed to fully align with both national and international climate commitments. From the analysis of the four indicators, it is evident that there has not been much progress in key action areas such as renewable energy adoption, reduction of fossil fuel dependency on GDP, energy efficiency improvement and reduction in emissions intensity. A few AP countries have recently initiated policies and mechanisms related to emissions trading, but these are voluntary carbon markets. Individuals, companies and organisations buy and sell carbon credits on a voluntary basis and there are no legal bindings to trade carbon. For meaningful emissions reduction, a compliant carbon market along with the voluntary market is crucial. High-emitting sectors such as electricity, transport, waste and industrial processes need to be covered by the compliance market to ensure a large and rapid emissions reduction in the region. As most of the carbon markets in the AP region are newly established, they are experiencing structural challenges, including disclosure issues, limited availability of emissions data and human and institutional capacity (Middle East Council on Global Affairs, 2024). A unified carbon market for the AP region, like the European Union Emissions Trading Scheme, would foster collaborations and strengthen technical experiences. In addition, a unified carbon market would improve the quality of carbon credits and data transparency, which is crucial to attracting international, regional and local carbon traders in the region.

The rapid large-scale adoption of renewable energy is crucial to achieving international and national climate goals set by each AP country. However, the urgency of adopting renewable energy at a large scale is not evident in most AP countries. Bahrain's target is to achieve only 10% renewable energy share in the energy mix by 2035, while Qatar's 2030 renewable energy target is only 18%. The 2050 renewable energy targets for the UAE and Kuwait are 44 and 50%, respectively. Given the urgency required to address climate change, these renewable energy targets are inadequate. Delaying meaningful climate action could leave the AP region vulnerable to volatile fossil fuel prices and increased carbon costs. In addition, the slow transition to renewable energy sources could lead to missed economic opportunities in the green energy industries and could make it difficult to meet international and domestic climate commitments. The current emissions trends and also the climate change policies and targets set by the AP countries suggest that global warming will continue in the region as well as worldwide for decades. This makes climate change adaptation not just necessary but urgent for the AP countries. Community-based adaptation techniques for sustainable agriculture, water and waste management and renewable energy are some of the crucial adaptation strategies that must be prioritised now. It is also important to improve data collection and the analysis of climate impacts on various regions and sectors within the AP region to address the climate impacts better. This study highlighted the climate change policies and strategies of the AP countries so that countries can learn from each other's climate actions and harmonised strategies by avoiding economic disadvantages and policy clashes. This book chapter is likely to support policymakers as well as the public in understanding policy gaps and the urgency of implementing various climate initiatives. In addition, this chapter of the book will inform policymakers and the public about the efforts being made by the countries within the AP to reduce emissions and address climate impacts across various sectors.

References

Abbas, S. Y. (2022). Feasibility of anaerobic digestion as an option for biodegradable and sewage sludge waste management in the Kingdom of Bahrain. *Desalination and Water Treatment, 263*, 160–167.

Akasha, H., Ghaffarpasand, O., & Pope, F. D. (2023). Climate change, air pollution and the associated burden of disease in the Arabian Peninsula and neighbouring regions: A critical review of the literature. *Sustainability, 15*(4), 3766.

Al-Dailami, A., Ahmad, I., Kamyab, H., Abdullah, N., Koji, I., Ashokkumar, V., & Zabara, B. (2022). Sustainable solid waste management in Yemen: Environmental, social aspects, and challenges. *Biomass Conversion and Biorefinery, 15*, 1–27.

Aleisa, E., & Alsaleh, A. (2024). Upcycling food waste into animal feed: An environmental assessment based on food waste quantities in different district types in Kuwait. *Journal of Engineering Research, 15*, 1134.

Alingal, S. (2024). *MoT highlights public bus infrastructure programme.* Gulf Times. https://www.gulf-times.com/article/681653/qatar/mot-highlights-public-bus-infrastructure-programme

AlKheder, S. (2021). Promoting public transport as a strategy to reduce GHG emissions from private vehicles in Kuwait. *Environmental Challenges, 3*, 100075.

Almazroui, M., Islam, M. N., Saeed, F., Alkhalaf, A. K., & Dambul, R. (2017). Assessing the robustness and uncertainties of projected changes in temperature and precipitation in AR5 global climate models over the Arabian Peninsula. *Atmospheric Research, 194*, 202–213.

Almazroui, M., Saeed, F., Islam, M. N., & Alkhalaf, A. K. (2016). Assessing the robustness and uncertainties of projected changes in temperature and precipitation in AR4 global climate models over the Arabian Peninsula. *Atmospheric Research, 182*, 163–175.

Almulhim, A. I., & Cobbinah, P. B. (2024). Framing resilience in Saudi Arabian cities: On climate change and urban policy. *Sustainable Cities and Society, 101*, 105172.
AlSabbagh, M., Siu, Y. L., Guehnemann, A., & Barrett, J. (2017). Integrated approach to the assessment of CO_2e-mitigation measures for the road passenger transport sector in Bahrain. *Renewable and Sustainable Energy Reviews, 71*, 203–215.
Al-Sakkaf, A. S., Zhang, J., Yao, F., Hamed, M. M., Simbi, C. H., Ahmed, A., & Shahid, S. (2024). Assessing exposure to climate extremes over the Arabian Peninsula using ERA5 reanalysis data: Spatial distribution and temporal trends. *Atmospheric Research, 300*, 107224.
Al-Salem, S. M. (2015). Carbon dioxide (CO_2) emission sources in Kuwait from the downstream industry: Critical analysis with a current and futuristic view. *Energy, 81*, 575–587.
Al-Sarihi, A. (2023). *The GCC and the road to net zero*. Middle East Institute. https://www.mei.edu/publications/gcc-and-road-net-zero
Al-Shaiba, A., Wilson, A., & Cochrane, L. (2023). The enablers, opportunities and challenges of electric vehicle adoption in Qatar: A systematic review of the literature and assessment of progress toward transportation transformation targets. *PLoS Climate, 2*(9), e0000271.
Alvarez, E., Korkeakoski, M., & Luukkanen, J. (2021). Long-Range integrated development analysis: The Cuban Isla de la Juventud study case. *Energies, 14*(10), 2865.
Arab News. (2023). *Bahrain expands EV charging network through KPS-ABB deal*. https://www.arabnews.com/node/2399506/business-economy
Arab Urban Development Institute. (2023). *Bahrain metro project*. https://araburban.org/en/infohub/projects/
Arab Times. (2024). *Sustainable future: Kuwait's vision for waste management by 2040*. https://www.arabtimesonline.com/news/sustainable-future-kuwaits-vision-for-waste-management-by-2040
Astute Analytica. (2024). *Kuwait electric vehicle market*. https://www.astuteanalytica.com/industry-report/kuwait-electric-vehicle-market
Atack, P. R. (2024). *UAE-Oman rail link contract awarded*. Railway Technology. https://www.railway-technology.com/news/uae-oman-rail-link-contract-awarded/?cf-view
Attia, A. S. (2020). Traditional multi-story house (Tower House) in Sana'a City, Yemen: An example of sustainable architecture. *Alexandria Engineering Journal, 59*(1), 381–387.
Authority for Public Services Regulation. (2024). *Homeowner's guide to energy efficiency*. https://www.apsr.om/pdfs/ee/guide-en.pdf
Babonneau, F., Benlahrech, M., & Haurie, A. (2022). Transition to zero-net emissions for Qatar: A policy based on Hydrogen and CO_2 capture and storage development. *Energy Policy, 170*, 113256.
Bahrain News Agency. (2019). *National waste management strategy discussed*. https://www.bna.bh/en/NationalWasteManagementStrategydiscussed.aspx?cms=q8FmFJgiscL2fwIzON1%2BDlQcyUX9XAmaasNuMsgd7EY%3D
Balushi, A. A. (2023). *Biofuel use set to redefine transport dynamics*. Oman Observer. https://www.omanobserver.om/article/1130616/oman/biofuel-use-set-to-redefine-transport-dynamics
Banhidarah, A. K., Al-Sumaiti, A. S., Wescoat, J. L., Jr., & Nguyen, H. T. (2020). Electricity-water usage for sustainable development: An analysis of United Arab Emirates farms. *Energy Policy, 147*, 111823.
Bell, J. (2025). *Etihad Rail's proposed 30-min high-speed rail service between Abu Dhabi, Dubai unveiled*. AlArabia News. https://english.alarabiya.net/News/gulf/2025/01/24/etihad-rail-s-30-min-high-speed-rail-service-between-abu-dhabi-dubai-unveiled
Bureau Veritas. (2023). *VOC Kuwait: New energy efficiency regulations are to be implemented on 8th October 2024*. https://verigates.bureauveritas.com/en/news/voc-kuwait-new-energy-efficiency-regulations-be-implemented-8th-october-2024
Carnegie Endowment for International Peace. (2024). *Assessing climate adaptation plans in the Middle East and North Africa*. https://carnegieendowment.org/research/2024/04/assessing-climate-adaptation-plans-in-the-middle-east-and-north-africa?lang=en
Clean Development Mechanism. (2025). *Adaptation*. https://cdmdna.gov.sa/page/34

Climate Change Laws. (2009). *National strategy for renewable energy and energy efficiency.* https://climate-laws.org/documents/national-strategy-for-renewable-energy-and-energy-efficiency_ae49?id=national-strategy-for-renewable-energy-and-energy-efficiency_3748

Climate Change Laws. (2017a). *Bahrain's national energy efficiency action plan (NEEAP).* https://climate-laws.org/documents/bahrain-s-national-energy-efficiency-action-plan-neeap_1051

Climate Change Laws. (2017b). *Bahrain's national renewable energy action plan (NREAP).* https://climate-laws.org/documents/bahrain-s-national-renewable-energy-action-plan-nreap_7e16?id=bahrain-s-national-renewable-energy-action-plan-nreap_c0c4

Climate Vulnerable Forum. (2016). *Climate vulnerable forum commit to stronger climate action at COP22.* https://web.archive.org/web/20180713050510/thecvf.org/cvf-2016-forum-press-release/

Climate Action Tracker. (2023). *UAE: Policies and action.* https://climateactiontracker.org/countries/uae/policies-action/

Department of Energy. (2019). *UAE capital unveils ambitious energy efficiency strategy on day two of 24th world energy congress.* https://www.doe.gov.ae/en/Media-Centre/News/UAE-Capital-Unveils-Ambitious-Energy-Efficiency-Strategy-on-Day-Two-of-24th-World-Energy-Congress

Deutsche Bahn. (2025). *Abu Dhabi metro, UAE.* https://db-engineering-consulting.com/en/projects/abu-dhabi-metro-uae/

Environment Authority. (2018). *National strategy for adaptation and mitigation to climate change for the Sultanate of Oman (2020–2040).* https://ea.gov.om

EPA. (2021). *Nationally determined contributions: Updating the first NDC.* https://epa.gov.kw/Portals/0/PDF/KuwaitNDCEN.pdf

EVreporter. (2023). *Saudi Arabia EV landscape.* https://evreporter.com/saudi-arabia-ev-landscape-2023/

Food and Agriculture Organization. (2009). *National strategy for renewable energy and energy efficiency.* https://faolex.fao.org/docs/pdf/yem217398E.pdf

Generis Global Legal Services. (2024). *Climate change initiatives in Kuwait: Government policies and renewable energy targets.* https://generisonline.com/climate-change-initiatives-in-kuwait-government-policies-and-renewable-energy-targets/

GIZ. (2014). *Country report on the solid waste management in Yemen.* https://www.retech-germany.net/fileadmin/retech/05_mediathek/laenderinformationen/Jemen_RA_ANG_WEB_Laenderprofile_sweep_net.pdf

Government Communications Office. (2024). *Ministry of environment and climate change launches 2024–2030 strategy.* https://www.gco.gov.qa/en/media-centre/top-news/ministry-of-environment-and-climate-change-launches-2024-2030-strategy

Gulf Business. (2024). *Here's what Dubai's metro stations will look like in 2030.* https://gulfbusiness.com/what-dubai-metro-stations-will-look-like-in-2030/

Haddad, M., & Cordahi, G. A. (2022). *How sustainable mobility can transform Qatar.* Gulf Times. https://www.gulf-times.com/story/716017/How-sustainable-mobility-can-transform-Qatar

Hamed, M. M., Al-Hasani, A. A. J., Nashwan, M. S., Sa'adi, Z., & Shahid, S. (2024). Assessing the growing threat of heat stress in the North Africa and Arabian Peninsula region connected to climate change. *Journal of Cleaner Production, 447*, 141639.

Hasan, M. A., Chapman, R., & Frame, D. J. (2022). Understanding road transport emissions reduction policies using multi-criteria analysis. In *Handbook of climate change mitigation and adaptation* (pp. 3203–3223). Springer.

Hasan, M. A., Abubakar, I. R., Rahman, S. M., Aina, Y. A., Chowdhury, M. M. I., & Khondaker, A. N. (2020). The synergy between climate change policies and national development goals: Implications for sustainability. *Journal of Cleaner Production, 249*, 119369.

Hasan, M. A., Frame, D. J., Chapman, R., & Archie, K. M. (2019). Emissions from the road transport sector of New Zealand: Key drivers and challenges. *Environmental Science and Pollution Research, 26*, 23937–23957.

International Energy Agency. (2024). *Saudi Arabia corporate average fuel economy (CAFE) standards.* https://www.iea.org/policies/8395-saudi-arabia-corporate-average-fuel-economy-cafe-standards

International Energy Agency. (2025). *Countries and regions.* https://www.iea.org/countries

Invest Oman. (2025). *Circular economy.* https://investoman.om/circular-economy

Jamal, H. F., & El-Fattah, A. A. (2023). An overview of solid waste management and privatisation in Kingdom of Bahrain. *Frontiers in Environmental Science, 11*, 1302711.

Kameyama, Y., & Kawamoto, A. (2018). Four intermediate goals: A methodology for evaluation of climate mitigation policy packages. *Climate Policy, 18*(2), 210–220.

Kanglebao. (2024). *Oman's path to sustainable waste management: A closer look at the country's initiatives.* https://kanglebao.com/blogs/omans-path-to-sustainable-waste-management-a-closer-look-at-the-countrys-initiatives

Khan, Z. A., Salami, B. A., Hussain, S. A., Hasan, M. A., Al-Ramadan, B. M., & Rahman, S. M. (2023). Dynamics of greenhouse gas emissions from cement industries in Saudi Arabia—Challenges and opportunities. *IEEE Access, 11*, 125631–125647.

Khondaker, A. N., Hasan, M. A., Rahman, S. M., Malik, K., Shafiullah, M., & Muhyedeen, M. A. (2016). Greenhouse gas emissions from energy sector in the United Arab Emirates: An overview. *Renewable and Sustainable Energy Reviews, 59*, 1317–1325.

KISR. (2025). *Energy efficiency technologies.* https://www.kisr.edu.kw/en/program/18/

Kotwicki, V., & Al Sulaimani, Z. (2009). Climates of the Arabian Peninsula–past, present, future. *International Journal of Climate Change Strategies and Management, 1*(3), 297–310.

Kuwait Government Online. (2024). *Kuwait announces modernising renewable energy goals.* https://e.gov.kw/sites/kgoenglish/Pages/ApplicationPages/NewsDetail.aspx?nid=26565794

Lahlou, F. Z., AlNouss, A., Govindan, R., Hazrat, B., Mackey, H. R., & Al-Ansari, T. (2023). Water and sludge resource planning for sustainable agriculture: An energy-water-food-waste nexus approach. *Sustainable Production and Consumption, 38*, 130–148.

Lahlou, F. Z., Mackey, H. R., McKay, G., Onwusogh, U., & Al-Ansari, T. (2020). Water planning framework for alfalfa fields using treated wastewater fertigation in Qatar: An energy-water-food nexus approach. *Computers and Chemical Engineering, 141*, 106999.

Mahmoud, M. (2024). *Building a more resilient Bahrain: An integrated approach to climate change, socioeconomic, and governance challenges.* Middle East Institute. https://www.mei.edu/publications/building-more-resilient-bahrain-integrated-approach-climate-change-socioeconomic-and

Malik, K., Rahman, S. M., Khondaker, A. N., Abubakar, I. R., Aina, Y. A., & Hasan, M. A. (2019). Renewable energy utilisation to promote sustainability in GCC countries: Policies, drivers, and barriers. *Environmental Science and Pollution Research, 26*, 20798–20814.

Marcenaro, F. (2024). *Thirsty sands: Climate change and water stress in the Arabian Peninsula.* https://iari.site/2024/05/11/thirsty-sands-climate-change-and-water-stress-in-the-arabian-peninsula/

Masoumi, A. H., Esmaeili, H. R., Khosravi, R., Gholamhosseini, A., Korkmaz, M., & Jeppesen, E. (2024). Species on the move: Impacts of climate change on the spatial range of endemic fishes of the eco-sensitive semi-arid area of the Arabian Peninsula. *Science of the Total Environment, 947*, 174095.

Matar, W., Mansouri, N. Y., & Umeozor, E. (2024). Energy policy pathways to inform climate policy in Saudi Arabia. *Journal of Cleaner Production, 445*, 141190.

Mehryar, S. & AlSahli, M. (2025). *Sustainable climate adaptation strategies for Kuwait: A multi-hazard approach integrating extreme heat and flood adaptation solutions.* The London School of Economics and Political Science. https://www.lse.ac.uk/middleeastcentre/research/kuwait-programme/kuwait-academic-collaborations/2024-25/sustainable-climate-adaptation-strategies-kuwait

Mendez, C., Contestabile, M., & Bicer, Y. (2023). Hydrogen fuel cell vehicles as a sustainable transportation solution in Qatar and the Gulf Cooperation Council: A review. *International Journal of Hydrogen Energy, 48*(99), 38953–38975.

Metro Rail Today. (2024). *Mecca metro project: A transformative urban mobility initiative in Saudi Arabia.* https://metrorailtoday.com/article/mecca-metro-project-a-transformative-urban-mobility-initiative-in-saudi-arabia

Middle East Institute. (2023). *Carbon trading in the MENA region: Opportunities and challenges.* https://www.mei.edu/publications/carbon-trading-mena-region-opportunities-and-challenges

Middle East Council on Global Affairs. (2024). *Gulf transition to low-carbon economies: The role of carbon markets.* https://mecouncil.org/publication/gulf-transition-to-low-carbon-economies-the-role-of-carbon-markets

Ministry of Transport. (2016). *Hamad international airport announces long-term commitment to reduce carbon emissions.* https://www.mot.gov.qa/en/news/hamad-international-airport-announces-long-term-commitment-reduce-carbon-emissions

Ministry of Information. (2020). *Kuwait national development plan 2020–2025.* https://media.gov.kw/assets/img/Ommah22_Awareness/PDF/NewKuwait/Revised%20KNDP%20-%20EN.pdf

Ministry of Media. (2021). *Waste management law in Saudi Arabia.* https://saudipedia.com/en/article/2898/government-and-politics/systems/waste-management-law-in-saudi-arabia

Ministry of Municipality and Environment. (2021). *Nationally determined contribution (NDC).* https://www.mme.gov.qa/pdocs/cview?siteID=2&docID=23348&year=2021

Ministry of Energy. (2023). *Saudi Arabia launches groundbreaking greenhouse gas crediting and offsetting mechanism to advance global climate goals.* https://moenergy.gov.sa/en/MediaCenter/ClimateWeek/Pages/KSA-Launches-GHG-Crediting-and-Offsetting-Mechanism-GCOM.aspx

Ministry of Energy and Infrastructure. (2023). *Minister of energy and infrastructure outlines details of national electric vehicles policy.* https://www.moei.gov.ae/en/media-center/news/13/7/2023/minister-of-energy-and-infrastructure-outlines-details-of-national-electric-vehicles-policy

Ministry of Energy. (2024a). *Renewable energy.* https://www.moenergy.gov.sa/en/OurPrograms/RenewableEnergy/Pages/default.aspx

Ministry of Energy. (2024b). *Renewable energy: Completed work.* https://www.moenergy.gov.sa/en/OurPrograms/RenewableEnergy/Pages/CompletedWork.aspx

Ministry of Climate Change and Environment. (2024a). *National climate adaptation action plan.* https://u.ae/en/about-the-uae/strategies-initiatives-and-awards/strategies-plans-and-visions/environment-and-energy

Ministry of Climate Change and Environment. (2024b). *National climate change plan of the UAE 2017–2050.* https://u.ae/en/about-the-uae/strategies-initiatives-and-awards/strategies-plans-and-visions/environment-and-energy/national-climate-change-plan-of-the-uae

Ministry of Climate Change and Environment. (2024c). *The UAE's net zero 2050 strategy.* https://u.ae/en/about-the-uae/strategies-initiatives-and-awards/strategies-plans-and-visions/environment-and-energy/the-uae-net-zero-2050-strategy

Ministry of Energy and Minerals. (2024). *The ministry of energy and minerals issues a ministerial decision to establish the Oman net zero centre.* https://mem.gov.om/en-us/Media-Center/News/ArtMID/608/ArticleID/1415

Ministry of Transport. (2024). *Transportation master plan for Qatar—2050.* https://www.mot.gov.qa/en/land-transport/transportation-master-plan-for-qatar-2050

Ministry of Environment and Climate Change. (2024). *"Environment" and "GGGI" prepare to develop Qatar's national adaptation plan.* https://www.mecc.gov.qa/en/environment-and-gggi-prepare-to-develop-qatars-national-adaptation-plan/

Ministry of Energy. (2025). *Saudi energy efficiency program.* https://www.moenergy.gov.sa/en/OurPrograms/SPFEE/Pages/default.aspx

Ministry of Media. (2025). *NEOM project.* https://saudipedia.com/en/article/229/economy-and-business/projects/neom-project

Ministry of Energy and Infrastructure. (2025). *The UAE energy strategy 2050.* https://moei.gov.ae/en/about-ministry/energy-strategies-to-achieve-net-zero

Ministry of Energy and Minerals. (2025). *The ministry of energy and minerals unveils a policy for self-generation, direct sale, and wheeling of electricity.* https://mem.gov.om/Media-Center/News/ArtMID/608/ArticleID/1422/The-Ministry-of-Energy-and-Minerals-Unveils-a-Policy-for-Self-Generation-Direct-Sale-and-Wheeling-of-Electricity

Muscat Daily. (2024). *Oman prepares a national plan for climate change adaptation.* https://www.muscatdaily.com/2024/09/04/oman-prepares-national-plan-for-climate-change-adaptation

Nasser, A. (2024). *Climate change: A new battlefield in Yemen's ongoing conflict.* https://arabcenterdc.org/resource/climate-change-a-new-battlefield-in-yemens-ongoing-conflict/

National Centre for Waste Management. (2023). *Strategic master plan.* https://mwan.gov.sa/en/strategic-plan

National Platform GOV.SA. (2025). *Environment.* https://www.my.gov.sa/wps/portal/snp/aboutksa/environmentalProtection

OHCHR. (2021). *Frequently asked questions on human rights and climate change.* https://www.ohchr.org/en/publications/fact-sheets/fact-sheet-no-38-frequently-asked-questions-human-rights-and-climate

Oman Observer. (2021). *Oman to regulate 11 waste streams to support circular economy.* https://www.omanobserver.om/article/1105427/business/economy/oman-to-regulate-11-waste-streams-to-support-circular-economy

Oman News Agency. (2023). *Oman launches initiatives to reduce carbon emissions.* https://omannews.gov.om/topics/en/80/show/119076

Oman Observer. (2024a). *Oman charts path for energy transformation.* https://www.omanobserver.om/article/1159427/business/oman-charts-path-for-energy-transformation

Oman Observer. (2024b). *Oman to roll out new policy on recycling, waste management.* https://www.omanobserver.om/article/1156524/business/economy/oman-to-roll-out-new-policy-on-recycling-waste-management

Oman Observer. (2024c). *Oman to unveil new law on waste sector privatisation this year.* https://www.omanobserver.om/article/1148081/business/oman-to-unveil-new-law-on-waste-sector-privatisation-this-year

Oman Petroleum and Energy Show. (2025). *OETC unveils a green financing framework to drive a low-carbon economy.* https://omanpetroleumandenergyshow.com/newfront/news/oetc-unveils-green-financing-framework-to-drive-low-carbon-economy

Oxford Business Group. (2025). *Qatar's increasing renewable energy generation capacity and sustainability.* https://oxfordbusinessgroup.com/reports/qatar/2024-report/energy-utilities/renewing-interest-increasing-generation-capacity-in-renewable-energy-solutions-aligns-with-the-countrys-long-term-sustainability-drive-overview/

Phillips, C. (2024). *Transitioning to sustainable transport in Saudi Arabia.* https://activetravelscore.us/wp-content/themes/jobify/theme-images/reports/transitioning-to-sustainable-transport-in-saudi-arabia.pdf

PwC. (2024). *The future is electric: A strategy for EV adoption in the UAE.* https://www.pwc.com/m1/en/publications/documents/2024/future-is-electric-strategy-ev-adoption-uae.pdf

Qarjouli, A. (2023). *Qatar launches its first electric vehicle company with intellectual property.* Doha News. https://dohanews.co/qatar-launches-first-electric-vehicle-company-with-intellectual-property

Qatar Free Zones Authority. (2020). *GAME tests the first e-vehicles "Made in Qatar" at QFZ.* https://qfz.gov.qa/game-tests-the-first-e-vehicles-made-in-qatar-at-qfz

Qatar General Electricity and Water Corporation. (2020). *Qatar national renewable energy policy.* https://www.km.qa/RenewableEnergy/Documents/Renewable%20energy%20policy-En.pdf

Qatar General Electricity and Water Corporation. (2024). *Qatar national renewable energy strategy.* https://km.qa/RenewableEnergy/Documents/QNRES_Strategy_EN.pdf

Qatar National Master Plan. (2025). *Climate change strategy (CCS) for urban planning and urban development sector in the state of Qatar.* https://www.mme.gov.qa/QatarMasterPlan/English/strategicplans.aspx?panel=ccs

Rahman, M. M., Akter, R., Abdul-Bari, J. B., Hasan, M. A., Rahman, M. S., Abu-Shoaib, S., Shatnawi, Z. N., Alshayeb, A. F., Shalabi, F. I., Rahman, A., Alsanad, M. A., & Rahman, S. M. (2022a). Analysis of climate change impacts on the food system security of Saudi Arabia. *Sustainability, 14*(21), 14482.

Rahman, M. M., Rahman, S. M., Shafiullah, M., Hasan, M. A., Gazder, U., Al Mamun, A., Mansoor, U., Kashifi, M. T., Reshi, O., Arifuzzaman, M., Islam, M. K., & Al-Ismail, F. S. (2022b). Energy demand of the road transport sector of Saudi Arabia: Application of a causality-based machine learning model to ensure sustainable environment. *Sustainability, 14*(23), 16064.

Rahman, M. M., Rahman, S. M., Rahman, M. S., Hasan, M. A., Shoaib, S. A., & Rushd, S. (2021). Greenhouse gas emissions from solid waste management in Saudi Arabia: Analysis of growth dynamics and mitigation opportunities. *Applied Sciences, 11*(4), 1737.

Rahman, S. M., Khondaker, A. N., Hasan, M. A., & Reza, I. (2017). Greenhouse gas emissions from road transportation in Saudi Arabia-a challenging frontier. *Renewable and Sustainable Energy Reviews, 69*, 812–821.

Railway Supply. (2024). *GCC railways project: Connecting Gulf nations with a unified railway network by 2030.* https://www.railway.supply/en/gcc-railways-project-connecting-gulf-nations-with-a-unified-railway-network-by-2030/

Rao, K. K., Al Mandous, A., Al Ebri, M., Al Hameli, N., Rakib, M., & Al Kaabi, S. (2024). Future changes in the precipitation regime over the Arabian Peninsula with special emphasis on UAE: Insights from NEX-GDDP CMIP6 model simulations. *Scientific Reports, 14*(1), 151.

ReliefWeb. (2023). *Climate change impacts on Yemen and adaptation strategies.* https://reliefweb.int/report/yemen/climate-change-impacts-yemen-and-adaptation-strategies

SAIS Observer. (2024). *Saudi Arabia's strategic embrace of electric vehicle and renewable energy.* https://saisobserver.org/2024/04/08/saudi-arabias-strategic-embrace-of-electric-vehicle-and-renewable-energy/

Salim, A. M., & Alsyouf, I. (2020). Renewable energy in the United Arab Emirates: Status and potential. In *Proceedings of the 2020 advances in science and engineering technology international conferences (ASET)* (pp. 1–5). IEEE.

Saudi and Middle East Green Initiatives. (2024a). *SGI target: Reduce carbon emissions by 278 MTPA by 2030.* https://www.greeninitiatives.gov.sa/about-sgi/sgi-targets/reduce-carbon-emissions

Saudi and Middle East Green Initiatives. (2024b). *MGI target: Grow 50 billion trees across the Middle East.* https://www.sgi.gov.sa/about-mgi/mgi-targets/planting-trees/?csrt=9846642149537293834

Saudi and Middle East Green Initiatives. (2024c). *SGI target: Grow 10 billion trees across Saudi Arabia.* https://www.sgi.gov.sa/about-sgi/sgi-targets/greening-saudi/?csrt=9846642149537293834

Saudi Arabia Transport and Mobility. (2025). *Redefining public transportation: Saudi Arabia's investment in metro systems and high-speed rail.* https://saudimobilityconsulting.com/redefining-public-transportation-saudi-arabias-investment-in-metro-systems-and-high-speed-rail

Serdar, M. Z., Lahlou, F. Z., & Al-Ansari, T. (2024). Enhancing resilience in the face of climate change: Addressing cascading risks within the EWF Nexus. *Progress in Disaster Science, 15*, 100390.

Sharma, R., Wahbeh, S., Sundarakani, B., Manikas, I., & Pachayappan, M. (2024). Enhancing domestic food supply in the UAE: A framework for technology-driven urban farming systems. *Journal of Cleaner Production, 434*, 139823.

Sial, N. R., Al-Abri, H., Qyyum, M. A., Al-Abri, R., & Ala'a, H. (2024). Techno-economic assessment of green hydrogen production in the unexploited landlocked northern-western region of Oman to realise the national sustainable strategy. *International Journal of Hydrogen Energy, 54*, 1–13.

Swiss Saudi Tech. (2024). *Electric vehicles promotion in KSA: Steering towards a greener future*. https://swisssaudi.com/2024/04/15/electric-vehicles-promotion-in-ksa-steering-towards-a-greener-future/

Symms, A. & Kler, A. S. (2024). *Waste management: A Qatari perspective*. https://www.tamimi.com/law-update-articles/waste-management-a-qatari-perspective/

Tamkeen. (2024). *Transport and logistics sector*. https://www.tamkeen.bh/wp-content/uploads/2024/09/Transport-Logistics-Report-English.pdf

The Peninsula. (2024). *Sustainable waste management*. https://thepeninsulaqatar.com/editorial/16/09/2024/sustainable-waste-management

The World Bank. (2024). *World development indicators*. https://databank.worldbank.org/source/world-development-indicators

Think Hazard. (2024). *Bahrain*. https://thinkhazard.org/en/report/21-bahrain

Times of Oman. (2023). *Transport ministry launches special programme to cut down carbon emissions*. https://timesofoman.com/article/130892-transport-ministry-launches-special-programme-to-cut-down-carbon-emissions

Times of Oman. (2024). *First electric public bus expected to roll out soon in Oman*. https://timesofoman.com/article/147601-first-electric-public-bus-expected-to-roll-out-soon-in-oman

TÜV SÜD. (2025). *Yemen energy efficiency labelling scheme*. https://www.tuvsud.com/en-id/services/product-certification/yemen-energy-efficiency-labelling-scheme

UAE Legislation. (2018). *Federal law no. (12) of 2018 on integrated waste management*. https://uaelegislation.gov.ae/en/legislations/1164/download

Umar, T. (2021). Sustainable energy production from municipal solid waste in Oman. In *Proceedings of the institution of civil engineers-engineering sustainability* (Vol. 175, pp. 3–11). Thomas Telford Ltd.

UNDP. (2025). *National energy efficiency programme phase 2*. https://www.undp.org/saudi-arabia/projects/national-energy-efficiency-programme-phase-2

UNEP. (2024a). *Bus rapid transit system (BRT) for Yemen*. https://tech-action.unepccc.org/wp-content/uploads/sites/2/2024/12/tna-policy-brief-mitigation-yemen-brt.pdf

UNEP. (2024b). *Hybrid electric vehicle (HEV) technology for Yemen*. https://tech-action.unepccc.org/wp-content/uploads/sites/2/2024/12/tna-policy-brief-mitigation-yemen-hev.pdf

UNFCCC. (2018). *Third national communication to the conference of the parties of United Nations framework convention on climate change: Republic of Yemen*. https://unfccc.int/sites/default/files/resource/Yemen_TNC_%202018.pdf

UNFCCC. (2019). *Kuwait national adaptation plan 2019–2030*. https://unfccc.int/sites/default/files/resource/Kuwait-NAP-2019-2030.pdf

UNFCCC. (2020a). *Bahrain's third national communication*. https://unfccc.int/documents/256562

UNFCCC. (2020b). *Second nationally determined contribution of the United Arab Emirates*. https://unfccc.int/sites/default/files/NDC/2022-06/UAE%20Second%20NDC%20-%20UNFCCC%20Submission%20-%20English%20-%20FINAL.pdf

UNFCCC. (2021). *Second nationally determined contribution*. https://unfccc.int/sites/default/files/NDC/2022-06/Second%20NDC%20Report%20Oman.pdf

UNFCCC. (2024). *About RCC MENA and South Asia*. https://unfccc.int/about-us/regional-collaboration-centres/rcc-for-the-middle-east-north-africa-and-south-asia/about-rcc-mena-and-south-asia

United Arab Emirates' Government portal. (2025). *Energy efficiency and conservation*. https://u.ae/en/information-and-services/environment-and-energy/water-and-energy/energy-efficiency/energy-efficiency-and-conservation

URB. (2025a). *93km sustainable urban highway*. https://urb.ae/projects/theloop/

URB. (2025b). *Dubai Cycle City 2040*. https://urb.ae/projects/dubaicyclecity/

Vinodhkumar, B., Ullah, S., Kumar, T. L., & Al-Ghamdi, S. G. (2024). Amplification of temperature extremes in Arabian Peninsula under warmer worlds. *Scientific Reports, 14*(1), 16604.

Waked, A., & Afif, C. (2012). Emissions of air pollutants from road transport in Lebanon and other countries in the Middle East region. *Atmospheric Environment, 61*, 446–452.

Waste and Recycling. (2025). *The national agenda for integrated waste management was launched in the UAE.* https://www.wasterecyclingmea.com/news/the-national-agenda-for-integrated-waste-management-launched-in-the-uae

Watts, N., Amann, M., Arnell, N., Ayeb-Karlsson, S., Belesova, K., Berry, H., et al. (2018). The 2018 report of the Lancet countdown on health and climate change: Shaping the health of nations for centuries to come. *The Lancet, 392*(10163), 2479–2514.

Wiebelt, M., Breisinger, C., Ecker, O., Al-Riffai, P., Robertson, R., & Thiele, R. (2013). Compounding food and income insecurity in Yemen: Challenges from climate change. *Food Policy, 43*, 77–89.

World Green Economy Organization. (2024). *Regional collaboration centre for the Middle East, North Africa and South Asia.* https://worldgreeneconomy.org/regional-collaboration-centre-for-the-middle-east-north-africa-and-south-asia/

Zawya. (2025). *Oman aims for metro project by 2032, minister says.* https://www.zawya.com/en/business/transport-and-logistics/oman-aims-for-metro-project-by-2032-minister-says-p8lm89db

Chapter 11
Saudi Arabia's Climate Leadership: National Initiatives, Institutional Frameworks and Alignment with the SDGs

Fayma Mushtaq, Luai Muhammad Alhems, and Majid Farooq

11.1 Introduction

The Arabian Peninsula is one of the most climatically vulnerable regions in the world, marked by high temperatures, limited freshwater availability, expanding urbanization and growing pressure on ecosystems (AlSarmi & Washington, 2011; Hamed et al., 2024). Within this context, the Kingdom of Saudi Arabia (KSA) the largest country in the region faces considerable environmental challenges (Khan et al., 2020). These include rising greenhouse gas emissions, land degradation, water scarcity, desertification, biodiversity loss and increasing energy demands, all of which have been exacerbated by climate change and rapid economic development (DeNicola et al., 2015). Historically reliant on its vast oil reserves, Saudi Arabia is undergoing a significant transformation in how it perceives and manages environmental issues. The country is reorienting its development trajectory under Vision 2030, a strategic roadmap that seeks to diversify the economy, enhance environmental sustainability and improve the quality of life for its citizens (Moshashai et al., 2020). Climate action has thus become a critical pillar of this broader vision, which aligns national development with global climate and sustainability objectives.

To respond effectively to emerging climate risks, KSA has introduced a wide array of national strategies, institutional frameworks and project-based interventions aimed at enhancing environmental governance and sustainability. These efforts include the

F. Mushtaq · L. M. Alhems
Applied Research Centre for Environment and Marine Studies, King Fahd University of Petroleum and Minerals, Dhahran, Saudi Arabia

M. Farooq (✉)
Department of Ecology, Environment and Remote Sensing, Government of Jammu and Kashmir, Srinagar, Jammu and Kashmir, India
e-mail: majid.farooq@jk.gov.in

F. Mushtaq et al. (eds.), *Climate Change in the Arabian Peninsula*,
https://doi.org/10.1007/978-3-032-02481-7_11

launch of the Saudi Green Initiative (SGI) and the Middle East Green Initiative (MGI), as well as the establishment of specialized institutions such as the National Center for Vegetation Cover and Combating Desertification (NCVC), the National Water Strategy (NWS), the National Center for Waste Management (MWAN) and the National Meteorological Center (NMC). Together, these bodies are strengthening the regulatory and scientific foundation for long-term environmental planning and climate resilience.

This chapter explores the structure, objectives and progress of key Saudi environmental and climate initiatives and demonstrates how they align with several Sustainable Development Goals (SDGs) mainly SDG 6 (Clean Water and Sanitation), SDG 7 (Affordable and Clean Energy), SDG 11 (Sustainable Cities and Communities), SDG 12 (Responsible Consumption and Production), SDG 13 (Climate Action) and SDG 15 (Life on Land). By presenting a comprehensive account of the country's institutional and policy landscape, this chapter highlights Saudi Arabia's evolving leadership in climate governance and its commitment to a sustainable and resilient future.

11.2 Saudi Arabia's Vision 2030: A Pathway Toward Sustainable Transformation

Saudi Arabia's National Vision 2030 is an ambitious and comprehensive strategic framework that seeks to transform the Kingdom's economic, social and environmental landscape. Launched in 2016, this forward-looking initiative is designed to reduce Saudi Arabia's long-standing dependence on oil revenues by diversifying the economy and promoting development across key public service sectors, including health, education, infrastructure, tourism and recreation. Vision 2030 is structured around three central pillars: a vibrant society, a thriving economy and an ambitious nation each serving as a foundation for national renewal and sustainable development. These guiding themes are supported by a wide array of programs and institutional mechanisms aimed at fostering economic innovation, social empowerment, cultural enrichment and environmental stewardship.

A defining feature of Vision 2030 is its strategic integration of climate action and sustainability objectives, making environmental responsibility an essential component of Saudi Arabia's national development agenda. The Vision sets forth ambitious targets, including achieving net-zero greenhouse gas (GHG) emissions by 2060 and sourcing 50% of the Kingdom's electricity from renewable energy by the year 2030 (Kamboj et al., 2023). This shift signifies a major policy realignment and reflects a commitment to combat climate change through decarbonization, green infrastructure investment and the promotion of sustainable economic models.

Renewable Energy Initiatives

Central to this vision is the Kingdom's concerted effort to harness its abundant solar radiation and desert topography to scale up renewable energy production. Through a suite of innovative projects and national strategies, Saudi Arabia aims to position itself as a global leader in renewable energy, particularly in solar and wind power (Al-Gahtani, 2024). These efforts are not only aligned with environmental imperatives but also with economic diversification and long-term energy security goals. The renewable energy roadmap under Vision 2030 identifies renewables as a key pillar for achieving sustainable development. It prioritizes the expansion of renewable energy infrastructure while addressing existing barriers such as market limitations, policy gaps and technological constraints. Strategic policy frameworks have been introduced to overcome these challenges and facilitate widespread deployment of solar photovoltaic (PV) systems, wind farms and utility-scale energy storage solutions (Al Zohbi & AlAmri, 2020). Recent statistics and project updates reveal tangible progress, with growing capacity installations and the adoption of emerging technologies such as smart grid systems, hybrid energy models and decentralized power generation. These initiatives not only reduce current pollution levels but also set a foundation for sustainable urbanization and economic resilience.

Financial and Regulatory Frameworks

In parallel with technological and infrastructural advancements, Vision 2030 integrates sustainability and climate risk considerations into the national financial and regulatory architecture. The Kingdom is actively embedding sustainability into its financial sector, advancing green finance mechanisms and promoting responsible investment practices to catalyze climate-aligned growth (Alhejaili, 2024). Regulatory reforms are underway to align capital markets and investment strategies with the goals of Vision 2030, fostering greater transparency, accountability and environmental performance. To support these transitions, Saudi Arabia is investing in the development of comprehensive sustainability assessment mechanisms. These frameworks are essential for tracking and evaluating progress toward national climate targets and Sustainable Development Goals (SDGs). Moreover, Vision 2030 emphasizes the active participation of stakeholders at all levels governmental bodies, private sector actors, civil society organizations and local communities to ensure inclusive and equitable implementation of climate policies (Alshuwaikhat & Mohammed, 2017). This multi-level engagement strategy ensures that the vision is not only top-down but also community-driven and contextually responsive.

The Saudi Green Initiative (SGI) and Middle East Green Initiative (MGI)

A cornerstone of Saudi Arabia's Vision 2030 is the Saudi Green Initiative (SGI), launched in 2021 as a transformative national strategy to enhance environmental protection, reduce greenhouse gas emissions and accelerate the transition to clean energy. The SGI is grounded in three primary objectives: expanding vegetation cover, increasing renewable energy deployment and protecting natural ecosystems. Key milestones of the SGI include the planting of 10 billion trees across Saudi Arabia,

Fig. 11.1 Key sustainable development goals (SDGs) targeted by the Saudi green initiative (SGI) and middle east green initiative (MGI), illustrating their integrated approach to climate action, ecosystem restoration, renewable energy and sustainable urban development. *Source* (United Nations, Sustainable Development Goals, https://www.un.org/sustainabledevelopment/news/communications-material/)

the rehabilitation of degraded lands and the expansion of protected areas to encompass more than 30% of the Kingdom's total landmass (Metwally & Ibrahim, 2022). Additionally, the initiative aims to increase the share of renewables to 50% of the electricity mix by 2030, thereby reinforcing Saudi Arabia's energy diversification goals. In parallel, the Middle East Green Initiative (MGI) reflects the Kingdom's commitment to fostering regional cooperation and collective climate solutions. The MGI aims to plant 50 billion trees across the Middle East, promote regional knowledge sharing and support sustainable land and ecosystem management beyond national borders (Onyeaka & Akinsemolu, 2024).

These two flagship programs collectively target several Sustainable Development Goals (SDGs), most notably (Fig. 11.1):

1. **SDG 13**: Climate Action
2. **SDG 15**: Life on Land
3. **SDG 7**: Affordable and Clean Energy

They also generate co-benefits for:

4. **SDG 3**: Good Health and Well-being
5. **SDG 11**: Sustainable Cities and Communities

Nationally Determined Contributions (NDCs) Under the Paris Agreement

In its commitment to global climate governance, Saudi Arabia has articulated ambitious Nationally Determined Contributions (NDCs) under the framework of the Paris Agreement. These updated commitments reflect a marked escalation in ambition compared to the Kingdom's initial targets and underscore its proactive stance in addressing climate change. Saudi Arabia's strategy for reducing greenhouse gas (GHG) emissions is established in a balanced approach that combines policy measures, technological innovation and institutional frameworks. The updated NDCs aim to achieve an annual reduction of 278 million tonnes of CO_2 equivalent by 2030, a goal that more than doubles the earlier target of 130 million tonnes (Hamieh et al., 2022). This significant enhancement reaffirms the Kingdom's commitment to environmental stewardship while sustaining economic development. To accomplish these

Fig. 11.2 Alignment of Saudi Arabia's updated nationally determined contributions (NDCs) with sustainable development goals (SDGs). *Source* (United Nations, Sustainable Development Goals, https://www.un.org/sustainabledevelopment/news/communications-material/)

ambitious reductions, Saudi Arabia is diversifying its energy mix, promoting energy efficiency and advancing carbon capture, utilization and storage (CCUS) technologies (Hamieh et al., 2022). These measures are reinforced by the development of detailed implementation roadmaps, with several components already articulated in recent official submissions (Luomi et al., 2022). Key sectors such as power and water, which together contribute to more than 40% of the nation's GHG emissions, are prioritized for transformative policy actions (Wogan et al., 2019). These sectors are central to the Kingdom's mitigation strategy, ensuring both environmental and economic returns. To operationalize its climate goals, Saudi Arabia is actively deploying a comprehensive suite of mitigation strategies, including:

1. Expanding energy efficiency programs
2. Scaling up renewable energy investments
3. Enhancing waste management systems
4. Promoting land restoration and afforestation
5. Strengthening water conservation mechanisms

Furthermore, the NDCs integrate adaptive capacity building in climate-vulnerable sectors such as agriculture, urban development and water resources. These measures directly contribute to (Fig. 11.2):

1. **SDG 6**: Clean Water and Sanitation
2. **SDG 12**: Responsible Consumption and Production

Building on the institutional foundations, Saudi Arabia's specialized environmental bodies have initiated impactful programs that reflect operational progress toward Vision 2030 and the SDGs. For instance, the National Center for Vegetation Cover and Combating Desertification (NCVC) has led large-scale afforestation campaigns to restore degraded lands, contributing directly to SDG 15 (Life on Land). The National Center for Waste Management (MWAN) has implemented comprehensive recycling and circular economy frameworks in alignment with SDG 12 (Responsible Consumption and Production). Similarly, the National Meteorological Center (NMC) supports SDG 13 (Climate Action) through enhanced early warning systems and climate forecasting services, critical for climate resilience in an arid region. Collectively, these initiatives illustrate a growing recognition within the Kingdom of the need to transition toward a low-carbon, climate-resilient development pathway. The strategic alignment of national policy frameworks with global sustainability targets reinforces Saudi Arabia's emerging role as a regional leader in climate action.

11.3 National Strategies and Frameworks

Saudi Arabia has launched a series of robust national strategies and institutional frameworks aimed at transforming its environmental governance and aligning national development with global sustainability agendas. These initiatives represent a strategic integration of climate mitigation, adaptation, conservation and public engagement, designed to deliver long-term environmental resilience. Many of these frameworks are directly aligned with the United Nations Sustainable Development Goals (SDGs), particularly SDGs 6 (Clean Water and Sanitation), 7 (Affordable and Clean Energy), 11 (Sustainable Cities and Communities), 13 (Climate Action) and 15 (Life on Land).

Saudi Green Initiative (SGI) Governance and Implementation

The Saudi Green Initiative (SGI) governance represents a multifaceted approach that integrates organizational, corporate and governmental strategies to achieve environmental sustainability in line with Saudi Vision 2030. The initiative is centered on increasing green cover, reducing carbon emissions and promoting sustainable practices across various sectors. SGI's governance architecture incorporates corporate governance mechanisms, internal governance structures and supportive public policies collectively enhancing environmental performance and long-term sustainability outcomes. At the core of SGI's governance lies an emphasis on institutional structures conducive to environmental sustainability. These include nature-based organizational systems that align with Vision 2030's long-term objectives (Mostafa & Nafady, 2022). The integration of modern technological solutions and innovative design practices further strengthens environmental management and enhances the effectiveness of sustainability planning (Mostafa & Nafady, 2022). To operationalize its ambitious goals, the SGI is supported by a comprehensive implementation framework that translates strategic governance principles into measurable actions across key sectors.

1. *SGI Monitoring and Reporting Mechanisms*: A robust system of ecological performance indicators has been established to assess afforestation progress, tree survival rates, carbon sequestration and ecosystem health. Technologies such as satellite imaging, drone surveillance and AI-powered dashboards play a central role in monitoring and reporting.
2. *Cross-Sectoral Integration*: SGI is integrally linked with broader national strategies in water conservation, renewable energy and urban resilience. For instance, afforestation contributes to urban cooling, dust storm reduction and public health

improvement, aligning with national goals in health, energy and environmental security.

3. *Legal and Regulatory Frameworks*: New legislation supports SGI by regulating land use, protecting newly afforested zones, banning overgrazing and promoting the use of native plant species. Environmental Impact Assessments (EIAs) are now required for major reforestation projects to ensure ecological integrity and sustainability.
4. *Public Engagement*: Nationwide campaigns such as *"Let's Make it Green"* engage communities, schools and the private sector in afforestation and conservation efforts. Youth participation is institutionalized through school-based environmental clubs and national volunteer programs focused on ecological restoration.
5. *Climate Resilience Co-Benefits*: Beyond carbon mitigation, SGI delivers significant adaptation benefits. These include reduced sandstorm frequency, improved air quality, stabilized soil systems and the regeneration of biodiversity corridors, reinforcing the initiative's value in addressing both environmental and public health challenges.

While the SGI governance framework is comprehensive, certain challenges persist. These include the need for stronger regulatory enforcement, broader stakeholder engagement and mainstreaming sustainability practices across educational institutions and civil society (Sharabi et al., 2023). Moreover, the effectiveness of governance mechanisms in reducing emissions and promoting sustainability requires regular evaluation and adaptation to keep pace with evolving climate goals (Sarwar & Alsaggaf, 2021).

Middle East Green Initiative (MGI): Advancing Regional Climate Cooperation

Launched alongside SGI, the Middle East Green Initiative (MGI) represents Saudi Arabia's regional extension of its environmental leadership. With an ambitious target of planting 50 billion trees across the Middle East including 10 billion within the Kingdom MGI is envisioned as a multi-country platform for collective environmental action and climate diplomacy. The initiative aims to significantly increase green cover in the Middle East and North Africa (MENA) region, which currently has less than 1% of the world's forests and only 3% of its land area covered by forests (Croitoru & Liagre, 2013). By planting trees, the initiative seeks to reduce carbon emissions and enhance carbon sequestration, contributing meaningfully to climate change mitigation efforts (Kouyate et al., 2022). This ambitious afforestation effort is a key part of a broader strategy to combat climate change, reduce desertification and enhance biodiversity. The initiative is particularly significant given the region's arid climate and limited forest cover, which presents unique challenges and opportunities for sustainable development. MGI is driven by regional cooperation agreements, technical collaboration and capacity-building programs. It emphasizes shared research on drought-resistant species, smart irrigation and cross-border conservation strategies. By promoting collaborative land rehabilitation and climate action, MGI enhances ecosystem integrity across arid and semi-arid zones of the Arabian Peninsula and beyond. Institutionally, MGI fosters:

1. *Bilateral and Multilateral Agreements*: Formal partnerships have been initiated between Saudi Arabia and neighbouring countries for joint afforestation and desertification control programs.
2. *Regional Climate Science and Data Sharing*: MGI promotes the establishment of regional climate data centres and joint research platforms for carbon accounting, reforestation metrics and climate risk monitoring.
3. *Capacity Development and Technical Training*: Saudi Arabia hosts training workshops for regional experts on reforestation, biodiversity protection and environmental governance.
4. *Sustainability Diplomacy*: MGI acts as a soft power tool that positions the Kingdom as a regional convener of climate dialogue, aligning the initiative with SDG 13, SDG 15 and SDG 17.

MGI's scale and ambition make it a pioneering model for transboundary climate cooperation. It amplifies Saudi Arabia's leadership in the Global South and strengthens regional resilience against the impacts of desertification, extreme heat and water scarcity. Table 11.1 presents a detailed comparative overview of SGI and MGI, highlighting their core objectives, implementation frameworks, progress and alignment with the United Nations Sustainable Development Goals (SDGs).

Vision 2030 EnvironmZental Pillars

Saudi Vision 2030, the Kingdom's overarching transformation blueprint, places environmental sustainability at the heart of national development. Its environmental pillars focus on enhancing the quality of life, preserving natural ecosystems and promoting renewable energy. The Vision emphasizes improved waste management, increased reliance on clean technologies and the conservation of marine and desert biodiversity. As a cross-cutting agenda, it integrates SDGs 7, 11, 12 (Responsible Consumption and Production) and 13.

The National Environment Strategy (NES)

The NES provides a comprehensive framework for sustainable environmental governance. It aims to restructure the environmental sector through decentralization, transparency and performance-based regulation. The strategy establishes five specialized national centers: the National Center for Waste Management (MWAN), the National Center for Vegetation Cover and Combating Desertification (NCVC), the National Center for Wildlife (NCW), the National Center for Environmental Compliance (NCEC) and the National Meteorological Center (NMC). These institutions coordinate sectoral policies, conduct research and monitoring and enforce environmental standards across the Kingdom advancing SDG 13 and SDG 15.

National Renewable Energy Program (NREP)

The NREP under the Ministry of Energy aims to diversify the Kingdom's energy mix by significantly increasing the share of renewable energy. The program targets 50%

Table 11.1 Comparative overview of SGI and MGI

Element	Saudi green initiative (SGI)	Middle east green initiative (MGI)
Launch and governance	Announced by His Royal Highness the Crown Prince Mohammed bin Salman in March 2021 as the Kingdom's domestic climate-action umbrella. Led by the Ministry of Energy and Ministry of Environment, Water and Agriculture (MEWA), with implementation support from NCVC, NCW, NCEC, NCM, MWAN and the Public Investment Fund (PIF). Embedded in Vision 2030 and Saudi Arabia's updated NDC under the Paris Agreement	Launched simultaneously (2021) as the region-wide counterpart to SGI. Convened through an annual MGI Summit hosted by Riyadh, bringing heads of state, multilateral banks and UN agencies together. Saudi Arabia provides the secretariat and seed financing; execution rests on voluntary country-led contributions across the Middle East and North Africa (MENA)
Headline targets	*Afforestation and Land Restoration*: Plant 10 billion trees and rehabilitate 40 million ha of degraded land by 2060. *Conservation*: Increase legally protected terrestrial and marine areas to $\geq$30% of national territory by 2030. *Emissions*: Reduce $\geq$278 Mt CO_2-e year^{-1} by 2030 through renewables, Carbon Capture, Utilization and Storage (CCUS), energy efficiency and methane abatement. *Energy*: Reach 50% renewable electricity share by 2030 under the National Renewable Energy Program (NREP)	*Afforestation*: Coordinate the planting of 40 billion trees across MENA (10 billion inside KSA + 30 billion in partner states). *Land Rehabilitation*: Restore $\approx$ 200 million ha of degraded rangelands, mangroves and agricultural soils. *Regional Emissions*: Target a collective reduction of ~670 Mt CO_2-e year^{-1} and 60% methane abatement from hydrocarbon activities. *Sand and Dust Storm Mitigation*: Reduce SDS outbreaks through green belts and early-warning cooperation
Key delivery pillars	*Nature-based Solutions*: Mangrove restoration, wadis and mountain forest recovery, urban green corridors. *Low-Carbon Energy*: Utility-scale solar and wind (e.g., Sakaka, Sudair, Dumat Al-Jandal), green hydrogen at NEOM, grid interconnections. *Circular Carbon Economy (CCE)*: CCUS hubs (1.5 kt CO_2 day^{-1} pilot, scaling to 9 Mt CO_2 year^{-1} by 2030), waste-to-energy, advanced recycling. *Community and Private-Sector Participation*: School greening programmes, corporate tree-credits, ESG-linked sukuk/green bonds via Public Investment Fund (PIF)	*Regional Afforestation Fund*: USD 10 bn pledge (Saudi share $\approx$ 15%) to co-finance large-scale tree-planting and rangeland regeneration. *Knowledge-Sharing and Capacity-Building*: Joint research with ICARDA, FAO, UNEP on arid-land agro-forestry and drought-resilient species. *Cross-Border Climate Early-Warning Systems*: Data exchange on sandstorms, heatwaves, flood risks (co-led by NCM and WMO). *Green Finance Catalyst*: Regional green-bond framework and blended-finance windows with the Green Climate Fund (GCF), World Bank, IsDB

(continued)

Table 11.1 (continued)

Element	Saudi green initiative (SGI)	Middle east green initiative (MGI)
Institutional architecture	*Steering Committee*: Ministers of Energy, Environment/Water and Agriculture, Investment, Finance. *Technical Task-Forces*: Carbon Markets, Renewable Procurement, Smart Agriculture, Protected-Area Expansion. *Monitoring and Verification*: Remote-sensing (NCVC/NCM), national GHG inventory (MEWA and GaStat), annual performance report aligned with *Global Stocktake*	*MGI Summit Secretariat:* Housed within MEWA; rotates chairmanship with partner states. *Regional Working Groups:* Afforestation, Methane, Marine Conservation, Climate Finance. *Tracking Platform*: An open GIS dashboard showing tree survival rates, land-rehabilitation progress and emission reductions
Primary SDG linkages	*SDG 13 Climate Action*: Large-scale emission reduction and resilience. *SDG 15 Life on Land*: Massive afforestation, desertification control, biodiversity corridors. *SDG 6 Clean Water and Sanitation*: Watershed restoration, soil-moisture retention, aquifer recharge. *SDG 7 Affordable and Clean Energy*: 50% renewables target, green hydrogen roll-out. *SDG 14 Life Below Water*: Coral-reef and mangrove protection. *SDG 17 Partnerships*: Domestic PPPs and link-up with GCF, UNEP, WEF, WRI	*SDG 13, 15, 17* as above, scaled regionally. *SDG 2 Zero Hunger*: Rangeland rehabilitation boosts pastoral and agro-pastoral livelihoods. *SDG 11 Sustainable Cities*: Regional sand-storm reduction, urban heat-island mitigation. *SDG 14*: Safeguards Red Sea and Gulf ecosystems via cross-border MPAs
Current progress (2021–mid 2024)	~600 million trees planted/naturally regenerated ($\approx$ 6% of 10 bn target). Protected-area coverage expanded from 16% $\rightarrow$ ~21%. Renewable capacity in operation/under contract >20 GW (on path to 58.7 GW by 2030). First sovereign and PIF green bonds (~USD 11 bn) fully subscribed	Pledges from 17 MENA states covering $\approx$14 billion of the 40 billion-tree goal. USD 2.5 bn Saudi-backed *MGI Climate Fund* launched for adaptation and mitigation. Regional methane-reduction roadmap endorsed at MGI Summit 2023. Joint WMO-Arab League sand-storm early-warning pilot operational in KSA, UAE, Kuwait

of electricity generation from renewables by 2030. Major solar and wind projects are being implemented across regions, such as the Sakaka PV Solar Plant and Dumat Al Jandal Wind Farm. These efforts align with SDG 7 and support decarbonization efforts under SDG 13.

Circular Carbon Economy (CCE) Framework

Endorsed by the G20 under Saudi Arabia's presidency, the Circular Carbon Economy (CCE) framework promotes a holistic approach to managing carbon emissions through the principles of reduce, reuse, recycle and remove. The CCE strategy supports the Kingdom's broader goals for carbon neutrality and energy security. It includes investment in carbon capture, utilization and storage (CCUS) and clean hydrogen development, with strong ties to SDG 9 (Industry, Innovation and Infrastructure) and SDG 13.

National Strategy for Water (NSW) 2030

The NSW 2030 addresses Saudi Arabia's water scarcity challenge through an integrated water resources management approach. It focuses on demand-side efficiency, reduction of groundwater dependency, desalination innovation, wastewater reuse and institutional restructuring of the water sector. Massive investments have been made in wastewater treatment and recycled water infrastructure, particularly in Riyadh and Jeddah. Technologies like treated sewage effluent (TSE) reuse for agriculture and landscaping and solar-powered desalination plants such as the Al Khafji facility, are pivotal components. In addition, the Saline Water Conversion Corporation (SWCC), one of the world's largest desalination operators, is transitioning towards renewable-powered desalination and low-carbon technologies. Water conservation campaigns and smart irrigation systems are also being implemented under the guidance of the Ministry of Environment, Water and Agriculture (MEWA), targeting behavioral change among consumers. This strategy advances SDG 6 and supports adaptation to climate-induced water stress.

NEOM: A Model for Sustainable Urbanism

New Future for the Middle East (NEOM) is a $500 billion mega-city development in northwest Saudi Arabia that aspires to become a global hub for innovation, sustainable living and climate-resilient infrastructure. Designed as a zero-carbon, fully renewable-powered urban ecosystem, NEOM integrates smart mobility, desalination powered by clean energy and regenerative design principles. Its flagship project, *The Line*, is a 170-km car-free linear city powered by 100% renewable energy and designed to have no carbon emissions. NEOM directly supports SDG 11 (Sustainable Cities and Communities), SDG 9 (Industry, Innovation and Infrastructure) and SDG 13 (Climate Action) and represents a model for future urban resilience in arid environments.

The Red Sea Project

Led by the Red Sea Global (RSG), this regenerative tourism initiative covers 28,000 km^2 along Saudi Arabia's western coastline and includes over 90 pristine islands. The project aims to set a new global benchmark in sustainable tourism by achieving a net-positive environmental impact. It involves zero-waste-to-landfill policies, 100% renewable energy, carbon neutrality and conservation of coral reefs, mangroves and endangered species. The Red Sea Project contributes to SDG 12 (Responsible Consumption and Production), SDG 14 (Life Below Water) and SDG 15 (Life on Land), while enhancing socio-economic development in line with SDG 8 (Decent Work and Economic Growth).

International and Regional Collaboration

Saudi Arabia recognizes that climate change is a transboundary challenge requiring cooperative solutions. The Kingdom has thus intensified its participation in international frameworks, directly contributing to SDG 17 (Partnerships for the Goals). At the global level, Saudi Arabia is an active participant in the United Nations Framework Convention on Climate Change (UNFCCC), the Paris Agreement and the G20 Climate and Energy Working Groups. It has played a visible role in recent global climate summits, including COP27 (Sharm El-Sheikh, 2022) and COP28 (Dubai, 2023), where it showcased its national efforts under Vision 2030 and supported climate finance and adaptation priorities. Collaborations with organizations such as the United Nations Environment Programme (UNEP), the United Nations Convention to Combat Desertification (UNCCD) and regional partners have also facilitated technology transfer, research initiatives and capacity-building programs. Notably, Saudi Arabia has contributed to the Arab Framework Action Plan on Climate Change, strengthening collective action across Arab states. In parallel, the Kingdom has supported humanitarian climate adaptation initiatives in vulnerable countries, reflecting its commitment to climate equity and resilience in the Global South. These efforts reflect the Kingdom's role in promoting a shared environmental agenda across the Arabian Peninsula and the broader Middle East.

Financing Climate Action

Mobilizing adequate financial resources is essential for achieving Saudi Arabia's climate and sustainability goals. Aligned with SDG 13 (Climate Action), SDG 8 (Decent Work and Economic Growth) and SDG 17 (Partnerships for the Goals), the Kingdom is advancing climate finance through public and private channels. The Public Investment Fund (PIF) plays a central role by investing in green infrastructure, renewable energy and nature-based solutions. Flagship projects like NEOM, The Line and the Red Sea Project serve as models for sustainable urban development. Saudi Arabia collaborates with the Green Climate Fund (GCF) as both a donor and recipient to co-finance projects in clean energy, land restoration and water security. The Kingdom is also building institutional capacity to access GCF resources directly. Complementing these efforts, Saudi Arabia has introduced a regional Voluntary Carbon Market, promoting emission offsets and financing verified environmental projects. Financial tools such as green bonds, sustainability-linked loans and ESG-compliant investments are increasingly being adopted by domestic institutions. These strategies reflect Saudi Arabia's shift from an oil-based economy toward a diversified, low-carbon and climate-resilient development model.

Collectively, these initiatives reflect Saudi Arabia's integrated approach to sustainable development, showcasing how national strategies are systematically aligned with the United Nations Sustainable Development Goals (SDGs) to advance climate resilience, economic diversification and ecological stewardship (Table 11.2).

Table 11.2 Overview of national strategies and institutions supporting Saudi Arabia's sustainability goals

Initiative/institution/strategy	Targeted sector/domain	Aligned SDGs
Saudi vision 2030	Economy, society, environment	SDG 8, 9, 11, 13, 17
The Saudi green initiative (SGI)	Afforestation, emissions reduction	SDG 13, 15
Middle east green initiative (MGI)	Regional climate cooperation	SDG 13, 17
National environment strategy	Environmental policy and governance	SDG 13, 16
National renewable energy program	Energy and power sector	SDG 7, 9, 13
NEOM	Urban development, smart cities	SDG 11, 13
Red sea project	Tourism, marine ecosystems	SDG 14, 12, 13
National center for vegetation cover (NCVC)	Land restoration, biodiversity	SDG 13, 15
National center for environmental compliance (NCEC)	Environmental monitoring, compliance and enforcement	SDG 13, 16
National center for wildlife (NCW)	Wildlife conservation, protected areas	SDG 14, 15
National center for waste management (MWAN)	Solid waste and circular economy	SDG 12, 13
National meteorological center (NMC)	Climate monitoring and early warning	SDG 13
Green Saudi financing (PIF, green bonds, ESG funds)	Climate finance and green investment	SDG 8, 9, 13, 17
Green climate fund (GCF) engagement	Adaptation and capacity building	SDG 13, 17
Regional voluntary carbon market	Carbon trading and offsetting	SDG 13
Desalination and water reuse programs	Water security, infrastructure	SDG 6, 9, 13
International partnerships (e.g., UNFCCC, GCF)	Global and regional climate policy	SDG 13, 17

11.4 Key Environmental Agencies in Saudi Arabia

Saudi Arabia has instituted a comprehensive environmental governance framework supported by specialized national agencies that operate under the Ministry of Environment, Water and Agriculture (MEWA). These institutions are essential for translating strategic visions into actionable outcomes across the Kingdom.

1. *National Center for Waste Management (MWAN)*: MWAN is tasked with implementing sustainable waste management systems, advancing recycling practices and minimizing landfill use. It supports circular economy principles and

addresses SDG 12 (Responsible Consumption and Production) and SDG 11 (Sustainable Cities and Communities).

2. *National Center for Vegetation Cover and Combating Desertification (NCVC)*: NCVC oversees efforts to rehabilitate degraded land, combat desertification and promote afforestation projects aligned with the Saudi Green Initiative. It plays a critical role in ecosystem restoration and aligns with SDG 15 (Life on Land) and SDG 13 (Climate Action).
3. *National Meteorological Center (NMC)*: The NMC is responsible for monitoring climate variables, issuing early warnings for extreme weather events and supporting climate science research. It contributes to strengthening resilience to climate-related hazards (SDG 13) and enhancing weather-based decision-making in agriculture, health and urban planning.
4. *National Center for Environmental Compliance (NCEC)*: The National Center for Environmental Compliance (NCEC) plays a pivotal role in regulating and enforcing environmental standards across Saudi Arabia. By conducting inspections, monitoring industrial activities and ensuring adherence to national environmental regulations, the NCEC helps safeguard ecosystems from pollution and unsustainable practices. It employs advanced monitoring tools, including remote sensing and environmental impact assessments, to oversee compliance across public and private sectors. These efforts directly align with SDG 12 (Responsible Consumption and Production) and SDG 13 (Climate Action), reinforcing the Kingdom's commitment to sustainable industrial growth and environmental protection.
5. *National Center for Wildlife (NCW)*: The National Center for Wildlife (NCW) is the lead institution for biodiversity conservation in Saudi Arabia. It manages a network of protected areas, oversees species conservation programs and implements strategies to preserve terrestrial and marine ecosystems. The NCW also engages in ecological research, education and international collaboration to enhance habitat restoration and species resilience. Its work contributes significantly to SDG 15 (Life on Land) through efforts to combat desertification and preserve endemic wildlife and to SDG 14 (Life Below Water) through the protection of marine biodiversity and coastal habitats.
6. *National Center for Meteorology (NCM)*: The National Center for Meteorology (NCM) is responsible for providing accurate weather forecasts, climate monitoring and early warning services. The center supports national disaster preparedness and climate adaptation strategies by generating data critical for water resource planning, agriculture, urban resilience and public health. Its role has grown increasingly vital in the face of rising climate variability, enabling data-driven decision-making and international cooperation in climate science. Through its services, the NCM strengthens national capacity to respond to extreme weather and climate risks, directly advancing SDG 13 (Climate Action) and supporting SDG 11 (Sustainable Cities and Communities).

11.5 Challenges and Future Directions

While Saudi Arabia has made remarkable progress in embedding climate change into its national agenda, several persistent challenges continue to influence the pace and effectiveness of its climate action. Chief among these is the ongoing dependence on fossil fuels, which complicates economic diversification efforts despite substantial investments in renewable energy and sustainable infrastructure. The transition toward a green economy necessitates a more aggressive phase-out of carbon-intensive industries and stronger incentives for low-carbon innovation across sectors. Another challenge lies in the integration of advanced climate technologies. While the Kingdom has demonstrated leadership through projects like NEOM and the Red Sea Development, scaling these technologies across all regions and sectors remains a complex undertaking. Technological adoption is often hindered by gaps in institutional capacity, limited access to global best practices and challenges in research and development localization. Community engagement and public awareness are also critical areas requiring greater focus. Achieving sustainable climate resilience demands grassroots participation, behavioral change and a shared understanding of climate risks and responsibilities. Expanding environmental education in schools, launching nation-wide climate awareness campaigns and enhancing the role of civil society organizations are essential steps toward fostering a culture of environmental stewardship. Future directions must therefore emphasize holistic strategies that combine robust regulatory frameworks, educational outreach and integrated climate governance. Strengthening cross-sectoral collaboration, enabling decentralized climate action at local government levels and reinforcing transparency in implementation and reporting will be key to advancing Saudi Arabia's climate leadership in the decades ahead.

11.6 Conclusion

Saudi Arabia has redefined its future by transitioning from a traditional, oil-dependent economy to a pioneer in climate leadership in the Middle East. This remarkable change is firmly framed in the ambitious Vision 2030 plan which has guided national strategies towards a more diversified and sustainable future. The emphasis on reducing fossil fuel dependence and investing in renewable energy, such as solar and wind power, symbolizes the Kingdom's commitment to mitigating the effects of climate change while driving economic growth. The outcome of these efforts is evident in the integrated approach combining robust climate adaptation, mitigation and sustainable development initiatives. Major programs like the Saudi SGI and MGI exemplify the Kingdom's resolve to reduce carbon emissions drastically by planting billions of trees and rehabilitating millions of hectares of degraded land. These afforestation and reforestation projects are designed not only to enhance carbon sequestration but also to improve air quality and promote biodiversity across the region. Concurrently, the National Renewable Energy Program (NREP) plays a pivotal role by targeting the generation of 50% of Saudi Arabia's electricity from renewable sources by 2030. Investments in renewable energy infrastructure such as the Sakaka solar PV plant and wind energy projects are tangible steps towards diversifying the energy mix and ensuring energy security for the future. This approach offers a scalable model that other developing countries can look to as they balance economic growth with environmental sustainability. Alongside clean energy initiatives, Saudi Arabia is also addressing pressing water security concerns through innovative desalination technologies powered by renewable energy. Given the Kingdom's arid environment, significant investments in desalination, wastewater treatment and recycling programs help reduce reliance on increasingly scarce freshwater resources while safeguarding the agricultural sector. These measures illustrate a holistic approach to managing natural resources efficiently amid climate stress. Saudi Arabia's commitment to sustainability extends to enhancing climate resilience across sectors. In urban settings, projects such as NEOM and the Red Sea Development Project integrate climate-smart designs, resilient infrastructure and sophisticated water management systems to mitigate the adverse effects of extreme weather events, ranging from severe heatwaves to potential flooding and sea-level rise. Adaptation strategies include promoting drought-resistant crops, modernizing irrigation technologies and employing smart urban planning practices to create environments that are better prepared to face climate challenges. At the international level, the Kingdom's role is equally significant. By actively participating in global climate forums like COP27 and COP28 and supporting initiatives such as the Humanitarian and Arab Framework Action Plan on Climate Change, Saudi Arabia is not only shaping climate policy but also fostering regional and global collaboration. The integration of international partnerships into national climate strategies exemplifies the Kingdom's leadership in mobilizing financial resources and advancing technological innovation to achieve long-term sustainability goals. In conclusion, the collective efforts across multiple domains from renewable energy to water security and from afforestation initiatives to advanced urban planning underscore Saudi Arabia's determination to create a resilient, sustainable future. Through a well-balanced mix of adaptation and mitigation, supported by international cooperation and robust policy frameworks, the Kingdom demonstrates that economic prosperity and environmental stewardship can indeed go hand in hand, forging a pathway that not only addresses current challenges but also paves the way for a sustainable and inclusive future for generations to come.

References

Al Zohbi, G., & AlAmri, F. G. (2020). Current situation of renewable energy in Saudi Arabia: Opportunities and challenges. *Journal of Sustainable Development, 13*(2), 98–106. https://doi.org/10.5539/jsd.v13n2p98

Al-Gahtani, S. F. (2024). Saudi Arabia's journey toward a renewable future. *Energies, 17*(11), 2444. https://doi.org/10.3390/en17112444

Alhejaili, M. O. (2024). Integrating climate change risks and sustainability goals into Saudi Arabia's financial regulation: Pathways to green finance. *Sustainability, 16*(10), 4159. https://doi.org/10.3390/su16104159

AlSarmi, S., & Washington, R. (2011). Recent observed climate change over the Arabian Peninsula. *Journal of Geophysical Research: Atmospheres, 116*(D11), 459. https://doi.org/10.1029/2010JD015459

Alshuwaikhat, H. M., & Mohammed, I. (2017). Sustainability matters in national development visions—Evidence from Saudi Arabia's vision for 2030. *Sustainability, 9*(3), 408. https://doi.org/10.3390/su9030408

Croitoru, L., & Liagre, L. (2013). *Contribution of forests to a green economy in the Middle East and North Africa region.* Plan Bleu. http://documents.irevues.inist.fr/handle/2042/53607

DeNicola, E., Aburizaiza, O. S., Siddique, A., Khwaja, H., & Carpenter, D. O. (2015). Climate change and water scarcity: The case of Saudi Arabia. *Annals of Global Health, 81*(3), 342–353.

Hamed, M. M., Al-Hasani, A. A. J., Nashwan, M. S., Saadi, Z., & Shahid, S. (2024). Assessing the growing threat of heat stress in the North Africa and Arabian Peninsula region connected to climate change. *Journal of Cleaner Production, 441*, 141639. https://doi.org/10.1016/j.jclepro.2024.141639

Hamieh, A., Rowaihy, F., Al-Juaied, M., Abo-Khatwa, A. N., Afifi, A. M., & Hoteit, H. (2022). Quantification and analysis of CO_2 footprint from industrial facilities in Saudi Arabia. *Environmental Challenges, 16*, 100299. https://doi.org/10.1016/j.ecmx.2022.100299

Kamboj, P., Hejazi, M., Alhadhrami, K., Qiu, Y., Kyle, P., & Iyer, G. (2023). *Saudi Arabia net zero GHG emissions by 2060: Transformation of the electricity sector* (KAPSARC Discussion Paper No. KS-2023-DP31). https://doi.org/10.30573/ks--2023-dp31

Khan, M. B., Hameed, I., & Khan, K. (2020). Kingdom of Saudi Arabia: The land of opportunities. In *Research, innovation and entrepreneurship in Saudi Arabia* (pp. 1–16). Routledge.

Kouyate, D., Abohassan, R. A. A., & Elfeel, A. A. (2022). Does arid land agroforestry contribute to achieving the Saudi Green Initiative goals? A review. *International Journal of Environment and Climate Change, 12*(11), 2707–2719. https://doi.org/10.9734/ijecc/2022/v12i1131259

Luomi, M., Alshehri, T., & Yılmaz, F. (2022). *The gulf cooperation council and the circular carbon economy: Progress and potential* (KAPSARC Discussion Paper KS--2022-DP06). https://doi.org/10.30573/ks--2022-dp06

Metwally, W. M., & Ibrahim, V. A. R. (2022). The green areas in the city and sustainable approach: Analytical study of Saudi Crown Prince Mohammed bin Salman's tree-planting program. *IOP Conference Series: Earth and Environmental Science, 1026*(1), 012045. https://doi.org/10.1088/1755-1315/1026/1/012045

Moshashai, D., Leber, A. M., & Savage, J. D. (2020). Saudi Arabia plans for its economic future: Vision 2030, the national transformation plan and Saudi fiscal reform. *British Journal of Middle Eastern Studies, 47*(3), 381–401.

Mostafa, M., & Nafady, D. A. (2022). Nature organizational structure systems in support of "Saudi Arabia's green initiative" utilized to keep up with 2030's vision. *International Design Journal, 12*(5), 39–50.

Onyeaka, H., & Akinsemolu, A. A. (2024). Advancing green education in MENA region: Challenges, opportunities and best practices. *Sustainable Development, 12*, 3182. https://doi.org/10.1002/sd.3182

Sarwar, S., & Alsaggaf, M. I. (2021). The role of governance indicators to minimize the carbon emission: A study of Saudi Arabia. *Management of Environmental Quality: An International Journal, 32*(5), 970–988. https://doi.org/10.1108/MEQ-11-2020-0275

Sharabi, W. A., Faraj, A. O. K., & Mazyu, M. A. (2023). *A proposal to enhance sustainability in Saudi universities in light of the Saudi Green Initiative.* https://doi.org/10.20944/preprints202307.0157.v2

Wogan, D., Carey, E., & Cooke, D. (2019). *Policy pathways to meet Saudi Arabia's contribution to the Paris Agreement* (KAPSARC Discussion Paper KS--2018-DP49). King Abdullah Petroleum Studies and Research Center. https://ideas.repec.org/p/prc/dpaper/ks--2018-dp49.html

MIX
Papier aus verantwortungsvollen Quellen
Paper from responsible sources
FSC® C105338

If you have any concerns about our products,
you can contact us on
ProductSafety@springernature.com

In case Publisher is established outside the EU,
the EU authorized representative is:
Springer Nature Customer Service Center GmbH
Europaplatz 3, 69115 Heidelberg, Germany

Printed by Libri Plureos GmbH
in Hamburg, Germany